Multifragmentation in Heavy-Ion Reactions

Multifragmentation in Heavy-Ion Reactions
Theory and Experiments

edited by

Rajeev K. Puri
Yu-Gang Ma
Arun Sharma

JENNY STANFORD
PUBLISHING

Published by

Jenny Stanford Publishing Pte. Ltd.
101 Thomson Road
#06-01, United Square
Singapore 307591

Email: editorial@jennystanford.com
Web: www.jennystanford.com

British Library Cataloguing-in-Publication Data
A catalogue record for this book is available from the British Library.

Multifragmentation in Heavy-Ion Reactions: Theory and Experiments

ISBN 978-981-4968-69-0 (Hardcover)
ISBN 978-1-003-38513-4 (eBook)

Contents

Preface

The journey of this book started in 2017, intending to bring together different active researchers across the world working in the field of intermediate-energy heavy-ion collisions.

The first step was to collect information about the research being conducted by various experts across the world working in this field. In this context, an in-depth search was carried out, and we could gather information about hundreds of researchers from more than 30 countries.

The next step was to contact some of the prominent researchers regarding their research contributions for the book. In response to our requests and the reminders sent to them, we received 15 research contributions, each corresponding to a chapter of the book. These chapters cover a wide range of theoretical and experimental findings. For the readers to make the most out of these and for consistency, the chapters have been reviewed, edited, processed, and placed in a logical order. The book will enable readers to understand the recent developments and current status of the field. It brings together the diverse perspectives of what we have learned so far and presents some open questions.

Our journey became challenging when work had to be stopped for a long period due to the COVID-19 pandemic. But now, when the journey is ending, we hope the chapters by authors from different parts of the world will be valuable for the readers. Moreover, this book will also further open up avenues and opportunities for future research in the area of intermediate-energy heavy-ion collisions.

This book would have not been possible without the dedicated and intellectual contributions of many researchers. We would like to thank all the people associated directly or indirectly with its completion. We also appreciate the trust Jenny Stanford Publishing showed in us and their help in bringing out the book.

We hope the book will be helpful to the community working in the field of intermediate-energy heavy-ion collisions. We also plan to bring out another volume on similar lines if we receive sufficient contributions from researchers.

Rajeev K. Puri
Yu-Gang Ma
Arun Sharma
March 2023

Chapter 1

Multifragmentation and Associated Phenomenon: Recent Progress and New Challenges

Arun Sharma[a] **and Rajeev K. Puri**[b]

[a]*Department of Physics, Govt. Degree College Billawar, Jammu 184204, India*
[b]*Department of Physics, Panjab University, Chandigarh 160014, India*
arungaur.ju@gmail.com

1.1 Introduction

The ultimate goal of this edited volume is to explore the recent progress and new challenges in intermediate energy heavy-ion collisions, particularly, the multifragmentation and associated phenomenon. The central topics to be reviewed in this volume are nuclear multifragmentation, nuclear phase transition, nuclear symmetry energy, etc. The present edited volume gives only a report of the momentary status of the field of multifragmentation and associated phenomenon and does not mean to be a definitive and unique issue. It is worth mentioning that this field has not yet revealed its secrets, and this motivated us to work on the current issue for the nuclear physics community. Now, before reviewing

Multifragmentation in Heavy-Ion Reactions: Theory and Experiments
Edited by Rajeev K. Puri, Yu-Gang Ma, and Arun Sharma

ISBN 978-981-4968-69-0 (Hardcover), 978-1-003-38513-4 (eBook)
www.jennystanford.com

the recent progress in the field of nuclear multifragmentation and associated phenomenon, let us first put some light on the concept of the equation of state of nuclear matter.

1.2 Nuclear Equation of State

One of the main research topics in the field of nuclear physics is to access the properties of nuclear matter, which is made up of an infinite number of protons and neutrons and these fermions interact via van der Waals–like forces. To understand this, one generally discusses various aspects of nuclear matter equation of state [Li (2008), Baran (2005), Li (2007), Youngblood (1999), Reisdorf (2012), Verde (2014), Li (1997)]. The nuclear matter equation of state (NEOS) is a non-trivial relation between thermodynamical variables characterizing a medium. Now, let us discuss why do we study nuclear matter equation of state: (1) The study of nuclear matter equation of state is important because bulk nuclear properties are mainly determined by NEOS. (2) Nuclear matter equation of state helps to understand the formation of compact astrophysical objects such as Neutron Star, Supernova, and Black Holes. (3) It also contributes toward the knowledge of the signal of phase transition in nuclear collisions and to understand a phenomenon like the formation of atomic elements. The NEOS is characterized by nuclear incompressibility and depending on the values of incompressibility, there are two possible equations of state, i.e., soft and hard [Aichelin (1991)]. The nuclear equation of state of isospin symmetric nuclear matter ($n = p$) behaves in a different manner and relatively well known when compared with the equation of state of isospin asymmetric nuclear matter [Li (2008), Baran (2005), Youngblood (1999), Li (2007), Reisdorf (2012), Verde (2014), Li (1997)]. Thus, the equation of state of isospin asymmetric nuclear matter is a long-standing issue, and thus it has attracted the nuclear physics community [Lopez (2012), Baran (2002), Chen (2005), Brown (2000), Tsang (2012), Dai (2014), Colonna (2014)]. In the whole world, a large number of radioactive beam facilities have been constructed or planned to study the isospin asymmetric nuclear matter (having large n to p ratio) and the structure of rare isotopes. With the help of these radioactive beam facilities, one can probe isospin physics, especially heavy-ion collisions induced by neutron-rich nuclei at intermediate energies. The ultimate aim of isospin physics is to determine the density dependence term of nuclear symmetry energy in the EOS of asymmetric nuclear matter. A large number of studies have been carried out in this context [Li (2008), Li (1998), Li (2001), Danielewicz (2002), Lattimer (2004), Baran (2005), Steiner (2005), Li (2019)]. At intermediate energies, a large number of phenomena such as multifragmentation, flow can affect the equation of state of nuclear matter.

1.3 Nuclear Multifragmentation: Theory and Experiments

The nuclear multifragmentation phenomenon was predicted for the first time by Bondrof in 1976 [Bondorf (1976), Bondorf (1979)] and studied since the early 1980s [Bondorf (1995), Borderie (2008)]. It occurs if the excitation energy of an excited matter (due to the collisions of projectile and target) is comparable with the binding energy of nucleus [Bondorf (1976), Bondorf (1979), Borderie (2008), Jakobsson (1982), Jakobsson (1990)]. This phenomenon results in the formation of many fragments of different masses. During the nuclear collisions, fragmenting systems are formed, and to get necessary information about them, several experimental facilities have been installed so far in the whole world.

The real advances in the field of multifragmentation were only made with the advent of efficient and more powerful 4π detectors [Souza (2006)].

During the initial years, the study of nuclear physics was confined to fusion, fission, and particle transfer. Due to the non-availability of accelerator facilities, it was not possible to accelerate heavier projectiles. At that time, the field of heavy-ion physics focussed on shooting light particles on heavy targets.

The Lawrence Berkely laboratory witnessed the installation of the first accelerator in the series known as BEVALAC accelerator [Moretto (1993), Phair (1995), Sangster (1992), Bowman (1991)]. In 1974, this facility started accelerating nuclei such as ^{56}Fe up to very high incident energy, i.e., 2.1 GeV/nucleon. This was just the starting of a new era in accelerator-based heavy-ion physics. With the passage of time and following the path of the first Berkley experiments, several high-energy accelerator facilities were installed at other parts of the world. In the recent years, multifragmentation studies have been performed with high-energy accelerators at Michigan State University (MSU), USA; Superconducting Super Collider (SSC) at BNL (USA); Grand Accelerator National D'ions Lourels (GANIL), France; NSF-Arizon Accelerator at the University of Arizona (USA); Relativistic Heavy-Ion Collisions (RHIC) (USA); Vivitron Accelerator in Straburg, France; Charged Heavy-Ion Mass and Resolving Array (CHIMERA) detector at Laboratori Nazionali del Sud in INFN, Catania; Superconduting Cyclotron (SC) at Texas, USA; heavy-ion Synchrotron SIS accelerator at GSI (Germany); Cooler Storage Ring (CSR) facility in Lanzhou, China [Xia (1998)] and Facility for Rare Isotope Beams (FRIB) [Bollen (2010)]. Also, an Isotope Separator On Line (ISOL) post-accelerate type of radioactive ion beams facility was proposed at VEEC in 1998. Many of the above-mentioned facilities have been upgraded to study the effects of isospin physics.

Over recent years, numerous collaborations have been constituted to study the physics of intermediate energy heavy-ion collisions. The collaborations like (1) the identification of Noyaux et Detection avec Resolution Array (INDRA) group, (2) A Large Acceptance Dipole magNet (ALADiN) group [Bohnet (2009)], (3) the FOur (FOPI) group, (4) the NIMROD Collaboration, (5) the KaoS collaboration, (6) the EOS Collaboration, (7) MULTICS Collaboration [Agostino (2000, 2003)], (8) ISIS Collaboration, (9) TAPS and the LAND collaborations [Venema (1993), Lambrecht (1994), Leifels (1993)], etc. have made significant advancements in the area of multifragmentation and associated phenomenon.

With the help of experimental facilities, early attempts were carried out to put some light on the complex picture of heavy-ion collisions at intermediate energies by performing emulsion experiments [Jakobsson (1990)]. These experiments also provide valuable information about the phenomenon like multifragmentation as well as vaporization (because the incident energy range in these experiments was between 25 MeV/nucleon to 200 MeV/nucleon). On the other hand, the Berkly group and the National Superconducting

Cyclotron Laboratory (NSCL) studied the symmetric and asymmetric reactions and emphasized to describe various features like angular distribution, excitation energy, cross section as well as velocity distribution of fragments, etc. [Moretto (1993), Phair (1995), Sangster (1992), Bowman (1991)], average multiplicity, and mass of the heaviest fragment [Williams (1997), Llope (1995), Stone (1995)], the neutron–proton double ratio, isospin diffusion, binding energy, and rapidity distribution [Zhang (2008), Tsang (2007), Liu (2007)]. Also, the INDRA group has made significant contribution in the field of multifragmentation. This group has focussed on investigating a variety of features such as kinetic energy spectra, Coulomb instability of fragments, velocity correlation, system size effects, and role of system size in entrance channel [Marie (2007), Loveland (1999), Frankland (2005)]. In addition to this, the ALADiN and FOur PI (FOPI) groups at GSI carried out experimental observations to shed light on the multifragmentation emission and total disassembly and shattering of nuclear matter after the nuclear reaction. The most renowned phenomenon of rise and fall behavior of intermediate mass fragments (IMFs) (when IMFs are plotted against incident energy and impact parameter) for the reaction of ${}^{197}_{79}Au + {}^{197}_{79}Au$ was reported by the ALADiN group [Bondorf (1995), Borderie (2008), Williams (1997), Schttauf (1996), Ogilvie (1991), 86, Botvina (1995), Blaich (1993), Peaslee (1988)]. However, the NIMROD group studied the heavy-ion collisions in the Fermi-energy domain. The phenomena like charge distribution, velocity spectrum, multiplicities, equation of state of isospin asymmetric nuclear matter, critical behavior in the disassembly of nuclei, average n/p ratio, symmetry energy, isotopic and isobaric yields were investigated by this group [Ma (2004), Wada (2004), Shetty (2009)]. In 2006, the RiKEN Radioactive Isotope Beam Factory (RIBF) was operated to generate elements from ${}^{4}He$ to ${}^{238}U$ [Yano (2007), Sakurai (2010)]. This facility helped to accelerate light ions such as ${}^{4}He$ (heavy ion such as ${}^{238}U$) up to 400(350) MeV/nucleon. A recently installed facility, namely FAIR (Facility for Antiproton and Ion Research) [GSI (2021), CBM (2011)], consists of interlinked machines for accelerating and storing particle beams of high quality.

As far as theoretical understanding of multifragmentation is concerned, several theoretical models have been developed (or still developing) and these models provide an opportunity to extract knowledge from the experimental observables. In some cases, using different models makes the situation much more complicated. Theoretically, one can study the multifragmentation at intermediate energy heavy-ion collisions with the help of two broad approaches: statistical and dynamical models. A wide variety of theoretical models have been proposed for nuclear multifragmentation [Bondorf (1995), Botvina (1995), Mekijan (1978), Randrup (1981), Hahn (1988), Gross (1990), Das (1993)]. The widely used statistical models are the percolation model [Jakobsson (1982), Jakobsson (1990), Li (1994)], Berlin multifragmentation [Jakobsson (1982), Jakobsson (1990), Sangster (1992), Colonna (1992)], statistical multifragmentation model [Randrup (1981), Hahn (1988)], lattice gas model [Souza (1994)], quantum statistical model (QSM), etc. In nuclear physics, the support for the statistical models came from the idea of compound nucleus (given by Niels Bohr in 1936) [Bohr (1936)], the Fong statistical fission model [Fong (1953)], the Landau theory of multifragmentation particle production [Landau (1953)], and Weiskopt evaporation models [Weisskopf (1937)]. The major limitations of statistical models are that they have limited predictive power and do not give time evolution of a reaction and have put forward the demand of dynamical models. Various dynamical approaches have been proposed keeping in view the consideration of mean field and nucleon–nucleon cross section on equal account for better understanding of intermediate energy heavy-ion collisions. The names of some of the dynamical approaches used in recent years are the Boltzman–Uehling–Uhlenback (BUU) model [Bertsch (1984), Aichelin (1985)], classical molecular dynamics (CMD) model [Wilets (1978), Bodmer (1980)], quantum molecular dynamics model [Aichelin (1991)], extended quantum molecular dynamics (EQMD) model [Maruyama (1996)], binding quantum molecular dynamics (BQMD) model [Maruyama (1996)], improved quantum molecular dynamics (ImQMD) model [Wang (2002)], antisymmeterized molecular dynamics (AMD) model [Ono (2004)], constrained molecular dynamics (CoMD) approach [Papa (2005)], fermionic molecular dynamics (FMD) model [Feldmeier (1997)], temperature-dependent quantum molecular dynamics (TMD) model [Puri (1994)], Pauli quantum molecular dynamics (PQMD) model [Peilert (1992)], relativistic quantum molecular dynamics (RQMD) model [Lehmann (1995)], ultra relativistic quantum molecular dynamics (UrQMD) model [Bass (1998), Abdel (2004)], modified quantum molecular dynamics (MQMD) model, and G-matrix quantum molecular dynamics (GQMD) model.

Some models have been upgraded to incorporate isospin degree of freedom to pin down the equation of state of neutron-rich matter along with the density dependence of nuclear symmetrical energy.

1.4 Nuclear Multifragmentation and Liquid–Gas Phase Transition

Due to the van der Waals behavior of nucleon–nucleon interaction, nuclear matter has been tentatively related with a liquid–gas phase transition [Bertsch (1983)]. At extremely different thermodynamical conditions, the nucleon–nucleon collisions at different incident energies (intermediate and relativistic energies) excite finite nuclear matter and subsequently, the excited nuclear matter disintegrates into intermediate mass fragments (IMFs) and hence signaling a liquid–gas phase transition. At higher excitation energies, the excited nuclear matter disintegrates into single nucleons [Bertsch (1983), Borderie (2008), Vient (2016)], and this phase is called nuclear vaporization. To investigate the phase diagram of a substance, one should have to probe the location of critical point and various critical exponents. The phenomenon of phase transition has been examined both experimentally and theoretically. Many experimental groups have observed signals of phase transition for system with different sizes and excitation energies. For example, the first effort to note this signal was carried out in the Fermi-Lab Purdue experiment [Finnet (1982)]. The research group of Bologna studied Fisher scaling critical exponents [Agostino (2003)] and also examined the occurrence of bimodal distribution [Bruno (2006)]. On the other hand, the INDRA collaboration predicted many signatures of phase transition for central collisions between Ni and Ni, Ni and Au, and Xe and Sn [Borderie (2006), Gulminelli (2002)]. A strong argument in favor of the first order liquid–gas phase transition was obtained with the evidence of experimental observation of spinodal decomposition in nuclear multifragmentation [Borderie (2001)]. Moreover, microcanonical heat capacity was experimentally observed in multifragmentation [Neindre (1999)], and it relates to liquid–gas phase transition [Chomaz (2000)]. In addition to the above studies, there are several other experimental studies performed to extract signatures of liquid–gas phase transition [Borderie (2008), Li (1994), Williams (1997), Agostino (2005)].

Meanwhile, several theoretical models such as statistical and dynamical models, percolation model, Fisher droplet model, lattice gas model, etc. have been developed to diagnose the nuclear liquid–gas phase transition [Borderie (2008), Li (1994), Richert (2001), Campi (1984), Stauffer (1994), Campi (2000), Berkenbusch (2002), Fisher (1967), Elliott (2002), Sood (2019), Sood (2021b), Ma (2005), Sharma (2016), Sharma (2016b)]. More recently, the theory of universal fluctuations [Gulminelli (2005), Botet (2001), Frankland (2005)] has been employed to understand critical behavior in finite systems.

1.5 Summary and Outlook

In conclusion of this introductory chapter of the present review volume, a few remarks are necessary. Undoubtedly, with the continuous advancements in the radioactive ion beam facilities (RIBs) around the world, a remarkable success has been achieved to minimize the secrets of the multifragmentation phenomenon at intermediate energy heavy-ion collisions. But still many exciting and challenging problems need to be solved. For example, the form of density dependence of the symmetry energy at supra-saturation density region is still uncertain. It is well known that several observables have been proposed and reported in the literature to pin down the density dependence of symmetry energy at sub-saturation and supra-saturation density regions, and in this direction, some success has been met to constrain the density dependence of symmetry energy at sub-saturation density region. The fast-growing field of isospin physics provides the opportunity to understand the high-density behavior of symmetry energy and equation of state of isospin asymmetric nuclear matter. On the other hand, a complete understanding of the phenomenon of liquid–gas phase transition is a challenging problem. The question of criticality remains elusive. It is worth mentioning that in the recent years, with enormous efforts, tremendous progress has been reported in the literature to understand multifragmentation and its relationship with the liquid–gas phase transition. However, putting forward a coherent picture of liquid–gas phase transition is still subject to debate.

In spite of all the challenges discussed in this chapter, there are great opportunities for the nuclear physics community to address all issues, especially with the continuous development and improvement in radioactive ion beam facilities. The present edited volume just gives review of some progress achieved in the perspective of various challenges mentioned above. The different articles reviewed in this volume are different in nature, reflecting the preferences of authors of the articles. Each chapter of the volume contributes most recent scientific knowledge in the field of heavy-ion collisions. We sincerely hope that to meet out the new challenges in this field and to progress further, the practice of writing present review volumes is very essential.

References

Abdel-Waged, K. (2004). Influence of medium modified angular distributions and mean field on the quasielastic and inelastic peaks of neutrons induced by 1.2 GeV protons on targets with $27 \leq A \leq 232$, *Phys. Rev. C* **70**, pp. 014605.

Aichelin, J. (1991). 'Quantum' molecular dynamics: A dynamical microscopic n-body approach to investigate fragment formation and the nuclear equation of state in heavy ion collisions, *Phys. Rep.* **202**, pp. 233–360.

Aichelin, J. and Bertsch, G. (1985). Numerical simulation of medium energy heavy ion reactions, *Phys. Rev. C* **31**, pp. 1730.

Baran, V. *et al.* (2020). Isospin effects in nuclear fragmentation, *Phys. Rep.* **703**, pp. 603.

Baran, V., Colonna, M., Greco, V., and Toro, M. D. (2005). Reaction dynamics with exotic nuclei, *Phys. Rep.* **410**, pp. 335.

Bass, S. A. *et al.* (1998). Microscopic models for ultrarelativistic heavy ion collisions, *Prog. Part. Nucl. Phys.* **41**, pp. 255.

Begemann-Blaich, M. *et al.* (1993). Quantum molecular dynamics simulation of multifragment production in heavy ion collisions at E/A=600 MeV, *Phys. Rev. C.* **48**, pp. 610.

Bertsch, G. and Siemens, P. J. (1983). Nuclear fragmentation, *Phys. Lett. B* **126**, pp. 9.

Bertsch, G. F., Kruse, H., and Gupta, S. D. (1984). Boltzmann equation for heavy ion collisions, *Phys. Rev. C* **29**, pp. R673.

Bodmer, A. R., Panos, C. N., and MacKellar, A. D. (1980). Classical-equations-of-motion calculations of high-energy heavy-ion collisions, *Nucl. Phys. A* **301**, pp. 359.

Bohnet, E. *et al.* (2009). Bimodal behavior of the heaviest fragment distribution in projectile fragmentation, *Phys. Rev. Lett.* **103**, pp. 072701.

Bohr, N. (1936). Neutron capture and nuclear constitution, *Nature* **137**, pp. 344.

Bollen, G. (2010). FRIB facility for rare isotope beams, *AIP Conf. Proc.* **1224**, pp. 432.

Bondorf, J. P. (1976). Heavy ion reactions between 30 and a few hundred MeV/nucleon, *Journal de Physique* **37**, pp. 5.

Bondorf, J. P. (1976). Proceeding of the EPS topical conference on large amplitude collective nuclear motions, Keszthely, Hungary, June 1979.

Bondrof, J. P., Botvina, A. S., Iljinov, A. S., Mishustin, I. N., and Sneppen, K. (1995). Statistical multifragmentation of nuclei, *Phys. Rep.* **257**, pp. 133–221.

Borderie, B. and Rivet, M. F. (2008). Nuclear multifragmentation and phase transition for hot nuclei, *Phys. Rep.* **61**, pp. 551–601.

Borderie, B. and Rivet, M. F. (2008). Nuclear multifragmentation and phase transition for hot nuclei, *Prog. Part. Nucl. Phys.* **61**, pp. 551.

Borderie, B. *et al.* (2001). Evidence for spinodal decomposition in nuclear multifragmentation, *Phys. Rev. Lett.* **86**, pp. 3252.

Borderie, B. *et al.* (2006). Many-fragment correlations and possible signature of spinodal fragmentation. In: *Dynamics and Thermodynamics with Nuclear Degrees of Freedom*, Chomaz, P. *et al.* (Eds.), Springer, Berlin, Heidelberg, pp. 243.

Botet, R. *et al.* (2001). Universal fluctuations in heavy-ion collisions in the Fermi energy domain, *Phys. Rev. Lett.* **86**, pp. 3514.

Botvina, A. S. *et al.* (1995). The effect of large angular momenta on multifragmentation of hot nuclei, *Nucl. Phys. A* **584**, pp. 737.

Bowman, D. R. *et al.* (1992). Equilibrium and non-equilibrium complex fragment emission in 50–100 MeV/u ^{139}La+ ^{12}C reactions, *Nucl. Phys. A* **523**, pp. 386.

Brown, B. A. (2000). Neutron radii in nuclei and the neutron equation of state, *Phys. Rep. Lett.* **85**, pp. 5296.

Bruno, M. *et al.* (2006). Bimodal pattern in the fragmentation of Au quasi-projectiles, *nucl-ex/0612030*.

Campi, X. *et al.* (1984). Sum rules study and a scaling property of fragmentation mass yield curves, *Phys. Lett. B* **138**, pp. 353.

Campi, X. *et al.* (2000). Analyzing fragmentation of simple fluids with percolation theory, *Eur. Phys. J. D* **11**, pp. 233.

CBM Physics Book, Springer, 2011; http://www.gsi.de/fair/experiment/CBM/PhysicsBook.html.

Chen, L. W., Ko, C. M., and Li, B. A. (2005). Determination of the stiffness of the nuclear symmetry energy from isospin diffusion, *Phys. Rep. Lett.* **94**, pp. 032701.

Chomaz, Ph, Duflot, V., and Gulminelli, F. (2000). Caloric curves and energy fluctuations in the microcanonical liquid-gas phase transition, *Phys. Rev. Lett.* **85**, pp. 3587.

Colonna, M. *et al.* (1994). Dynamics and statistics in multifragment production, *Phys. Lett. B* **283**, pp. 180.

Colonna, M., Baran, V., and Toro, M. D. (2014). Theoretical predictions of experimental observables sensitive to the symmetry energy, *Eur. Phys. J. A.* **50**, pp. 30.

D'Agostino, M. *et al.* (2000). Negative heat capacity in the critical region of nuclear fragmentation: An experimental evidence of the liquid-gas phase transition, *Phys. Rev. B.* **473**, pp. 219.

D'Agostino, M. *et al.* (2003). Critical-like behaviours in central and peripheral collisions: A comparative analysis, *Nucl. Phys.* **724**, pp. 455.

D'Agostino, M. *et al.* (2005). Nuclear liquid–gas phase transition: Experimental signals, *Nucl. Phys.* **749**, pp. 55.

Dai *et al.* (2014). Triton/^{3}He ratio as an observable for neutron-skin thickness, *Phys. Rev. C* **89**, pp. 014613.

Danielewicz, P., Lacey, R., and Lynch, W. G. (2002). Determination of the equation of state of dense matter, *Science* **298**, pp. 1592.

Das, C. B., Das, Mishra, A., Satpathy, M., and Satpathy, L. (1993). Statistical simultaneous multifragmentation model for heavy ion collisions with entrance channel characteristics, *J. Phys. G: Nucl. Part. Phys.* **19**, pp. 319.

Elliott, J. B. *et al.* (2002). Liquid to vapor phase transition in excited nuclei, *Phys. Rev. Lett.* **88**, pp. 04270.

Feldmeier, H. and Schnack, J. (1997). Fermionic molecular dynamics and short range correlations, *Prog. Part. Nucl. Phys.* **39**, pp. 393.

Finnet, J. E. *et al.* (1982). Nuclear fragment mass yields from high-energy proton-nucleus interactions, *Phys. Rev. Lett.* **49**, pp. 1321.

Fisher, M. E. (1967). The theory of condensation and the critical point, *Physics* **3**, pp. 255.

Fong, P. (1953). Neutron capture and nuclear constitution, *Phys. Rev* **89**, pp. 332.

Frankland, J. D. *et al.* (2005). Model-independent tracking of criticality signals in nuclear multifragmentation data, *Phys. Rev. C* **71**, pp. 034607.

Frankland, J. D. *et al.* (INDRA and ALADiN collaborations) (2005). Model-independent tracking of criticality signals in nuclear multifragmentation data, *Phys. Rev. C.* **71**, pp. 034607.

Gross, D. H. E. (1990). Statistical decay of very hot nuclei-the production of large clusters, *Prog. Rep. Phys.* **53**, pp. 605.

Gulminelli, F. *et al.* (2005). Distribution of the largest fragment in the lattice gas model, *Phys. Rev. C* **71**, pp. 054607.

Gulminelli, F. *et al.* (2006). Challenges in nuclear dynamics and thermodynamics. In: *Dynamics and Thermodynamics with Nuclear Degrees of Freedom*, Chomaz, P. *et al.* (Eds.), Springer, Berlin, Heidelberg, pp. 1–3.

Hahn, D. and Stocker, H. (1988). The quantum statistical model of fragment formation: Entropy and temperature extraction in heavy ion collisions, *Nucl. Phys. A* **476**, pp. 718.

http://www.gsi.de/fair/ and related sites therein.

Hubele, J. *et al.* (1992). Statistical fragmentation of Au projectiles at E/A= 600 MeV, *Phys. Rev. C.* **46**, pp. R1577.

Jakobsson, B. *et al.* (1982). The disintegration of nuclei in violent heavy ion interactions at 55AMeV-110AMeV, *Z. Phys. A.* **307**, pp. 293.

Jakobsson, B. *et al.* (1990). On the observation of a transition from fusion to multifragmentation in high multiplicity ^{16}O induced reactions, *Nucl. Phys. A* **509**, pp. 195.

Kleine Berkenbusch, M. *et al.* (2002). Event-by-event analysis of proton-induced nuclear multifragmentation: Determination of the phase transition universality class in a system with extreme finite-size constraints, *Phys. Rev. Lett.* **88**, pp. 022701.

Lambrecht, D. *et al.* (1994). Energy dependence of collective flow of neutrons and protons in ^{197}Au+^{197}Au collisions, *Z. Phys. A* **350**, pp. 115.

Landau, L. D. and Akad, I. (1953). On the multiple production of particles in high energy collisions, *Nauk SSSR, Ser. Fiz.* **17**, pp. 51.

Lattimer, J. M. and Prakash, M. (2004). The physics of neutron stars, *Science* **304**, pp. 536.

Lehmann, E. *et al.* (1995). Consequences of a covariant description of heavy-ion reactions at intermediate energies, *Phys. Rev. C* **51**, pp. 2113.

Leifels, Y. *et al.* (1993). Exclusive studies of neutron and charged particle emission in collisions of Au at 400 MeV/nucleon, *Phys. Rev. Lett.* **71**, pp. 963.

Li, B. A. and Ko, C. M. (1997). Equation of state of asymmetric nuclear matter and collisions of neutron-rich nuclei, *Phys. Rep. Lett.* **78**, pp. 1664.

Li, B. A. and Udo Schröder, W. (Eds.) (2001). *Isospin Physics in Heavy-Ion Collisions at Intermediate Energies*, Nova Science Publishers, Inc., New York, 2001.

Li, B. A., Chen, L. W., and Ko, C. M. (2008). Recent progress and new challenges in isospin physics with heavy-ion reactions, *Phys. Rep.* **464**, pp. 113.

Li, B. A., Ko, C. M., and Bauer, W. (1998). Isospin physics in heavy-ion collisions at intermediate energies, *Internat. J. Modern. Phys. E* **7**, pp. 147.

Li, B. A., Krastev, P. G., Wen, D. H., and Zhang, N. B. (2019). Towards understanding astrophysical effects of nuclear symmetry energy, *Eur. Phys. J. A.* **55**, pp. 117.

Li, T. *et al.* (1994). Mass dependence of critical behavior in nucleus-nucleus collisions, *Phys. Rev. C* **49**, pp. 1630.

Li, T. *et al.* (2007). Incompressibility of nuclear matter from the giant monopole resonance, *Phys. Rep. Lett.* **82**, pp. 691.

Liu, T. X. *et al.* (2007). Isospin diffusion observables in heavy-ion reactions, *Phys. Rev. C* **76**, pp. 034603.

Llope, W. J. *et al.* (1995). Autocorrelations and intermediate-mass-fragment multiplicities in central heavy-ion collisions, *Phys. Rev. C* **51**, pp. 1325.

Lopez-Quelle, M. *et al.* (2012). Asymmetric nuclear matter in the relativistic approach, *Nucl. Phys. A* **703**, pp. 603.

Loveland, W. *et al.* (1999). Heavy residue production in the interaction of Pb and Au 29 MeV/nucleon with, *Phys. Rev. C* **59**, pp. 1472.

Ma, Y. G. *et al.* (2004). Evidence of critical behavior in the disassembly of nuclei with A $\sim$ 36, *Phys. Rev. C.* **69**, pp. 031604.

Ma, Y. G. *et al.* (2005). Critical behavior in light nuclear systems: Experimental aspects, *Phys. Rev. C* **71**, pp. 054606.

Marie, N. *et al.* (2007). Experimental determination of fragment excitation energies in multifragmentation events, *Phys. Rev. C* **58**, pp. 256.

Maruyama, T., Niita, K., and Iwamoto, A. (1996). Extension of quantum molecular dynamics and its application to heavy-ion collisions, *Phys. Rev. C* **53**, pp. 297.

Mekijan, A. Z. (1978). Explosive nucleosynthesis, equilibrium thermodynamics, and relativistic heavy-ion collisions, *Phys. Rev. C* **17**, pp. 1051.

Moretto, L. G., Delis, D. N., and Wozniak, G. J. (1993). Experimental signature for statistical multifragmentation, *Phys. Rev. Lett.* **71**, pp. 3935.

Neindre N, Le, (1999). PhD Dissertation Universite de Caen, France.

Ogilvie, C. A. *et al.* (1991). Rise and fall of multifragment emission, *Phys. Rev. lett.* **67**, pp. 1214.

Ono, A. and Horiuchi, H. (2004). Antisymmetrized molecular dynamics for heavy ion collisions, *Prog. Part. Nucl. Phys.* **53**, pp. 501.

Papa, M., Giuliani, G., and Bonasera, A. (2005). Constrained molecular dynamics II: An N-body approach to nuclear systems, *J. Comp. Phys.* **208**, pp. 403.

Peaslee, G. F. *et al.* (1994). Energy dependence of multifragmentation in ^{84}Kr + ^{197}Au collisions, *Phys. Rev. C.* **49**, pp. R2271.

Peilert, G. *et al.* (1994). Dynamical treatment of Fermi motion in a microscopic description of heavy ion collisions, *Phys. Rev. C* **46**, pp. 1457.

Phair, L. *et al.* (1995). Reducibility and thermal scaling of charge distributions in multifragmentation, *Phys. Rev. Lett.* **75**, pp. 213.

Puri, R. K. *et al.* (1994). Temperature-dependent mean field and its effect on heavy-ion reactions, *Nucl. Phys. A* **575**, pp. 733.

Randrup, J. and Koonin, S. E. (1981). The disassembly of nuclear matter, *Nucl. Phys. A* **356**, pp. 223.

Reisdorf, W. *et al.* (2012). Systematics of azimuthal asymmetries in heavy ion collisions in the 1AGeV regime, *Nucl. Phys. A* **876**, pp. 1.

Richert, J. and Wagner, P. (2001). Microscopic model approaches to fragmentation of nuclei and phase transitions in nuclear matter, *Phys. Rep.* **350**, pp. 1.

Sakurai, H. (2010). RIKEN radioactive isotope beam factory project: Present status and perspectives, *Pram. J. Phys.* **75**, pp. 369.

Sangster, T. C. *et al.* (1992). Intermediate mass fragment emission in Fe+ Au collisions, *Phys. Rev. C* **46**, pp. 1404.

Schttauf, A. *et al.* (1996). Universality of spectator fragmentation at relativistic bombarding energies, *Nucl. Phys. A* **607**, pp. 457.

Sharma, A. *et al.* (2016). Isospin effects via Coulomb forces on the onset of multifragmentation in light and heavily charged systems, *Eur. Phys. J. A* **52**, pp. 42.

Sharma, A. *et al.* (2016). Multifragmentation of nearly symmetric and asymmetric reactions within a dynamical model, *Nucl. Phys. A* **945**, pp. 95–111.

Shetty, D. V., Souliotis, G. A., Galanopoulos, S., and Yennello, S. J. (2009). Effective nucleon mass and the nuclear caloric curve, *Phys. Rev. C.* **79**, pp. 034603.

Sood, S., Kumar, R., Sharma, A., and Puri, R. K. (2019). Cluster formation and phase transition in nuclear disassembly using a variety of clusterization algorithms, *Phys. Rev. C* **99**, pp. 054612.

Sood, S., Kumar, R., Sharma, A., Gautam, S., and Puri, R. K. (2021). Fragment emission and critical behavior in light and heavy charged systems, *Chin. Phys. C* **45**, 014101.

Souza de, R. T. *et al.* (2006). Detection. In: Chomaz, P. *et al.* (Eds.) *Dynamics and Thermodynamics with Nuclear Degrees of Freedom*, Springer, Berlin, Heidelberg, pp. 275–291.

Souza, S. R. *et al.* (1994). A dynamical model for multifragmentation of nuclei, *Nucl. Phys. A* **571**, pp. 159.

Stauffer, D. *et al.* (1994). *Introduction to Percolation Theory*, Taylor And Francis Publishers, London.

Steiner, A. W., Prakash, M., Lattimer, J. M., and Ellis, P. J. (2005). Isospin asymmetry in nuclei and neutron stars, *Phys. Rep.* **410**, pp. 335.

Stone, N. T. B., Llope, W. J., and Westfall, G. D. (1995). Transitions in nuclear disassembly as viewed by fragment charge correlations, *Phys. Rev. C* **51**, pp. 3157.

Tsang, M. B. *et al.* (2007). Fragmentation cross sections and binding energies of neutron-rich nuclei, *Phys. Rev. C* **76**, pp. R041302.

Tsang, M. B. *et al.* (2012). Constraints on the symmetry energy and neutron skins from experiments and theory, *Phys. Rev. C* **86**, pp. 015803.

Venema, L. *et al.* (1993). Azimuthal asymmetry of neutral pion emission in Au+ Au reactions at 1 GeV/nucleon, *Phys. Rev. Lett.* **71**, pp. 835.

Verde, G. (2014). Probing the nuclear equation-of-state and the symmetry energy with heavy-ion collisions, *EPJ Web of Conferences* **66**, pp. 01018.

Vient, E. *et al.* (2016). Is the caloric curve a robust signal of the phase transition in hot nuclei? *IL NUOVO CIMENTO:IWM-EC 2016* **39**, pp. 406.

Wada, R. *et al.* (2004). Reaction dynamics and multifragmentation in Fermi energy heavy ion reactions, *Phys. Rev. C.* **69**, pp. 044610.

Wang, N., Li, Z., and Wu, X. (2002). Improved quantum molecular dynamics model and its applications to fusion reaction near barrier, *Phys. Rev. C* **65**, pp. 064608.

Weisskopf, V. (1937). Statistics and nuclear reactions, *Phys. Rev* **52**, pp. 295.

Wilets, L., Yariv, Y., and Chestnut, R. (1978). Classical many-body model for heavy-ion collisions, *Nucl. Phys. A* **301**, pp. 359.

Williams, C. *et al.* (1997). Fragment distributions for highly charged systems, *Phys. Rev. C* **55**, pp. R2132.

Xia, J. W. *et al.* (1998). HIRFL STATUS and HIRFL-CSR project in Lanzhou, rerspectives, *Proceedings of the First Asian Particle Accelerators Conference, KEK, Japan* pp. 342.

Yano, Y. (2007). The RIKEN RI Beam Factory Project: A status report - INSPIRE, *Instrum. Methods Phys. Res. B* **261**, pp. 1009.

Youngblood, D. H., Clark, H. L., and Lui, Y. W. (1999). Isotopic dependence of the giant monopole resonance in the even-A ^{112,124}Sn isotopes and the asymmetry term in nuclear incompressibility, *Phys. Rep. Lett.* **99**, pp. 162503.

Zhang, Y. *et al.* (2008). The influence of cluster emission and the symmetry energy on neutron-proton spectral double ratios, *Phys. Lett. B* **664**, pp. 145.

Chapter 2

Statistical Multifragmentation in Heavy-Ion Reactions: Theory and Experiments

R. Ogul,[a] **N. Buyukcizmeci,**[a] **and A. S. Botvina**[b,c]

[a] *Selcuk University, Department of Physics, 42075 Konya, Turkey*

[b] *FIAS and ITP J.W. Goethe University, D-60438 Frankfurt am Main, Germany*

[c] *Institute for Nuclear Research, Russian Academy of Sciences, 117312 Moscow, Russia*

rogul@selcuk.edu.tr

2.1 Introduction

In this chapter, we shall discuss nuclear multifragmentation, which is a universal phenomenon occurring when a large amount of energy is deposited in a nucleus. It has been observed in nearly all types of high-energy nuclear reactions induced by hadrons, photons, and heavy ions (for reviews, see Refs. [1–3]). At low excitation energies, the produced nuclear system can be treated as a compound nucleus [4], which decays via evaporation of light particles or fission. However, at high excitation energy, possibly accompanied by compression during the initial dynamical stage of the reaction [5–7], the system will expand to subsaturation densities, thereby becoming

Multifragmentation in Heavy-Ion Reactions: Theory and Experiments
Edited by Rajeev K. Puri, Yu-Gang Ma, and Arun Sharma

ISBN 978-981-4968-69-0 (Hardcover), 978-1-003-38513-4 (eBook)
www.jennystanford.com

unstable, and will break up into many fragments. Multifragmentation is pronounced in case at least three intermediate mass fragments (IMFs) with $Z \geq 3$ are produced in one event.

Statistical models have proved to be very successful in nuclear physics. They are used for the description of nuclear decay when an equilibrated source can be identified in the reaction. The most famous example of such a source is the "compound nucleus" introduced by Niels Bohr in 1936 [4]. It was clearly seen in low-energy nuclear reactions leading to excitation energies of a few tens of MeV. It is remarkable that this concept works also for nuclear reactions induced by particles and ions of intermediate and high energies, when the nuclei break up into many fragments (multifragmentation). According to the statistical hypothesis, initial dynamical interactions between nucleons lead to the re-distribution of the available energy among many degrees of freedom, and the nuclear system evolves toward equilibrium. In the most general consideration, the process may be subdivided into several stages: (1) a dynamical stage leading to the formation of equilibrated nuclear system, (2) disassembly of the system into individual primary fragments, and (3) deexcitation of hot primary fragments. The dynamical models agree in that the character of the dynamical evolution changes after a few rescatterings of incident nucleons, when high-energy particles ("participants") leave the system. However, the time needed for equilibration and transition to the statistical description is uncertain and model dependent [8]. It is estimated to be around 100 fm/c for spectator matter, but the required parameters, i.e., the excitation energies, mass numbers, and charges of the predicted equilibrated sources, vary significantly even over longer times. To evade these ambiguities and other uncertainties related to the coupling with a particular dynamical model, alternative strategies have been developed [8–15] and are also followed here.

With the results of dynamical simulations as qualitative guidelines, the exact parameters of the thermalized sources are obtained from the analysis of the experimental fragmentation data. Usually, only a subgroup of observables is required for the determination of the source parameters (see, e.g., Refs. [9, 15]). The success of the statistical description is then judged from the quality of predictions for other observables and for the correlations between them.

2.2 Statistical Multifragmentation Model

As the model for multifragmentation, describing the reaction stages (2) and (3) defined above, the statistical multifragmentation model (SMM) is used (for a review see Ref. [2]). The SMM assumes statistical equilibrium of the excited nuclear system with mass number A_s, charge Z_s, and excitation energy E_x (above the ground state) within a low-density freeze-out volume. All breakup channels (partitions $\{p\}$) composed of nucleons and excited fragments are considered, and the conservations of baryon number, electric charge, and energy are taken into account. Besides the breakup channels, also the compound-nucleus channels are included, and the competition between all channels is permitted. In this way, the SMM covers the conventional evaporation and fission processes occurring at low excitation energy as well as the transition region between the low- and high-energy deexcitation regimes. In the thermodynamic limit, as demonstrated in Refs. [16–18], the SMM is consistent with the nuclear liquid–gas phase transition when the liquid phase is represented by an infinite nuclear cluster.

In the model, light nuclei with mass number $A \leq 4$ and charge $Z \leq 2$ are treated as elementary stable particles with masses and spins taken from the nuclear tables ("nuclear gas"). Only translational degrees of freedom of these particles contribute to the entropy of the system. Fragments with $A > 4$ are treated as heated nuclear liquid drops. Their individual free energies F_{AZ} are parametrized as a sum of the bulk, surface, Coulomb, and symmetry energy contributions

$$F_{AZ} = F^B_{AZ} + F^S_{AZ} + E^C_{AZ} + E^{sym}_{AZ} \qquad (2.1)$$

The standard expressions for these terms are $F^B_{AZ} = (-W_0 - T^2/\epsilon_0)A$, where T is the temperature, the parameter ϵ_0 is related to the level density, and $W_0 = 16$ MeV is the binding energy of infinite nuclear matter; $F^S_{AZ} = B_0 A^{2/3}((T_c^2 - T^2)/(T_c^2 + T^2))^{5/4}$, where $B_0 = 18$ MeV is the surface energy coefficient and $T_c = 18$ MeV is the critical temperature of infinite nuclear matter; $E^C_{AZ} = cZ^2/A^{1/3}$, where $c = (3/5)(e^2/r_0)(1 - (\rho/\rho_0)^{1/3})$ is the Coulomb parameter (obtained in the Wigner–Seitz approximation) with the charge unit e and

$r_0 = 1.17$ fm; $E_{AZ}^{sym} = \gamma(A-2Z)^2/A$, where $\gamma = 25$ MeV is the symmetry energy parameter. These parameters are those of the Bethe–Weizsäcker formula and correspond to the assumption of isolated fragments with normal density in the freeze-out configuration, an assumption found to be quite successful in many applications.

Within the microcanonical treatment [2, 19], the statistical weight of a partition p is calculated as

$$W_p \propto exp\, S_p, \tag{2.2}$$

where S_p is the corresponding entropy, depending on the fragments in this partition as well as on the parameters of the system.

In the grand-canonical treatment of the SMM [20], after integrating out translational degrees of freedom, the mean multiplicity of nuclear fragments with A and Z can be written as

$$\langle N_{AZ}\rangle = g_{AZ}\frac{V_f}{\lambda_T^3}A^{3/2}\exp\left[-\frac{1}{T}\left(F_{AZ}-\mu A-\nu Z\right)\right]. \tag{2.3}$$

Here g_{AZ} is the ground-state degeneracy factor of species (A, Z), $\lambda_T = \left(2\pi\hbar^2/m_N T\right)^{1/2}$ is the nucleon thermal wavelength, and $m_N \approx 939$ MeV/c^2 is the average nucleon mass. V_f is the free volume available for the translational motion of fragments. The free energy $F_{AZ} = F_{AZ}(T, \rho)$ of fragments with $A > 4$ and $Z > 2$, parametrized as given above, is a function of the temperature and density. The chemical potentials μ and ν are found from the mass and charge constraints:

$$\sum_{(A,Z)}\langle N_{AZ}\rangle A = A_s, \quad \sum_{(A,Z)}\langle N_{AZ}\rangle Z = Z_s \tag{2.4}$$

For the freeze-out density, one-third the normal nuclear density is assumed. This is a standard value, used previously in many successful applications and consistent with independent experimental determinations of the freeze-out density [21, 22]. The average density that corresponds to the freeze-out volume is usually taken in the range between 1/3 and 1/10 of the normal nuclear density $\rho_0 \approx 0.15$ fm^{-3}. In the case of thermal multifragmentation, the freeze-out density can be reliably estimated from experimental data on fragment velocities since they (to 80–90%) are determined by the Coulomb acceleration after the breakup. The experimental analyses of the kinetic energies, angle- and velocity-correlations of the fragments indeed point to values of (0.1–0.4) ρ_0 (Refs. [21, 22]).

As was established experimentally, an "ideal" picture of thermal multifragmentation begins to fail at excitation energies of about 5–6 MeV/nucleon [23]. At higher excitations, a part of the energy goes into a collective kinetic energy of the produced fragments, without thermalization. This energy is defined as the flow energy, and its share depends on the kind of reaction. For example, at the thermal excitation energy of $E_x \approx 6$ AMeV, the additional flow energy is around 0.2 AMeV in hadron-induced reactions, and it is around 1.0 AMeV in central heavy-ion collisions around Fermi energy. Since a dynamical flow itself can break matter into pieces, it is necessary to understand the limits of the statistical description in the case of a strong flow. This problem was addressed in the number of works within dynamical and lattice-gas models. Their conclusion is that a flow does not change statistical model predictions, if its energy is essentially smaller than the thermal energy. This justifies a receipt often used in statistical models, when the flow energy is included by increasing the velocities of fragments in the freeze-out volume according to the flow velocity profile [2]. This is in agreement with many experimental analyses. However, statistical models work surprisingly well even when the flow energy is comparable with the thermal energy, or even higher [24, 15]. This observation tells us about the phase space domination for the fragment formation in these reactions.

We especially stress two main achievements of statistical models in theory of nuclear reactions: First, a clear understanding has been reached that sequential decay via compound nucleus must give a way to nearly simultaneous breakup of nuclei at high excitation energies; and second, the character of this change can be interpreted as a liquid–gas type phase transition in finite nuclear systems. The results obtained in the nuclear multifragmentation studies can be applied in several other fields. First, the mathematical methods of the statistical multifragmentation can be used for developing thermodynamics of finite systems [11, 25]. These studies were stimulated by recent observation of extremely large fluctuations of energy of produced fragments, which can be interpreted as the negative heat capacity [26]. A very important advantage of the statistical approach to the cluster production is that the statistical equilibration is generally achieved in the astrophysical condition, and we can demonstrate the links for the nuclei production in both cases. As an illustration of this connection, we show in Fig. 2.1 examples of fragment mass distributions produced in multifragmentation reactions (see [27, 19]), and in astrophysical conditions associated with supernova type-II explosions. The calculations of nuclear composition of stellar matter at subnuclear densities were carried out with the SMM generalized for astrophysical conditions [28, 29]. One can see that the evolution of mass distributions with excitation energy is qualitatively the same for both the nuclear multifragmentation reactions and the supernova explosions. However, in the supernova environments, much heavier and neutron-rich nuclei can be produced because of screening of their charge by surrounding electrons. Here, we have

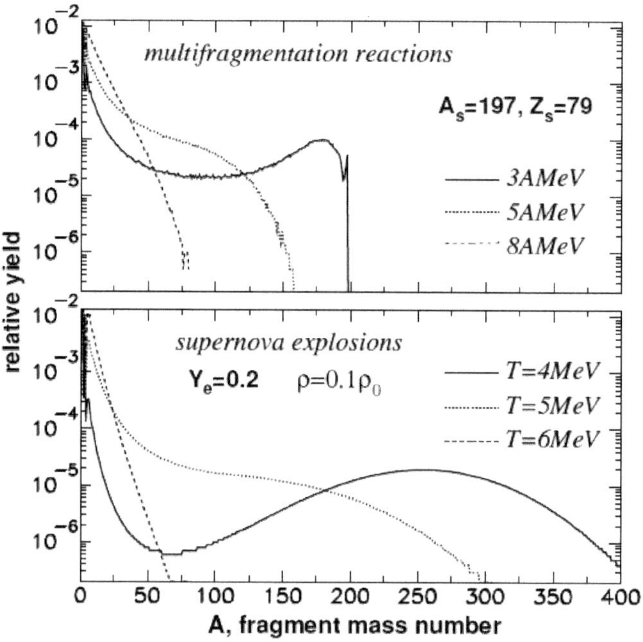

Figure 2.1 Mass distributions of hot fragments calculated with SMM: Top panel — for multifragmentation of Au^{197} nuclear sources at excitation energies of 3, 5, and 8 AMeV; and, bottom panel — for stellar matter at the baryon density ρ and electron fraction Y_e, typical for supernova explosions, at temperatures 4, 5, and 6 MeV.

applied the SMM to nuclear sources of different mass and isospin, which can easily be used in modern experiments. The excitation energy range is of $E_x = 2 - 20$ MeV/nucleon, and the freeze-out volume, where the intermediate mass fragments are located after expansion of the system, is $V = 3V_0$ ($V_0 = 4\pi A_0 r_0^3/3$ is the volume of a nucleus in its ground state). In other words, the fragment formation is described at a low density freeze-out ($\rho \approx \rho_0/3$), where the nuclear liquid and gas phases coexist. The SMM phase diagram has already been under intensive investigations (see, e.g., [17]) with the same parametrization of the surface tension of fragments versus the critical temperature T_c for the nuclear liquid–gas phase transition in infinite matter. Therefore, it is possible to extract T_c from the fragmentation data [30]. The symmetry energy contributes to the masses of fragments as well. Taking this contribution into account, we shall investigate the influence of the symmetry energy coefficient γ on fragment production in the multifragmentation of finite nuclei. In this section, we concentrate mainly on properties of hot fragments. However, their secondary deexcitation is important for final description of the data, which was included in SMM long ago, and its effects were already discussed somewhere else [2, 30, 12]. Here, we point out how the secondary deexcitation code can be modernized so that the change in the symmetry energy of nuclei during the evaporation is taken into account.

2.2.1 Isospin Dependence of Fragment Distributions

For our calculations, we consider Au^{197}, Sn^{124}, La^{124}, and Kr^{78} nuclei to see how isospin affects the multifragmentation phenomena with their neutron-to-proton ratios (N/Z), which read 1.49, 1.48, 1.18, and 1.17, respectively. The two sources Au^{197} and Sn^{124} have nearly the same N/Z ratios 1.49 and 1.48, respectively, but different mass numbers, whereas Sn^{124} and La^{124} have the same mass numbers, but very different N/Z ratios 1.48 and 1.18, respectively. To see the effect of the size of the nucleus in our study, we have also included Kr^{78}, which has the lowest mass number among the considered nuclei here, but its N/Z ratio 1.17 is very close to that of La^{124}. The relative yield of hot fragments produced after the breakup of Au^{197}, Sn^{124}, La^{124}, and Kr^{78} nuclei as a function of A/A_0 is given in Ref.[32] for $E_x = 3$, 4, 5, and 8 AMeV. As was shown in previous analysis of experimental data (see, e.g., [27, 31, 13]), these results are fully consistent with experimental observations. The effect of isospin of different sources in liquid–gas phase transition region can be clearly seen for different excitation energies. The mass distribution of fragments produced in the disintegration of various nuclei evolves with the excitation energy. For Au^{197} and Sn^{124} multifragmentation, onset is about 5 MeV/nucleon, and for La^{124} and Kr^{78}, it is about 4 AMeV at standard SMM parameters so that U-shape mass distributions disappear at these energies. We can also describe this evolution with the temperature (see the caloric curves below). At low temperatures ($T \leq 5$ MeV), there is a U-shape distribution corresponding to partitions with a few small fragments and one big residual fragment. At high temperatures ($T \geq 6$ MeV), the big fragments disappear and an exponential-like fall-off is observed. In the transition region ($T \simeq 5 - 6$ MeV), however, one observes a transition between these two regimes, which is rather smooth because of the finiteness of the systems. In Ref. [32], we demonstrate N/Z ratios of the hot fragments produced in the freeze-out volume of the same systems. One can see that the neutron content of the hot fragments is only slightly smaller than that of the whole system since most of the neutrons are still contained in fragments. It is also seen that the neutron richness of fragments is increasing with their mass number [19]. This is a general behavior for nuclear systems in equilibrium, where binding energy is influenced by interplay of the Coulomb and the symmetry energy. For some systems, one can observe specific isotopic effects such as the increase in the neutron richness of intermediate mass fragments (with Z=3–20) with the excitation energy (compare the results for 3 and 8 AMeV). This is because after removing heavy fragments by increasing the excitation energy, their neutrons are accumulated into the smaller fragments [19]. Usually a power-law fitting is performed with $Y(A) \propto A^{-\tau}$ and $Y(Z) \propto Z^{-\tau_Z}$, where $Y(A)$ and $Y(Z)$ denote the multifragmentation mass and charge yield as a function of fragment mass number A and charge number Z, respectively. The parameters τ (for mass distribution) and τ_Z (for charge distribution) can be considered in thermodynamical limit as critical exponents [33]. In calculations, we consider the fragments in the range $6 \leq A \leq 40$ for mass and $3 \leq Z \leq 18$ for charge number. The lighter fragments

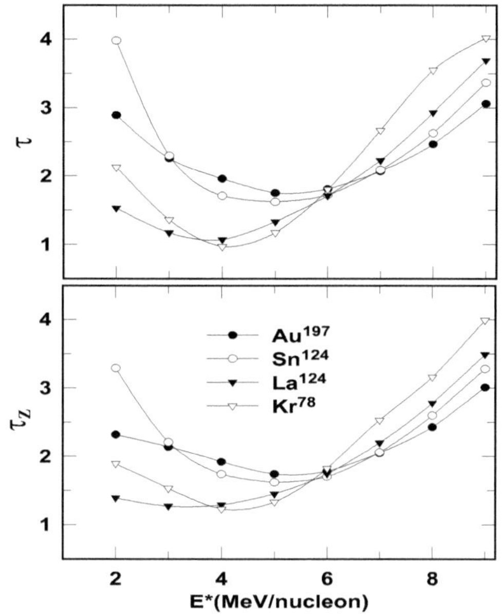

Figure 2.2 The values of critical exponents τ (top panel) and τ_Z (bottom panel), as a function of excitation energy E_x for Au^{197}, Sn^{124}, La^{124}, and Kr^{78}.

are considered as a nuclear gas. The obtained values of τ and τ_Z at $T_c = 18$ MeV as a function of E_x are given in Fig. 2.2 for cold fragmentation (with secondary deexcitation). There is a minimum at about $E_x = 5.3$, 5.2, 4, and 4.2 AMeV for Au^{197}, Sn^{124}, La^{124}, and Kr^{78}, respectively. The results for Au^{197} and Sn^{124} are very similar to each other and significantly different from those obtained for La^{124} and Kr^{78}. This means that intermediate mass fragment (IMF) distributions approximately scale with the size of the sources, and they depend on the neutron-to-proton ratios of the sources. This is because the symmetry energy still dominates over the Coulomb interaction energy for these intermediate-size sources. One may also see from these results that the lower N/Z ratio leads to smaller τ and τ_Z parameters with the flatter fragment distribution. A large N/Z ratio of the source favors the production of big clusters since they have a large isospin. This is the reason for the domination of partitions consisting of small IMFs with a big cluster in the transition

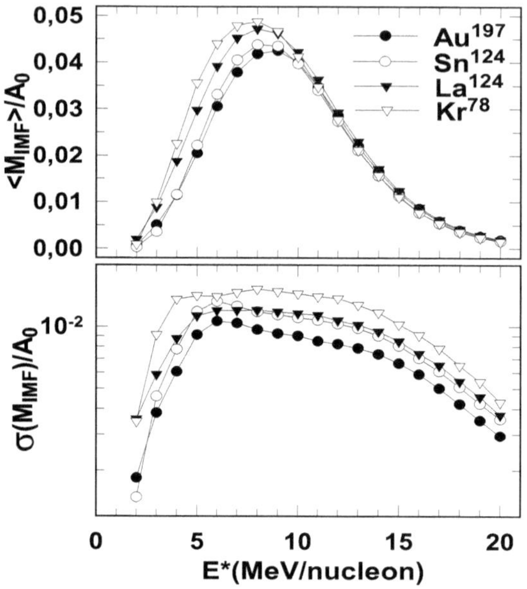

Figure 2.3 The average IMF multiplicity (top panel) and its variance (bottom panel) per nucleon versus E_x for various nuclei.

region (U-shape distribution). It is also seen from the bottom of Fig. 2.2 that the values of τ_Z are very close to the values of τ since the neutron-to-proton ratio of produced IMFs exhibits a small change within their narrow charge range (see also [13, 19]).

For completeness, we have calculated the average IMF multiplicity and its variance for the excitation energy range of 2–20 AMeV for all nuclei. To avoid the effect of the source size and leave only the isospin effect, IMF multiplicity is divided by A_0. In Fig. 2.3, we present the scaled average IMF multiplicity and its variance versus E_x. In this figure, one may clearly see an effect of the isospin and the source size on fragment multiplicities.

2.2.2 Largest Fragments in Fragment Partitions

In multifragmentation, the mass number of the largest fragment A_{max} may be used as an "order parameter" since it is directly related to the number of fragments produced during disintegration. We have analyzed its behavior with respect to the excitation energies. In order to remove the effect of source size and leave only an isospin effect, we have scaled it with the mass number of the source A_0. We have plotted A_{max}/A_0 as a function of excitation energy in Fig. 2.4 (top panel). The values of these quantities decrease first rapidly with E^* and then, beyond some value ($E^* \sim 10$ AMeV), decrease rather slowly. This suggests that the vaporization process becomes dominant around this point (see also an analysis in Ref. [33]). Another striking result of this behavior is that all curves almost coincide in the transition region as displayed on the top panel in Fig. 2.4 (i.e., a universality of the behavior of A_{max}). We have also calculated the average A_{max} variances in the multifragmentation stage for all nuclei at the same excitation energy range. The average variance of A_{max}/A_0 values is presented in the bottom panel of Fig. 2.4. We observe from this figure that the maximum variance values for all sources are seen at an excitation energy of 4–5 AMeV, which is exactly corresponding to the region of the transition from compound-like channels to the full multifragmentation. In the following paragraphs, we clarify why this kind of behavior takes place.

Besides the mean number of IMFs, the mean values of other charge correlations such as the relative asymmetries between the atomic numbers of the heaviest fragments and higher-order charge correlations can be reproduced as a function of bound charge Z_{bound} and maximum charge Z_{max}. These observables characterize the fragment formation in multifragmentation process in more details and can be described with SMM; see Refs. [2, 13]. Until now, there is no alternative description of such characteristics within dynamical fragment formation models.

2.2.3 Temperature and Bimodality

The temperature of fragments is an important ingredient for any description of nuclear multifragmentation. Besides statistical approaches, a temperature can even be introduced in dynamical (coalescence-like) processes of fragment production [24]. In SMM, the temperature of fragments T_f in separate partitions is defined from their energy balance according to the canonical prescription:

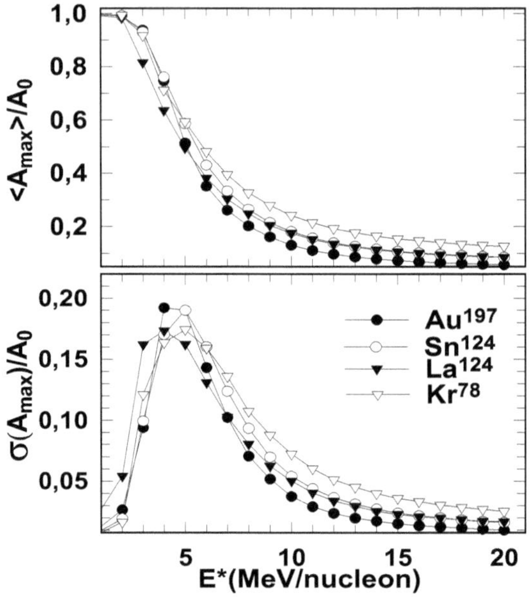

Figure 2.4 $< A_{max} >$ and its variance $\sigma(A_{max})$ divided by A_0 versus E^* for Au^{197}, Sn^{124}, La^{124}, and Kr^{78}.

$$m_0 + E^* = 1.5(n-1)T_f + \sum_i (m_i + E_i^*(T_f)) + E_{coul}, \tag{2.5}$$

where n is multiplicity of hot fragments and 'gas' particles, m_0 and m_i are ground masses of initial nucleus and the fragments, E_i^* are internal excitation energies of fragments, and E_{coul} is their Coulomb interaction energy. Masses and internal excitations are found according to their free energies (see Eq. (2.1) and Ref. [2]). The term with $(n-1)$ takes into account the center of mass constraint, which is important for finite systems [34]. In this approach, we allow for temperature to fluctuate from partition to partition, and we define the temperature of the system as the average one for the generated partitions. The methods involving fluctuations of temperature have been used in thermodynamics since long ago [35]. Our definition of temperature has another advantage: It is fully consistent with the

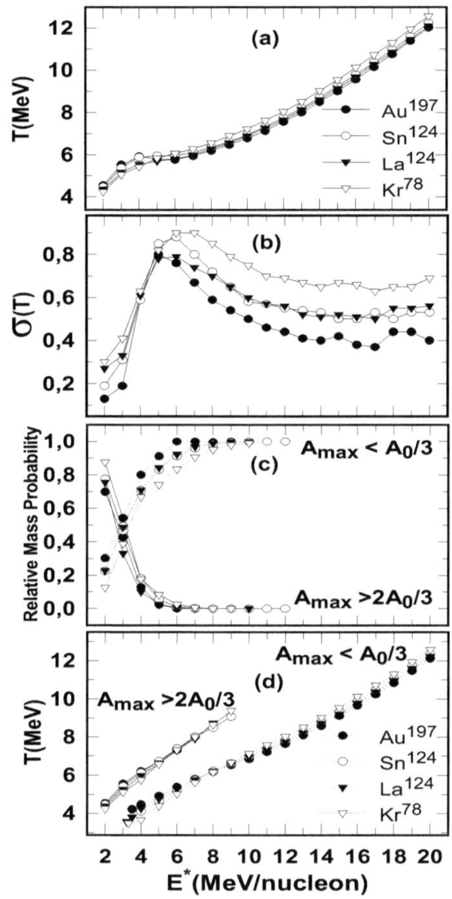

Figure 2.5 (a) Temperature, (b) its variance value, (c) probabilities of events with $A_{max} > 2A_0/3$ and $A_{max} < A_0/3$, and (d) average temperatures T versus E^* for Au^{197}, Sn^{124}, La^{124}, and Kr^{78}. The full and dotted lines in Fig. 2.5c and Fig. 2.5d represent the events with $A_{max} > 2A_0/3$ and $A_{max} < A_0/3$, respectively.

temperature commonly used in nuclear physics for isolated nuclei, which makes a natural connection with the physics of compound nucleus.

Figs. 2.5a and 2.5b show the results of the average temperature and its variance with respect to the excitation energy E_x for all nuclei under consideration. While one may see a plateau in the excitation energy range 3–7 AMeV in the upper panel of this figure, the variance of temperature exhibits a maximum value at about 5–6 MeV for standard SMM parameters, in agreement with the maximum fluctuations of A_{max}. The plateau-like caloric curve was reported long ago (e.g., [20, 2]). It was confirmed by experimental results [36], and by rather sophisticated calculations of Fermionic Molecular Dynamics [37] and Antisymmetrized Molecular Dynamics [38]. The reason for large fluctuations of the temperature and A_{max} can be clear from Figs. 2.5c and 2.5d. In these figures, we show relative mass probability and temperature values for two groups of fragment partitions, which are defined as $A_{max} > 2A_0/3$ (associated with compound-like channels) and $A_{max} < A_0/3$ (full multifragmentation). As can be seen in Fig. 2.5c, the probability of observing the first group events $A_{max} > 2A_0/3$ decreases rapidly in the excitation energy range 2–6 AMeV, and this probability for the second group $A_{max} < A_0/3$ increases within the same energy range. However, temperatures of these groups of fragments are essentially different, since the binding energy effect is different, and a disintegration into a large number of fragments takes more energy. The transition to the energy-consuming group $A_{max} < A_0/3$ occurs because of the phase space domination, despite a smaller temperature of this group. We will call it "bimodality" from now on (see also discussion for other models in Ref. [25]). It is an intrinsic feature of the phase space population in the SMM; however, it was only recently realized that it can be manifested in different branches of the caloric curve and in momenta of fragment distributions (see Ref. [27]). The bimodality provides an explanation why the caloric curve may have a plateau-like (or even a "back-bending") behavior and large fluctuations in the transition region.

2.2.4 Influence of the Critical Temperature

We have analyzed how the critical exponent τ depends on T_c, similar to Ref. [30], and on isospin of the sources. During multifragmentation, nuclei demonstrate critical behavior at considerably lower temperatures ($T^* \approx$ 5–6 MeV) than the critical temperature of nuclear matter T_c (see also Ref. [27, 31, 33]), because of Coulomb and finite size effects. As was shown in many papers (see, e.g., Ref. [33] and references in), the critical behavior can manifest as scaling of

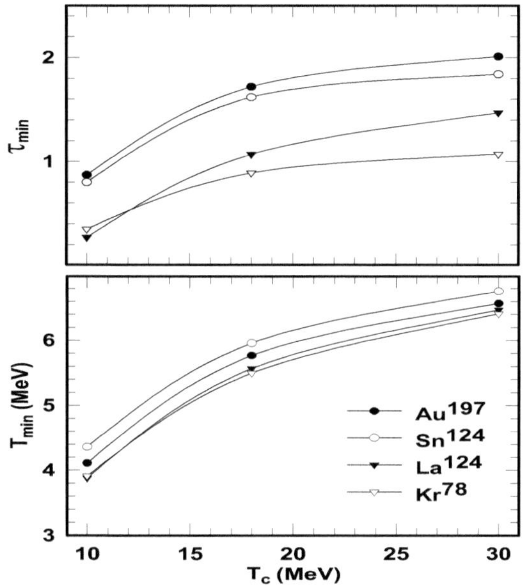

Figure 2.6 The values of τ_{min} and T_{min} as a function of the critical temperature T_c for various nuclei.

some observables around the critical point T^*. In the present work, we are interested in information about the critical temperature T_c, which is high enough to be extracted through the scaling in finite nuclei. However, this information can be obtained by looking at other features of fragment distributions, since the critical temperature influences the surface tension of fragments. We have extracted the minimum values of τ versus T_c and the corresponding temperatures of the systems, as shown in Fig. 2.6. A similar analysis of experimental data was carried out by Karnaukhov *et al.* [39] for residues produced after intranuclear cascade in Au target. One can see from Fig. 2.6 that both the τ_{min} and T_{min} temperatures corresponding to this minimum increase with T_c. However, the nuclei with lower isospin such as La^{124} and Kr^{78} exhibit much lower τ_{min} values. It is because the proton-rich nuclei are less stable, and they disappear rapidly with the excitation energy, which corresponds to a less pronounced U-shape mass distribution. This effect may be important for the explanation of some data, since sources with different isospin may manifest different τ_{min}. The decreasing T_{min} for lower T_c is explained by a very fast decrease in the surface tension with temperature in this case. Nevertheless, all parameters increase with T_c and tend to saturate at $T_c \to \infty$, corresponding to the case of the temperature-independent surface. In this case, only the translational and bulk entropies of fragments, but not the surface entropy, influence the probability of fragment formation.

2.2.5 Influence of the Symmetry Energy

Now, let us turn to the analysis of the symmetry energy of fragments, and its influence on fragment partitions. As we discussed, the symmetry energy for fragments is defined in SMM as $E^{sym} = \gamma(A - 2Z)^2/A$, where γ is a phenomenological coefficient. As an initial approximation for the SMM calculations, we assume γ=25 MeV, corresponding to the mass formula of cold nuclei. We have also performed the calculations for γ=8 MeV for Sn^{124} and La^{124} sources. In Fig. 2.7, we demonstrate the caloric curve, the mean maximum mass of fragments in partitions, and τ parameters for different assumptions on the symmetry energy. The results for γ=8 MeV are only slightly different from those obtained for the standard SMM assumption γ=25 MeV. From this figure, one can see a small decrease in temperature caused by the involvement of the fragments with unusual neutron and proton numbers into the dominating partitions. Therefore, the average characteristics of the produced hot fragments are not very sensitive to the symmetry energy, and special efforts should be taken in order to single out this effect.

The main effect of the symmetry energy is manifested in charge and mass variances of the hot fragments. In Fig. 2.8, we have shown the relative mass distributions of primary hot (top panels) and final cold fragments (after secondary deexcitation, middle and bottom panels) with $Z = 5$ and 10 at an excitation energy of 5 AMeV for $\gamma = 25$ and 8 MeV. The distribution of primary hot fragments becomes rather broad for $\gamma = 8$ MeV, and its maximum lies on the neutron-rich side.

2.2.6 Evaporation of Hot Fragments with Isospin Effects

The primary fragments are supposed to decay by the secondary evaporation of light particles, or by the Fermi breakup [12, 2].

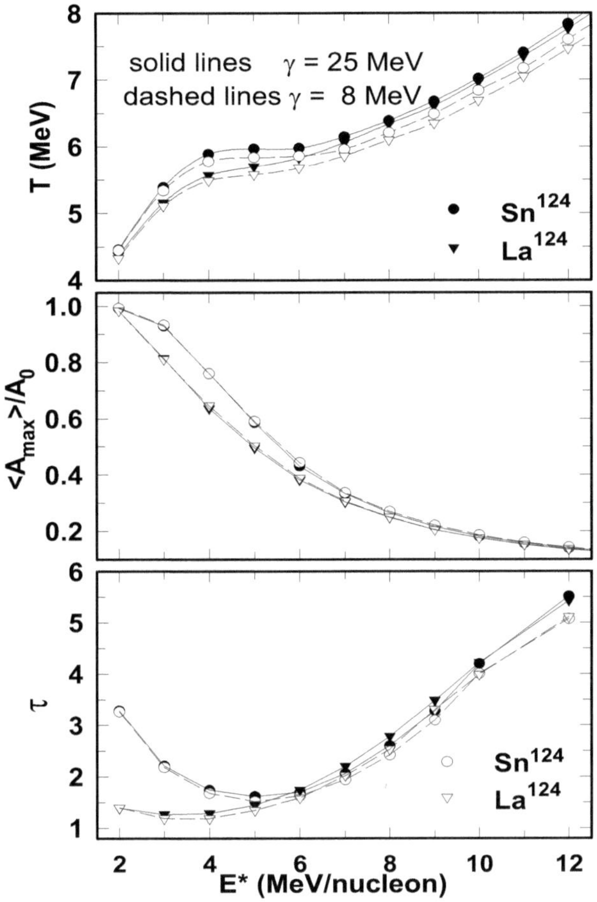

Figure 2.7 Variation in the caloric curves (top panel), A_{max}/A_0 (middle) and τ (bottom) of the nuclei Sn^{124} and La^{124} with symmetry energy versus excitation energy.

However, these codes use standard mass formulae obtained from fitting masses of cold isolated nuclei. If hot fragments in the freeze-out have smaller values of γ, their masses in the beginning of the secondary deexcitation will be different, and this effect should be taken into account in the evaporation process. Sequential evaporation is considered only for large nuclei ($A > 16$) evaporating lightest particles ($n,p,d,t,{}^3\text{He},\alpha$). We believe that we can estimate

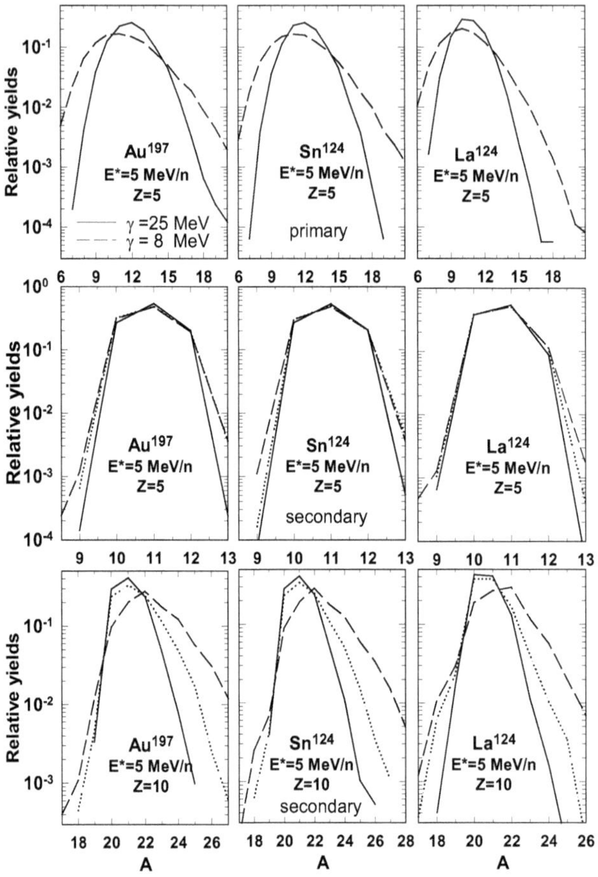

Figure 2.8 Isotopic distribution of primary and secondary fragments for $Z = 5$ and $Z = 10$ versus A at E^{*} = 5 AMeV for Au^{197}, Sn^{124} and La^{124}. Dotted lines for γ=8 MeV, and dashed lines for new evaporation depending on the symmetry energy (see the text).

the effect of the symmetry energy evolution during the sequential evaporation by including the following prescription. There are two different regimes in this process. First, in the case when the internal excitation energy of this nucleus is large enough ($E^{*}/A > 1$MeV), we take for hot fragments the standard liquid drop masses m_{ld} adopted in the SMM, as follows:

$$m_{AZ} = m_{ld}(\gamma) = m_n N + m_p Z - AW_0 + \gamma \frac{(A-2Z)^2}{A} + B_0 A^{2/3} + \frac{3e^2 Z^2}{5r_0 A^{1/3}}, \tag{2.6}$$

where m_n and m_p are masses of free neutron and proton. In the second regime corresponding to the lower excitation energies, we adopt a smooth transition to standard experimental masses with shell effects (m_{st}, taken from nuclear tables) with the following linear dependence,

$$m_{AZ} = m_{ld}(\gamma) \cdot x + m_{st} \cdot (1 - x), \tag{2.7}$$

where $x = \beta E^*/A$ (β= 1MeV^{-1}) and $x \leq 1$. The excitation energy E^* is always determined from the energy balance taking into account the mass m_{AZ} at the given excitation. This mass correction was included in a new evaporation code developed on the basis of the old model [12], taking into account the conservation of energy, momentum, mass, and charge number. We have checked that the new evaporation at γ=25 MeV leads to the results very close to those of the standard evaporation.

Generally, the secondary deexcitation pushes the isotopes toward the value of stability; however, the final distributions depend on the initial distributions for hot fragments. One can see that the results can also depend on whether the symmetry energy evolves during the evaporation or not. In Fig. 2.8, we demonstrate the results of two kinds of calculations. The first one is carried out according to the standard code [12], and the second one is with the above-described version taking into account the symmetry energy (mass) evolution during evaporation. The standard deexcitation leads to narrowing the distributions and to concentration of isotopes closer to the $\beta-$stability line by making the final distributions of the case of γ=8 MeV similar to the case of γ=25 MeV. However, a sensitivity to the initial values of γ remains, as one can see from the distributions of cold fragments with Z = 5 and 10. This difference in distributions is much more pronounced if we use the new evaporation. The final isotope distributions are considerably wider, and they are shifted to the neutron-rich side, i.e., the produced fragments are neutron rich. This effect has a simple explanation: By using the experimental masses at all steps of evaporation, we suppress the emission of charged particles by both the binding energy and the Coulomb barrier. Whereas, in the case of small γ in the beginning of evaporation, the binding energy essentially favors an emission of charged particles. When the nucleus is cooled sufficiently down to restore the normal symmetry energy, the remaining excitation is rather low ($E^*/A < 1$ MeV) for the nucleus to evaporate many neutrons.

The difference between the two kinds of evaporation calculations gives us a measure of uncertainty expected presently in the model. This uncertainty can be diminished by comparison with experiments. It is encouraging, however, that all final isotope distributions have some dependence on the γ parameter. This sensitivity gives us a chance to estimate the symmetry energy by looking at isotope characteristics. As was shown earlier, the actual γ parameter of the symmetry energy can be determined via an isoscaling analysis [40]. An analysis of the N/Z ratio of the produced fragments can also be used for this purpose. Recently, one may see some experimental evidences for an essential decrease in the symmetry energy of fragments with temperature in multifragmentation [41, 42]. The consequences of this decrease are very important for astrophysical processes [28].

In summary, we have shown the effects of isospin, critical temperature, and symmetry energy on the fragment distribution in multifragmentation. For this purpose, we have analyzed different nuclei with various neutron-to-proton ratio on the basis of SMM. The critical temperature of nuclear matter influences the fragment production in multifragmentation of nuclei through the surface energy, while the symmetry energy determines directly the neutron richness of the produced fragments. Effects of the critical temperature can be observed in the power law fitting parametrization of the fragment yields by finding the τ parameter. We have also shown the effect of the isospin and neutron excess of sources on the fragment distributions and on the τ parametrization. By selecting partitions according to the maximum fragment charge, we have demonstrated a bimodality as an essential feature of the phase transition in finite nuclear systems. This feature may allow for identification of this phase transition with the first-order one. It has been found out that the symmetry energy of the hot fragments produced in the statistical freeze-out is very important for isotope distributions, but its influence is not very large on the mean fragment mass distributions after multifragmentation. We have shown that the symmetry energy effect on isotope distributions can survive after secondary deexcitation. An extraction of this symmetry energy from the data is important for nuclear astrophysical studies.

2.2.7 Theory and Experiments at Fermi Energy Regime

In recent decades, it was shown by several groups that the SMM successfully reproduced the experimental data in the Fermi energy regime (see, e.g., [27, 43–47]. For example, multifragmentation yields, emerging from the central ^{197}Au + ^{197}Au collisions at the Fermi energy of E/A = 35 MeV, were analyzed with the SMM. Charge distributions, mean fragment energies, and two-fragment correlation functions are well reproduced by the statistical breakup of a large, diluted, and thermalized system slightly above the multifragmentation threshold. Fragment emission data observed in central ^{197}Au + ^{197}Au collisions at E/A = 35 MeV were in fully consistent with the statistical breakup of a single source at excitation energy per nucleon 5–6 MeV and at a low density freeze-out ($\rho \approx \rho_0/3 - \rho_0/6$) [43, 27]. The results of other analyses within SMM and other models were in close agreement with the TAMU experimental data measured in the Fermi energy regime (20–40 AMeV) for peripheral collisions given in Refs. [46, 48], and references therein. Particular isotopic effects, such as the odd–even staggering of the yield of final fragments, were studied by FAZIA collaboration [47] in peripheral collisions of ^{86}Kr on ^{124,112}Sn at a Fermi energy of E/A = 35 MeV. Similar statistical approaches were applied to obtain preliminary encouraging results with the help of the ensemble of residual sources [49]. SMM also well reproduced the data for isotopic and charge yields and IMFs measured in MSU experiments for central and peripheral collisions ^{124,112}Sn + ^{124,112}Sn in well above the Fermi energy regime at E/A = 50 MeV [50, 51].

In the next section, we shall discuss the breakup of projectiles at relativistic energies. We have opportunities to produce excited heavy hyper-residues in relativistic hadron-nucleus and peripheral heavy-ion collisions. The produced hypernuclei have a broad distribution in masses and isospin. They can even reach beyond the neutron and proton driplines and that opens a chance to investigate properties of exotic hypernuclei. One finds also the abundant production of multi-strange nuclei, of bound and unbound hypernuclear states with new decay modes. In addition, we can directly get an information on the hypermatter both at high and low temperatures. In this way, we make extension of nuclear reaction studies at low temperature into the **strange matter** sector.

2.3 Comparison with Experiments at High Energies

In this section, we shall extend our review to the multifragmentation of projectiles at relativistic energies. First, we summarize the recent analyses of ALADiN experimental data and theoretical predictions based on a statistical ensemble version of SMM (for details see Refs. [52, 53]). Then we summarize the other results in the literature obtained in a similar context. In the analysis of ALADiN experiments, we followed the procedure used by Botvina *et al.* [13] in describing the multifragmentation at relativistic energies. The ALADiN experiment S254, conducted in 2003 at the GSI Schwerionen Synchrotron (SIS), was designed to study isotopic effects in the fragmentation of projectiles ^{124}Sn, ^{107}Sn, and ^{124}La at relativistic energies (for details, see Ref. [52]). In Ref. [53], we also analyzed the production cross sections and isotopic distributions of projectile-like residues in the reactions ^{112}Sn + ^{112}Sn and ^{124}Sn + ^{124}Sn at an incident beam energy of 1 AGeV measured with the FRS fragment separator at the GSI laboratory [54].

Multifragmentation has been shown to be a fast process with characteristic times around 100 fm/c or less (see, e.g., Refs. [39, 55, 56]). Nevertheless, as evident from numerous analyses of experimental data, a high degree of equilibration can be reached in these reactions. In particular, chemical equilibrium among the produced fragments is very likely, while kinetic equilibration may not be reached [9]. Statistical models were found very suitable for describing the measured fragment yields [43, 45, 57–60]. Even though equilibration may not occur within the individual system

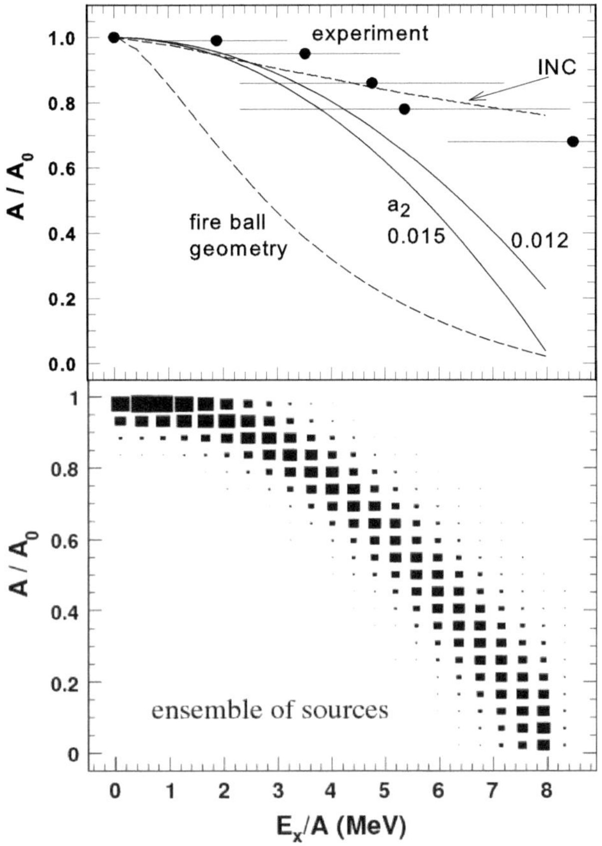

Figure 2.9 Top panel: Mean reduced mass number A/A_0 of relativistic projectile residuals as a function of their excitation energy E_x/A. Bottom panel: Ensemble of hot thermal sources represented in a scatter plot of reduced mass number A/A_0 versus excitation energy E_x/A.

during the primary reaction stage, the processes are, apparently, of such complexity that dynamical constraints beyond the conservation laws do not restrict the statistical population of the available partition space.

Taking multifragmentation into account is crucially important for a correct description of fragment production in high-energy reactions. Depending on the chosen projectile, it is responsible for 10 to 50% of the reaction cross section.

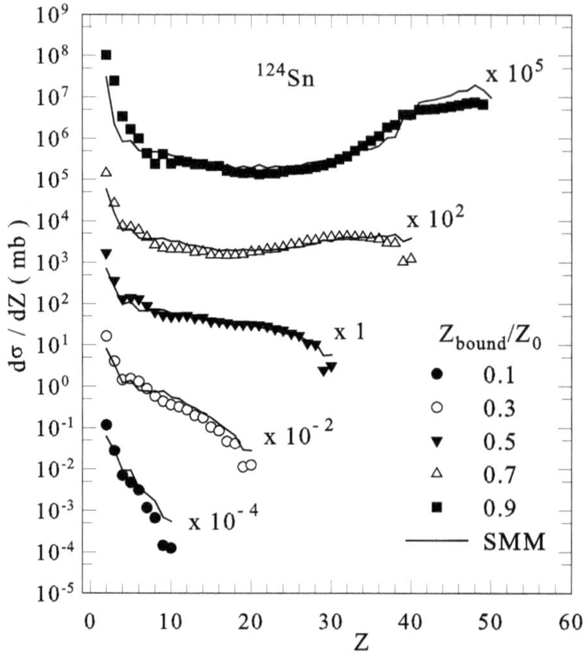

Figure 2.10 Experimental cross sections $d\sigma/dZ$ for the fragment production following collisions of ^{124}Sn projectiles sorted into five intervals of Z_{bound}/Z_0 with centers as indicated and width 0.2 (symbols) in comparison with normalized SMM calculations (lines). Different scale factors were used for displaying the cross sections as indicated.

After all, we have the possibility to investigate the phase diagram of nuclear matter at temperatures $T \approx 3$–8 MeV and densities around $\rho \approx 0.1$–$0.3\rho_0$ ($\rho_0 \approx 0.15$ fm^{-3} denotes the nuclear saturation density). These equilibrium conditions coincide with those of the nuclear liquid–gas coexistence region. Similar conditions are realized in stellar matter during the expansion phase of supernova explosions [61, 28, 29]. The study of multifragmentation thus receives a strong astrophysical motivation from its relation to stellar evolution and the formation of nuclei, occurring at densities near or below ρ_0 and over a wide range of isotopic asymmetries. The characteristics of multifragmentation as a function of the neutron-to-proton ratio N/Z of the disintegrating system are, therefore, of particular interest. We will also discuss the possible applications for astrophysics and supernova explosions at the end of this section.

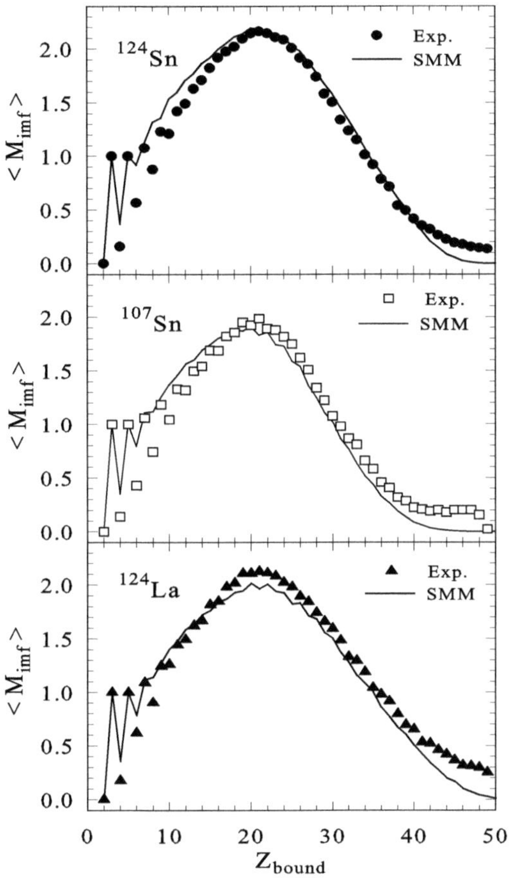

Figure 2.11 Comparison of SMM calculations (lines) with the experimental results (symbols) for the mean multiplicity $\langle M_{\text{imf}} \rangle$ of IMFs versus Z_{bound} for the three studied projectiles.

Previous ALADiN experimental data have provided extensive information on charged fragment production in multifragmentation reactions [62, 63]. In particular, they have demonstrated a "rise and fall" of multifragmentation with excitation energy [64, 65], and they have shown that the temperature remains nearly constant, around $T \sim 6$ MeV, during this process [9, 66, 67], as predicted within the statistical approach [68]. The observed large fluctuations of the

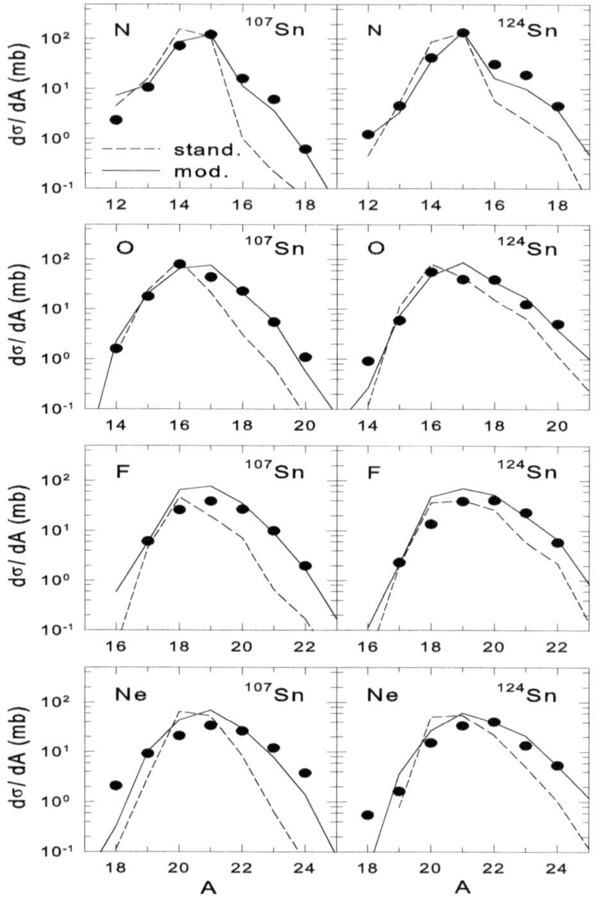

Figure 2.12 Isotopic distributions of fragments with $Z = 7$–10, for ^{107}Sn (left panels) and ^{124}Sn (right panels), integrated over $0.2 \leq Z_{\text{bound}}/Z_0 \leq 0.8$. The dots show experimental data. Solid and dashed lines for γ=25 and γ=14 MeV, respectively.

fragment multiplicity and of the size of the largest fragment in the transition region from a compound like decay to multifragmentation have been interpreted as manifestations of the nuclear liquid–gas phase transition in small systems (see, e.g., Refs. [27, 69–71] and references given therein). New data for projectile fragmentation at 600 AMeV are presented together with their statistical analysis

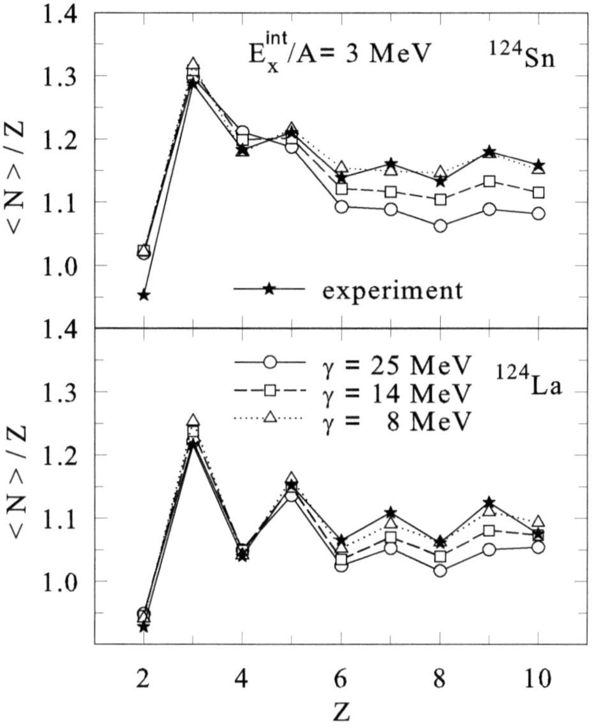

Figure 2.13 Mean neutron-to-proton ratio $\langle N \rangle/Z$ of fragments produced in the fragmentation of ^{124}Sn (top panel) and ^{124}La (bottom panel) projectiles in the range $0.4 \leq Z_{\text{bound}}/Z_0 < 0.6$. The experimental results (stars) are compared with SMM calculations using the three indicated values of the symmetry-term coefficient γ and $E_x^{\text{int}} = 3$ AMeV as the start value of the interpolation interval for the secondary-decay stage of the calculations. With this choice, $\gamma = 8$ MeV (triangles) gives the best agreement with the data.

within the framework of the SMM. The data were obtained during the ALADiN experiment S254 [72–74].

Now, let us concentrate on the analysis of ALADiN-S254 experiment, which was extensively studied in Ref. [52], and then briefly discuss the critical behavior of nuclear matter at high energies by giving some references in the literature. In the present work, we follow Ref. [13], which was rather effective in describing the multifragmentation of relativistic ^{197}Au projectiles, including their correlations and dispersions. There, the average masses of

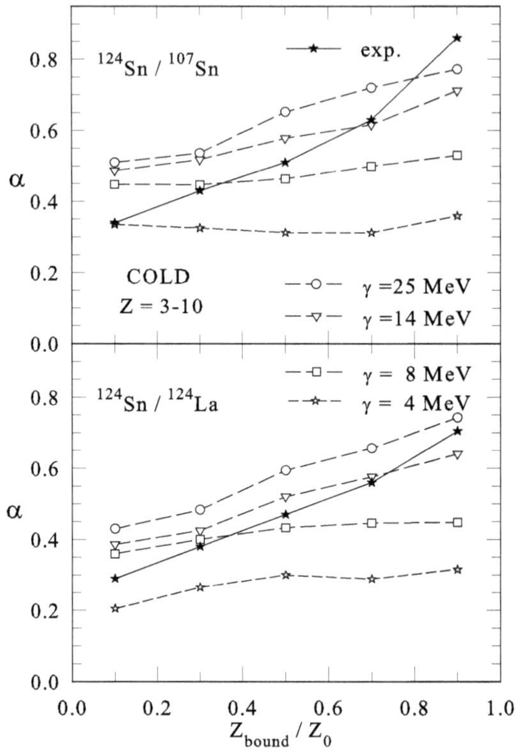

Figure 2.14 Experimental data (stars) and SMM ensemble calculations (open symbols) of isoscaling coefficients α extracted from fragment yield ratios ($3 \leq Z \leq 10$) for ^{124}Sn and ^{107}Sn projectiles (top panel) and for ^{124}Sn and ^{124}La projectiles (bottom panel) as a function of the reduced bound charge Z_{bound}/Z_0. Four different symmetry-term coefficients γ were used in the SMM calculations as indicated.

the equilibrated sources A_s were parameterized as $A_s/A_0 = 1 - a_1(E_x/A_s) - a_2(E_x/A_s)^2$, where E_x is the excitation energy of the sources in MeV, and A_0 is the projectile mass. The correlations obtained with parameters $a_1 = 0.001$ MeV^{-1} and $a_2 = 0.015$ MeV^{-2}, used in Ref. [13], as well as $a_2 = 0.012$ MeV^{-2} are shown in Fig. 2.9. It is interesting to note that a nearly identical correlation was obtained in Ref. [14] with a linear backtracing algorithm based on the SMM for ^{197}Au + Cu at 600 AMeV. As argued in Ref. [8], a dependence of this kind can be considered as consistent with BUU calculations. Recently, the realistic dynamical model calculations of nuclei collisions, see

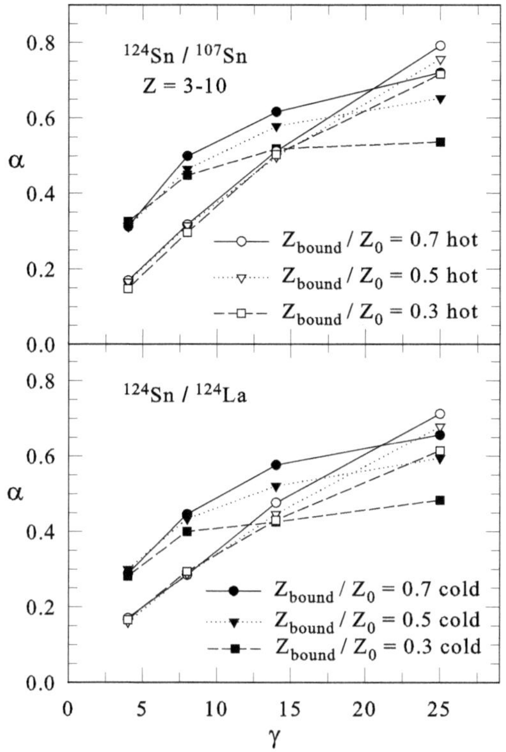

Figure 2.15 Theoretical isoscaling coefficients α calculated for three intervals in Z_{bound}/Z_0 and for both hot primary fragments (open symbols, legend in upper panel) and cold fragments after secondary deexcitation (closed symbols, legend in lower panel).

Fig. 11 in Ref. [75], have confirmed our assumptions on the ensemble of hot residue nuclear sources from the theory side.

In Fig. 2.9, the mean reduced mass number A/A_0 of relativistic projectile residuals is given as a function of their excitation energy E_x/A according to the fireball model of Gosset *et al.* [76], calculated for ^{107}Sn projectiles, and Intra-Nuclear Cascade calculations for ^{197}Au+C [77] (dashed lines), according to the experimental results for ^{197}Au nuclei reported by Pochodzalla *et al.* [66] and including the widths in E_x/A (dots), and according to the parametrization used in this work for $a_2 = 0.012$ and 0.015 MeV^{-2} (solid lines). In the bottom panel, the frequency of the individual sources is proportional to the area of the squares, as used in the SMM calculations.

The ensemble of thermal sources adopted for the present ALADiN analysis, for $a_2 = 0.015$ MeV^{-2}, is given in the bottom panel of Fig. 2.9 as a weighted distribution in the A_s/A_0 versus E_x/A_s plane. The widths are empirically adjusted, guided by the earlier results for ^{197}Au fragmentation [13]. The yield distributions as functions of the reduced mass A/A_0 and excitation energy per nucleon were also generated empirically, in a similar manner as described in the same reference.

An additional good example for the comparison of theory and experiments at relativistic energies has been given in our recent paper [78]. There, the experimental data measured by the Fragment Separator (FRS) of GSI [79], for the fragment mass, charge, and isotope yields emitted from the peripheral ^{136}Xe + Pb collisions at 1 GeV per nucleon, were analyzed within the SMM model. The ensemble of excited projectile-like sources was created in the framework of ALADiN analysis, which was mainly designed to calculate the data on multifragmentation, and to show the production of several nuclei, with $Z \leq 25$, is possible. For the completeness, it was demonstrated that the yield of near-projectile fragments (big residues) can also be reproduced within SMM by including the statistical weight of low excitations (compound nuclei, with $E^* < 1$ MeV per nucleon) in the whole weighting. To include the contribution of compound nuclei, the interval of excitation energy was chosen from 0 to 4 MeV per nucleon with the corresponding masses and charges of the nuclei. In this way, we developed an additional good example of the connection of the compound nucleus regime with the multifragmentation regime.

2.3.1 In-Medium Modification of Fragment Properties

The liquid-drop model for the description of the emerging fragments is an essential and very successful ingredient of the SMM [2]. It is based on the concept of an idealized chemical freeze-out, defined as the moment in time at which the nuclear forces have ceased to act between fragments. Their action within fragments can then be summarized with a liquid-drop description of the fragment properties with standard parameters. Only the Coulomb energy is reduced due to the presence of other fragments, an effect reasonably well accounted for within the Wigner–Seitz approximation [2]. However, in a more realistic treatment of the breakup stage, the excited primary fragments may have to be considered as not only excited but also having modified properties due to residual interactions with the environment. Their shapes and average densities may deviate from the equilibrium properties of isolated nuclei [80, 81]. Evidence for such effects has already been obtained from dynamical simulations [82] and from dynamical or statistical analyses of isotopic effects in a variety of reactions [41, 29, 48, 46]. Within the statistical approach, they can be accounted for in the fragment free energies by changing the corresponding liquid-drop parameters. In the following, we search for such effects by testing the sensitivity of fragment observables to the surface and symmetry energy terms in comparison with the present data.

At the last stage of the multifragmentation process, the hot primary fragments undergo deexcitation and propagate in the mutual Coulomb field. The realistic description of this stage is essential for obtaining reliable final fragment yields. To the extent that this can be achieved, the hot breakup stage will become accessible, which is the preeminent aim of this work. Possible modifications of the fragment properties must be taken into account in the first deexcitation steps, as the hot fragments are still surrounded by other species, while, at the end of the evaporation cascade, the standard properties must be restored. In the actual calculations, a linear interpolation between these two limiting cases has been introduced in the interval of excitation energies between zero and $E_x^{\rm int} = 1$ AMeV [32]. Conservations of energy and momentum are observed throughout this process.

Possible in-medium modifications of light clusters with $A \leq 4$, treated as elementary particles in the SMM, are not considered here. This seems justified because of their small primary multiplicities in the multifragmentation channels. For example, in the primary partitions of a single source ^{124}Sn at $E_x = 5$ AMeV, ^{2,3}H, and ^{3,4}He nuclei appear with mean multiplicities of 0.14, 0.07, 0.03, and 0.29, respectively. Together they amount to a fraction of less than 1.5% of the total mass of the system. The majority of light clusters observed in the final partitions stems from secondary decays [83]. Modifications of light clusters have been intensively studied with microscopic approaches, preferentially however in temperature–density domains at which larger fragments are of reduced importance (see, e.g., Refs. [84, 85] and references given therein).

2.3.2 Sensitivity to Symmetry Energy and Surface Term Parameters: Theory and Experiment

For event characterization, the total charge bound in fragments with $Z \geq 2$, Z_{bound}, has been introduced as in previous reports on ALADiN results (see, e.g., Refs. [9, 62, 63, 13]). Small Z_{bound} values correspond to high excitation energies of the sources that disintegrate predominantly into very light clusters ("fall" of multifragmentation). Large values of Z_{bound} correspond to low excitation energies, at which the decay changes its character from evaporation-like/fission-like processes to multifragmentation ("rise" of multifragmentation).

The measured cross sections $d\sigma/dZ$ for fragment production in reactions initiated by ^{124}Sn projectile are shown in Fig. 2.10, for example. The data have been sorted into five bins of the reduced bound charge Z_{bound}/Z_0 of width 0.2 spanning the range up to $Z_{\text{bound}} = Z_0$. As can be seen from Ref. [52], the evolution from U-shaped to exponential Z spectra appears without noticeable differences between all cases. The agreement between theory and experiment is overall very satisfactory and sufficient for performing the following analyses in individual intervals of Z_{bound}.

For the quantitative comparison of theory and experiment, the SMM ensemble calculations were globally normalized with respect to the measured Z_{bound} cross sections in the interval $10 < Z_{\text{bound}} \leq 30$ for which the agreement is best. The obtained factors are 0.00937 mb, 0.00804 mb, and 0.0106 mb per theoretical event for ^{107}Sn, ^{124}Sn, and ^{124}La projectiles, respectively, i.e., on average about 100 events per mb were calculated with the SMM.

As the first step, the sensitivity of the fragment–charge distributions and correlations to the intrinsic properties of these fragments was investigated with the SMM. Ensemble calculations were performed for different parameters of the liquid-drop description of the produced fragments and analyzed as a function of the obtained

Z_{bound}. To permit a more detailed comparison and to include the case of ^{124}La on the same scale, the reduced Z_{bound}/Z_0, normalized with respect to the atomic number Z_0 of the projectile, has been used for some observables and subdivided into five bins (0–0.2, 0.2–0.4, etc.), labeled by their central values (0.1, 0.3, etc.) in the legends of the figures.

The variation in the symmetry-term coefficient in the calculations, down to one third of its standard value, has a negligible effect on the resulting charge distributions except in the bin of the largest Z_{bound}.

As can be seen from the results for the multiplicities of intermediate-mass fragments (IMF, $3 \leq Z \leq 20$) shown in Ref. [52] as functions of Z_{bound} for the neutron-rich ^{124}Sn and the neutron-poor ^{124}La projectiles, symmetry energy term slightly influences the charge distribution. The effect of varying the symmetry-term coefficient is very small and visible only at the rise of multifragmentation on the right half of the figure. The fall part is expected to be universal because of the autocorrelation of the two quantities M_{imf} and Z_{bound} [86]. The autocorrelations dominate in the cases of Z_{bound} = 3 and 5, which can only be reached with partitions containing exactly one lithium fragment.

The surface energy of fragments is important for the charge yields [87]. As demonstrated in Ref. [52], the relatively small variation in the surface-energy coefficient B_0 from 18 to 20 MeV leads to considerable changes in the fragment production. The larger surface energy suppresses multifragmentation, leading to smaller IMF yields and multiplicities. And, vice versa, decreasing B_0 results in an intensive breakup of nuclei already at low excitation energy and to steeper Z spectra. Because of the mentioned autocorrelations, the influence of B_0 decreases with decreasing Z_{bound}.

The results of the SMM calculations, represented by the thin lines in the figures, were obtained with the normalized ensembles (parameter $a_2 = 0.015$ MeV^{-2}, cf. Fig. 2.9) and with liquid-drop parameters that provide an adequate description of the experimental observables and trends. The charge observables are mostly sensitive to the surface-term coefficient B_0, as discussed, previously. In order to maintain a consistent comparison, the evidence for a reduced symmetry term coefficient γ in the fragmentation channels, extracted from the analysis of isotope characteristics, has already been taken into account by choosing $\gamma = 14$ MeV for the three systems (Fig. 2.11).

For the surface-term coefficient, different values had to be chosen in order to obtain equivalent descriptions for the neutron-rich and neutron-poor systems. The good reproduction of the charge distributions shown in Ref. [52] has been achieved with the values $B_0 = 17.5$ MeV for ^{124}Sn and $B_0 = 19.0$ MeV for the neutron-poor ^{107}Sn and ^{124}La. They are determined with an accuracy of about 0.5 MeV for the chosen ensemble of excited sources (see Ref. [52]). It should be mentioned here again that the comparisons are made with the measured data and that the small inefficiencies of the spectrometer for very light fragments are taken into account by filtering the calculations. If corresponding corrections are applied to the data, as done in Ref. [73], the main effect is a shift in the maxima of the mean fragment multiplicity to about 10% larger values of Z_{bound} than shown in Ref. [52], while the increase in the maximum multiplicity itself is nearly negligible.

If modifications of the source parameters are simultaneously permitted, the effect of B_0 on fragment multiplicities can be compensated by a corresponding variation of a_2. The larger mean multiplicities $\langle M_{\text{imf}} \rangle$ and the shift in the peak of the multiplicity distribution to larger Z_{bound} observed by reducing the surface term (Ref. [52]) can also be obtained by increasing the excitation energies of the sources [87]. Equally satisfactory results for ^{124}Sn were, e.g., obtained with the parameter pair $a_2 = 0.015$ MeV^{-2} and $B_0 =$ 17.5 MeV used here as well as with $a_2 = 0.012$ MeV^{-2} and $B_0 =$ 19.0 MeV. The required N/Z dependence of B_0 is independent of this ambiguity, at least as long as the ensemble parameters remain the same and independent of the projectile N/Z. Evidence for a variation in the surface-term coefficient with the isotopic composition of the system has previously been derived from a τ analysis of fragment-charge yields $y(Z)$ measured for the fragmentation of ^{129}Xe, ^{197}Au, and ^{238}U at 600 AMeV [87].

2.3.3 Isotope Distributions

Cross sections measured for resolved isotopes and integrated over $0.2 \leq Z_{\text{bound}}/Z_0 \leq 0.8$ are presented in Fig. 2.12 for 7–10, where the heavier fragments with $Z \geq 7$ rather sensitive to the variation in γ.

The comparison with the data shows that the yields of neutron-rich isotopes are much better described if the reduced symmetry–energy coefficient is used. In particular, the heavier fragments with $Z \geq 7$ are rather sensitive to the variation in γ, as already evident from Fig. 2.12. The same conclusion can be drawn from the comparison of the mean neutron-to-proton ratios $\langle N \rangle / Z$ for elements of $Z \leq 10$ with the experimental results Ref. [88, 52]. Evidently, they depend very little on Z_{bound}, and the calculations are, therefore, only shown for the full range $0.2 \leq Z_{\text{bound}}/Z_0 < 0.8$. The modified calculations, with $\gamma = 14$ MeV, agree well with the data. Ensemble calculations performed with this choice for the ^{124}Sn and ^{124}La systems also provide an excellent description of the data (Fig. 2.13).

Summarizing this section, it may be stated that the isotopically resolved fragment yields provide evidence for a reduced symmetry energy in the hot environment at freeze-out, in accordance with previous findings [41, 29]. Similar observations were made in reaction studies performed at intermediate and relativistic energies [48, 46, 50].

2.3.4 Isoscaling and the Symmetry Term

Isoscaling concerns the production ratios R_{21} of fragments with neutron number N and proton number Z in reactions with different isospin asymmetry. It is defined as their exponential dependence on N and Z according to

$$R_{21}(N,Z) = Y_2(N,Z)/Y_1(N,Z) \propto exp(N \cdot \alpha + Z \cdot \beta) \qquad (2.8)$$

with parameters α and β. Here, Y_2 and Y_1 denote the yields from the more neutron-rich and the more neutron-poor reaction system, respectively. Isoscaling has the property that a comparison is made for the same isotopes produced by different sources, so that the effects of yield fluctuations due to nuclear structure effects may be reduced.

The comparison of the measured isotope yields from the fragmentation of ^{124}Sn and of the two neutron-poor systems has confirmed that isoscaling is observed [72]. The isoscaling parameter α, determined from the yields for $3 \leq Z \leq 10$ in different ranges of

Z_{bound}/Z_0, is shown in Fig. 2.14. It is seen to decrease rapidly as the disintegration of the spectator systems into fragments and light particles increases, as reported previously for the fragmentation of target spectators in reactions of ^{12}C on ^{112}Sn and ^{124}Sn at 300 and 600 AMeV [41]. Nearly identical results are obtained for the isotopic and isobaric pairs of reactions.

In the same figure, the results obtained from SMM ensemble calculations with different symmetry-term coefficients γ and determined for $3 \leq Z \leq 10$ are compared with the data. Smaller values of γ lead to smaller isoscaling parameters α because the isotope distributions become wider, and the variation in the yield ratios with N becomes correspondingly smaller. The SMM standard value $\gamma = 25$ MeV is applicable only in the bin of the largest Z_{bound}. Smaller values must be chosen to reproduce the rapidly decreasing parameter α in the fragmentation regime at smaller Z_{bound}.

The proportionality expected for the dependence of α on γ is demonstrated in Fig. 2.15. It is fulfilled in good approximation for the hot fragments at freeze-out (open symbols in the figure). Using the calculated values of α for $\gamma = 14$ MeV and the microscopic temperature for this interval, $T = 6.0$ MeV, it is even possible to test the widely used formula

$$\alpha \approx \frac{4\gamma}{T}\left(\frac{Z_1^2}{A_1^2} - \frac{Z_2^2}{A_2^2}\right), \tag{2.9}$$

where Z_1, A_1 and Z_2, A_2 are the atomic and mass numbers of the two thermalized systems. The numerical factor deduced from the ensemble calculations is 3.8, while the nominal coefficient, analytically derived in the zero-temperature limit, is 4 [40]. The validity of Eq. 2.9 at finite temperatures has recently also been confirmed by Souza *et al.* in a theoretical study with the SMM [89].

This relation was independently deduced from dynamical and statistical investigations [40, 90] and used in several analyses of experiments [41, 48, 46]. The SMM ensemble calculations show that, within some approximation, this relation remains valid for the cold fragment yields as long as γ is close to its conventional value of 25 MeV. As γ is lowered, the resulting α for the asymptotic fragments becomes larger and deviates increasingly from the value for hot fragments (Fig. 2.15). During the late stages of the secondary

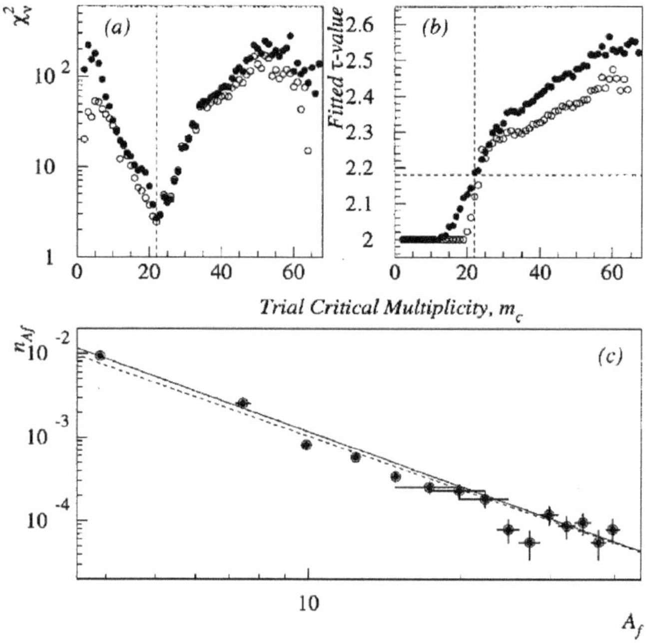

Figure 2.16 Determination of critical exponents τ and critical multiplicity m_c from fragment yield. (a) χ^2_ν values from the power law fit to the mass yield distribution for different m. (b) Values of τ as a function of m. (c) Power law fit to data point m=m_c, corresponding to the minimum value of χ^2_ν. The dashed line shows the fit to the open points, which exclude A=4 fragments. The black dot results include the A=4 fragments (see the text).

decays, the available daughter states are increasingly concentrated in the valley of stability, which narrows the initially wide isotope distributions. This effect has been demonstrated already and discussed in Ref. [41].

2.3.5 Phase Diagram and Critical Behaviour: High-Energy Experiments and Theory

Multifragmentation, furthermore, opens a unique possibility for investigating the phase diagram of nuclear matter at temperatures $T \approx$ 3–8 MeV and densities around $\rho \approx$ 0.1–0.3ρ_0 ($\rho_0 \approx$ 0.15 fm^{-3}

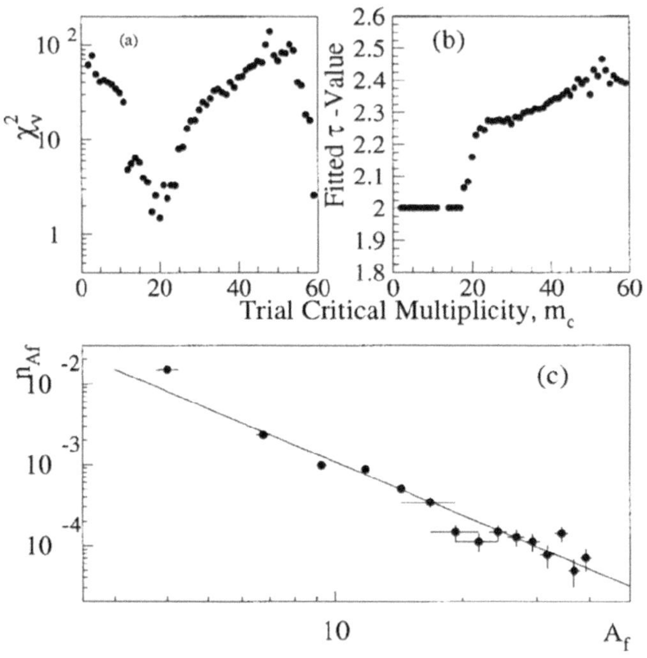

Figure 2.17 Determination of τ and m_c from SMM (see the text).

denotes the nuclear saturation density). These equilibrium conditions coincide with those of the nuclear liquid–gas coexistence region. Many statistical models have demonstrated that multifragmentation is a kind of a phase transition in highly excited nuclear systems. In the SMM, a link to the liquid–gas phase transition is especially strong. In particular, the surface energy of hot primary fragments is parametrized in such a way that it vanishes at a certain critical temperature. The SMM has predicted distinctive features of this phase transition in finite nuclei, such as the plateau-like anomaly in the caloric curve [68, 2], which have been later observed in experiments [66, 91]. Many other manifestations of the phase transition, such as large fluctuations and bimodality [52, 27, 32], critical behavior, and even values of critical exponents [27, 59], have been investigated within this model. The experimental data are usually in agreement with the predictions. Nevertheless, the properties of this phase transition are not yet fully understood.

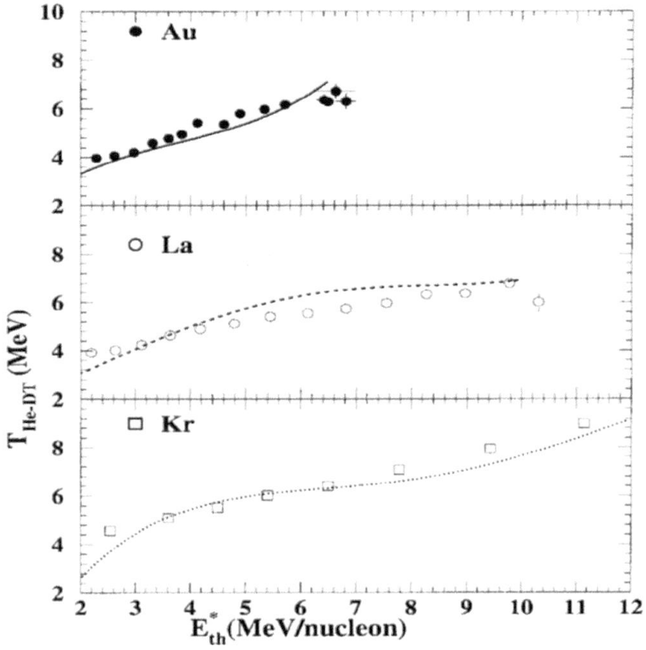

Figure 2.18 Caloric curves (T_f *vs.* E^*_{th}/A) for Kr, La, and Au. Points are experimental and curves are from SMM (see the text).

Table 2.1 Critical exponents from data and SMM_{cold} (experimental remnants), see the text.

Parameter	Data	SMM
m_c	22 ± 1	20 ± 2
τ	2.19 ± 0.02	2.17 ± 0.02
γ	1.4 ± 0.3	1.02 ± 0.23
σ	0.32 ± 0.05	0.63 ± 0.08
	0.54 ± 0.11	

The critical behavior observed in experimental data can also be explained within a percolation model [92], or a Fisher's droplet model [93], which corresponds to a second-order phase transition in the vicinity of the critical point. We must note, however, that the finiteness of the systems under investigation plays a crucial role. To

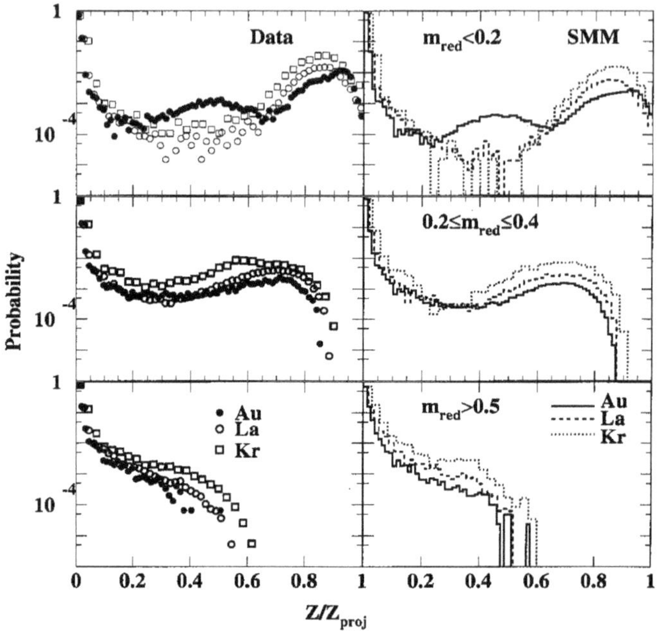

Figure 2.19 Second stage fragment charge distribution as a function of $Z/Z_{projectile}$. Results are shown for three reduced multiplicity intervals for both data and SMM (see the text).

connect this anomalous behavior with a real phase transition, one should study it in a thermodynamical limit. Within the SMM, this was done in Ref. [17], where multifragmentation of an equilibrated system was identified as a first-order phase transition. The mixed phase in this case consists of an infinite liquid condensate and gas of nuclear fragments of all masses. In a finite system, this mixed phase corresponds to U-shaped fragment distributions with the heaviest fragment representing the liquid phase. Thus, one can connect multifragmentation of finite nuclei with the fragmentation of a very big system. This is important for the application of statistical models in astrophysical environments (neutron stars, supernova explosions), where nuclear statistical equilibrium can also be expected [28].

Critical behaviour in nuclear multifragmentation was investigated in high-energy heavy-ion collisions by several groups (see, e.g.,

Refs. [31, 58, 59]). Multifragmentation results from 1 AGeV Au on C were compared with the SMM model in Ref. [58]. A large number of observables, including the fragment charge yield distributions, fragment multiplicity distributions, caloric curve, critical exponents, and the critical scaling function, are explored in this comparison. Figures 2.16 and 2.17 show the various critical exponents and the location of the critical point extracted from the data [58]. The comparison of the data and SMM calculations are given in Table 2.1. Very good agreement was obtained using SMM with the standard values of the SMM parameters. In a similar context, multifragmentation products emerging from 1 AGeV Au, La, and Kr and collisions with C were studied in Ref. [31, 59], and results were compared with intranuclear cascade and SMM model calculations (see Figs. 2.18 and 2.19). The results for first-stage particles and remnant were compared with INC simulations and those for second-stage emission were compared with SMM. Satisfactory agreement was obtained with the data.

Limits of thermal multifragmentation and angle correlations were studied in Refs. [24, 94, 3]. Time scales, critical exponents, and freeze-out volume in thermal multifragmentation were also investigated at relativistic energies [95–97]. Theoretical investigations within SMM on nuclear matter properties such as caloric curve, fragment correlation functions, time scales, critical behaviour, isoscaling and symmetry energy of fragments in the dense environment were done extensively (see, for example, Refs. [30, 41, 40]).

2.3.6 Possible Applications for Astrophysics and Supernova Explosions

In this section, we summarize the nuclear composition and the equation of state of the supernova matter at subnuclear densities expected during the collapse and subsequent explosions of massive stars within a statistical approach. Beside nuclear reactions, very similar conditions are realized in stellar matter during the expansion phase of type-II supernova explosions [61, 28, 29, 98]. When the cores of massive stars collapse in a burst of electron capture, high densities are reached; then repulsive NN interactions give rise to shockwaves, which is assumed to be responsible for supernova explosion. The study of multifragmentation thus receives a strong astrophysical motivation from its relation to stellar evolution and the formation of nuclei, occurring at densities near or below ρ_0 and over a wide range of isotopic asymmetries.

In the statistical approach, supernova matter is described as a mixture of nuclear species, electrons, photons, and perhaps neutrinos in thermal equilibrium. For the macroscopic scales, one can safely apply the grand-canonical approximation, which was previously used by many authors. We call this model the Statistical Model for Supernova Matter (SMSM). It was first introduced in Ref. [28] as generalization of the SMM [2]. In the statistical approach, it is assumed that protons, neutrons, electrons, photons, and neutrinos are in thermal equilibrium. In this mixture, we consider that (1) electrons balance positive charge, (2) photons change nuclear composition via photonuclear reactions, and (3) a strong neutrino wind from a protoneutron star. Neutron capture, photo-disintegration of nuclei, neutron evaporation, electron/positron and neutrino/antineutrino reactions are taken into account in the assumptions of statistical equilibrium.

For macroscopic scales, grand-canonical approximation of SMM model is used and. Each particle i with B_i baryon number, Q_i charge number, L_i lepton number, is characterized by a chemical potential μ_i. Three independent chemical potentials are determined from the conservations of the total number of baryons, charge, and leptons. Similarly, $\mu_{A,Z}$, μ_e, and μ_ν correspond conservation laws for Baryon, Charge, and Lepton numbers. The lepton number conservation is a valid concept only if neutrinos and antineutrinos are trapped in the system within the neutrinosphere; otherwise, μ_L= 0 (if they escape freely from stars then only conservation of B_i, Q_i, and Y_e is fixed, and the lepton number conservation is irrelevant). In this case, two remaining chemical potentials are determined from the conditions of baryon number conservation and electro-neutrality. Since the beta-equilibrium may not be achieved in a fast explosive process, we also often fix the electron fraction Y_e in the calculations.

Statistical ensemble with precise values of T, ρ_B, Y_e, and fixed value of A = 1000 baryons in a box. The density of fragments is fixed at ρ_0 = $0.15 fm^{-3}$, non-fixed composition of uncertain particle number. The system is divided into Wigner–Seitz cells containing one nucleus, neutrons, and electrons. We consider the fragments in the range $1 \leq$

$A \leq 1000$ for mass and $0 \leq Z \leq A$ for charge number. Since weak reactions are slow, we consider out of equilibrium lepton fraction ($Y_L = \rho_L/\rho_B$), or by the electron fraction ($Y_e = \rho_e/\rho_B$) with the lepton and electron (proton) densities. The baryon density is nearly the same as the density of matter. In Fig. 2.1, we show the results for nuclear multifragmentation and supernova explosions. Comparison of SMM findings with the results of other models can be found in Ref. [98] and references therein.

Consequently, one can see that the evolution of mass distributions with excitation energy is qualitatively the same for both the nuclear multifragmentation reactions and the supernova explosions. However, in the supernova environments, much heavier and neutron-rich nuclei can be produced because of the screening of their charge by surrounding electrons. This shows that studying the statistical multifragmentation reactions in the laboratory is important for understanding how heavy elements were synthesized in the universe.

2.4 Conclusion

In conclusion, a statistical approach is very effective and it has a solid theoretical justification: As demonstrated in kinetic theories, the system evolves toward the statistical equilibrium when the fragments form, i.e., when relative velocities of nucleons become low. This situation takes place after pre-equilibrium emission. Pre-equilibrium emission (multistep processes) takes place after the first stage (direct reaction processes) but long before the statistical equilibrium of the compound nucleus is reached. It was well justified that pre-equilibrium processes are an important part of the reaction cross section, at projectile energies starting from $\sim$10 AMeV up to $\sim$1 AGeV. At lower energies (when the excitation energy $E_x \lesssim 3$ MeV/nucleon), the standard compound nucleus picture is valid when sequential evaporation of light particles and fission are the dominant decay channels. Such evolution of nuclear decay mechanisms is predicted by all statistical models.

SMM provides very effective descriptions of experimental data, both at low and high energy ion reactions. For example, in our previous analysis with SMM at low- and high-energy regimes, an overall good agreement with the data was obtained. Similar justifications were also shown by various facilities with the analyses of the experiments such as Miniball/Miniwall (MSU), Multics (INFN), INDRA (GANIL), CHIMERA (LNS), NIMROD, MARS, FAUST (TAMU), LASSA, ISiS (IUCF), FASA (JINR), ALADiN, FRS (GSI), EOS (LBNL), and so on. The key theoretical point is the adequate selection of the equilibrium state.

SMM also gives a chance to investigate the phase diagram of nuclear matter at low temperatures and describes the liquid–gas type phase transition from nucleons to fragments. Also properties of fragments (e.g., their symmetry energy) during this phase transition can be investigated and extracted from the comparisons with experiment. Moreover, multifragmentation reactions are responsible for the production of some specific isotopes. Importance of multifragmentation reactions is now widely recognized, and in recent years, the interest to them has risen in several domains of research. Indeed, practical calculations of fragment production and transport in complex medium are needed for nuclear waste transmutation (environment protection), electro-nuclear breeding (new methods of energy production), proton and ion therapy (medical applications), and radiation protection of space missions (space research).

In addition, the critical temperature of nuclear matter influences the fragment production in the multifragmentation of nuclei through the surface energy, while the symmetry energy determines directly the neutron richness of the produced fragments. Effects of the critical temperature can be observed in the power law fitting parametrization of the fragment yields by finding τ parameter. One may investigate the effect of the isospin and neutron excess of sources on the fragment distributions and on the τ parametrization within SMM. By selecting partitions according to the maximum fragment charge, we have demonstrated a bimodality as an essential feature of the phase transition in finite nuclear systems. This feature may allow for the identification of this phase transition with the first-order one. It has been found out that the symmetry energy of the hot fragments produced in the statistical freeze-out is very important for isotope distributions, but its influence is not very large on the mean fragment mass distributions after multifragmentation.

The knowledge obtained after the analysis of nuclear reactions with statistical models can be directly used for other cases when the similar parameters of nuclear matter and fragments in equilibrium are expected. For example, in astrophysical applications, to evaluate the properties of fragments in supernova explosions and in crust of neutron stars.

Acknowledgments

The authors gratefully acknowledge the contributions of the distinguished colleagues through the concerning articles given in the references of this chapter. Several hundreds of papers concerning statistical multifragmentation have been published so far, and we offer an excuse that in this short chapter we cannot mention all works related to this field, and all the names of our colleagues contributed to the present results.

References

[1] Gross, D. H. E.(1990). *Rep. Prog. Phys.* **53**, pp. 605–658.

[2] Bondorf, J. P., Botvina, A. S., Iljinov, A. S., Mishustin, I. N., and Sneppen, K. (1995). *Phys. Rep.* **257**, pp. 133–221.

[3] Viola, V. E. *et al.* (2006). *Phys. Rep.* **434**, 1–2, pp. 1–46.

[4] Bohr, N. (1936). *Nature, (London)* **137**, 344.

[5] Bertsch, G. and Siemens, P. J. (1983). *Phys. Lett.* **126**, 1–2, pp. 9–12.

[6] Desbois, J., Boisgard, R., Ngo, C., and Nemeth, J. (1987). *Z. Phys. A* **328**, 1, pp. 101–113.

[7] Friedman, W. A. (1990). *Phys. Rev. C* **42**, 2, pp. 667–673.

[8] Barz, H. W. *et al.* (1993). *Nucl. Phys. A* **561**, 3, pp. 466–476.

[9] Xi, H.F. *et al.* (1997). *Z. Phys. A* **359**, 4, pp. 397–406.

[10] Botvina, A. S. and Mishustin, I. N. (1992). *Phys. Lett. B* **294**, 1, pp. 23–26.

[11] Li, B. A., DeAngelis, A. R., Gross, D. H. E. (1993). *Phys. Lett. B* **303**, 3–4, pp. 225–229.

[12] Botvina, A. S. *et al.* (1987). *Nucl. Phys. A* **475**, 4, pp. 663–686.

[13] Botvina, A. S. *et al.* (1995). *Nucl. Phys. A* **584**, 4, pp. 737–756.

[14] Désesquelles, P. *et al.* (1996). *Phys. Rev. C* **53**, 5, pp. 2252–2257.

[15] Le Fèvre A. *et al.* (2004). *Nucl. Phys. A* **735**, 1–2, pp. 219–247.

[16] Das Gupta, S. and Mekjian, A. Z. (1998). *Phys. Rev. C* **57**, 3, pp. 1361–1365

[17] Bugaev, K. A., Gorenstein, M. I., Mishustin, I. N., and Greiner, W. (2001). *Phys. Lett. B* **498**, 3–4, pp. 144–148.

[18] Chaudhuri, G. and Das Gupta, S. (2009). *Phys. Rev. C* **80**, 4, 044609.

[19] Botvina, A. S. and Mishustin, I. N. (2001). *Phys. Rev. C* **63**, 6, 061601(R).

[20] Botvina, A. S., Iljinov, A. S., and Mishustin, I. N. (1985). *Sov. J. Nucl. Phys.* **42**, 5, pp. 712–718.

[21] Fritz, S. *et al.* (1999). *Phys. Lett. B* **461**, 4, pp. 315–321.

[22] Viola, V. E., Kwiatkowski, K., Natowitz, J. B., and Yennello, S. J. (2004). *Phys. Rev. Lett.* **93**, 13, 132701.

[23] Lefort, T. *et al.* (2000). *Phys. Rev. C* **62**, 3, 031604.

[24] Neubert, W. and Botvina, A. S. (2003). *Eur. Phys. J. A* **17**, 4, pp. 559–571.

[25] Chomaz, P. *et al.* (2001). *Phys. Rev. E* **64**, 4, pp. 046114.

[26] D'Agostino, M. *et al.* (2000). *Phys. Lett. B* **473**, 3–4, pp. 219–225.

[27] D'Agostino, M. *et al.* (1999). *Nucl. Phys. A* **650**, 3, pp. 329–357.

[28] Botvina, A. S. and Mishustin, I. N. (2004). *Phys. Lett. B* **584**, 3–4, pp. 233–240.; Botvina, A. S. and Mishustin, I. N. (2010). *Nucl. Phys. A* **843**, pp. 98–132.

[29] Botvina, A. S. and Mishustin, I. N. (2005). *Phys. Rev. C* **72**, 4, 048801

[30] Ogul, R. and Botvina, A. S (2002). *Phys. Rev. C* **66**, 5, 051601(R).

[31] Hauger, J. A. *et al.* (2000). *Phys. Rev. C* **62**, 2, 024616.

[32] Buyukcizmeci, N., Ogul, R., and Botvina, A. S. (2005). *Eur. Phys. J. A* **25**, 1, pp. 57–64.

[33] Srivastava, B. K. *et al.* (2002). *Phys. Rev. C* **65**, 5, 054617.

[34] Botvina, A. S. and Mishustin, I. N. (2003). *Phys. Rev. Lett.* **90**, 17, 179201.

[35] Landau, L. D. and Lifshitz, E. M. (1980). *Statistical Physics*, Part 1, 3rd ed. Pergamon, New York, pp. 340.

[36] Trautmann, W. (2001). *Nucl. Phys. A* **685**, pp. 233C–245C.

[37] Schnack, J. and Feldmeier, H. (1997). *Phys. Lett. B* **409**, 1–4, pp. 6–10.

[38] Sugawa Y. and Horiuchi, H. (1999). *Phys. Rev. C* **60**, 6, 064607.

[39] Karnaukhov, V. *et al.* (2003). *Phys. At. Nucl.* **66**, 7, pp. 1242–1251.

[40] Botvina, A. S., Lozhkin, O. V., and Trautmann, W. (2002). *Phys. Rev. C* **65**, 4, 044610.

[41] Le Fèvre A. *et al.* (2005). *Phys. Rev. Lett.* **94**, 16, 162701.

[42] Shetty, D. V. *et al.* (2005). *Phys. Rev. C* **71**, 2, 024602.

[43] D'Agostino, M. *et al.* (1996). *Phys. Lett. B* **371**, 3–4, pp. 175–180.

[44] Desesquelles, P. *et al.* (1998). *Nucl. Phys. A* **633**, 3, pp. 547–562.

[45] Avdeyev, S. P. *et al.* (2002). *Nucl. Phys. A* **709**, pp. 392–414.

[46] Souliotis, G. A. *et al.* (2007). *Phys. Rev. C* **75**, 1, 011601(R).

[47] Piantelli, S. *et al.* (2013). *Phys. Rev. C* **88**, 6, 064607.

[48] Iglio, J. *et al.* (2006). *Phys. Rev. C* **74**, 2, 024605.

[49] Buyukcizmeci, N. *et al.* (2015). *Nucl. Sci. and Tech.* **26**, 2, S20507.

[50] Ogul, R. *et al.* (2009). *J. Phys. G* **36**, 11, 115106.

[51] Buyukcizmeci, N. *et al.* (2012). *J. Phys. G: Nucl. Part. Phys.* **39**, 11, 115102.

[52] Ogul, R. *et al.* (2011). *Phys. Rev. C* **83**, 024608.

[53] Imal, H. *et al.* (2015). *Phys. Rev. C* **91**, 034605.

[54] Föhr, V. *et al.* (2011). *Phys. Rev. C* **84**, 054605.

[55] Beaulieu, L. *et al.* (2000). *Phys. Rev. Lett.* **84**, 26, pp. 5971–5974.

[56] Vermani, Y. K. and Puri, R. K. (2009). *Euro. Phys. Lett.* **85**, 6, pp. 62001.

[57] Botvina, A. S., Iljinov, A. S., and Mishustin, I. N. (1990). *Nucl. Phys. A* **507**, 3–4, pp. 649–674.
[58] Scharenberg, R. P. *et al.* (2001). *Phys. Rev. C* **64**, 5, 054602.
[59] Srivastava, B. K. *et al.* (2001). *Phys. Rev. C* **64**, 0541605(R).
[60] Bellaize N. *et al.* (2002). *Nucl. Phys. A* **709**, pp. 367–391.
[61] Ishizuka, C., Ohnishi, A., and Sumiyoshi, K. (2003). *Nucl. Phys. A* **723**, 3–4, pp. 517–543.
[62] Kreutz, P. *et al.* (1993). *Nucl. Phys. A* **556**, 4, pp. 672–696.
[63] Schüttauf, A. *et al.* (1996). *Nucl. Phys. A* **607**, 4, pp. 457–486.
[64] Ogilvie, C. A. *et al.* (1991). *Phys. Rev. Lett.* **67**, 10, pp. 1214–1217.
[65] Hubele, J. *et al.* (1991). *Z. Phys. A* **340**, 3, pp. 263–270.
[66] Pochodzalla, J. *et al.* (1995). *Phys. Rev. Lett.* **75**, 6, pp. 1040–1043.
[67] Trautmann, W. *et al.* (2007). *Phys. Rev. C* **76**, 6, 064606
[68] Bondorf, J., Donangelo, R., Mishustin, I. N., and Schulz, H. (1985). *Nucl. Phys. A* **444**, 3, pp. 460–476
[69] Campi, X. (1988). *Phys. Lett. B* **208**, 3–4, pp. 351–354.
[70] Gulminelli, F. and D'Agostino, M. (2006). *Eur. Phys. J. A* **30**, 1, pp. 253–262.
[71] Chaudhuri, G. and Das Gupta, S. (2007). *Phys. Rev. C* **75**, 3, 034603.
[72] Trautmann, W. *et al.* (2008). *Int. J. Mod. Phys. E* **17**, 9, pp. 1838–1849.
[73] Sfienti, C. *et al.* (2009). *Phys. Rev. Lett.* **102**, 15, 152701.
[74] Turzo, K. *et al.* (2004). *Eur. Phys. J. A* **21**, 2, pp. 293–301
[75] Botvina, A. S. *et al.* (2017). *Phys. Rev. C* **95**, 1, 014902.
[76] Gosset, J. *et al.* (1977). *Phys. Rev. C* **16**, pp. 629.
[77] Botvina, A. S., Gudima, K. K., Iljinov, A. S., and Mishustin, I. N. (1994). *Phys. At. Nucl.* **57**, 4, pp. 628–635.
[78] Imal, H. and Ogul, R. (2021). *Nucl. Phys. A*, **1014**, pp. 122261.
[79] Henzlova, D. *et al.* (2008). *Phys. Rev. C* **78**, pp. 044616.
[80] Botvina, A. S., Larionov, A. B., and Mishustin, I. N. (1995). *Phys. At. Nucl.* **58**, 10, pp. 1703–1712.
[81] Baran, V. *et al.* (2002). *Nucl. Phys. A* **703**, 3–4, pp. 603–632.

[82] Ono, A., Danielewicz, P., Friedman, W. A., Lynch, W. G., and Tsang, M. B. (2004). *Phys. Rev. C* **70**, 041604(R).

[83] Buyukcizmeci, N., Botvina, A. S., Mishustin, I. N., and Ogul, R. (2008). *Phys. Rev. C* **77**, 3, 034608.

[84] Typel, S., Röpke, G., Klähn, T., Blaschke, D., and Wolter, H. H. (2010). *Phys. Rev. C* **81**, 1, 015803.

[85] Natowitz, J. B. *et al.* (2010). *Phys. Rev. Lett.* **104**, 20, 202501.

[86] Hubele, J. *et al.* (1992). *Phys. Rev. C* **46**, 5, pp. R1577–R1581.

[87] Botvina, A. S. *et al.* (2006). *Phys. Rev. C* **74**, 4, 044609.

[88] Łukasik, J. *et al.* (2008). *Nucl. Instr. and Meth. A* **587**, 2–3, pp. 413–419.

[89] Souza, S. R., Tsang, M. B., Carlson, B. V., Donangelo, R., Lynch, W. G., and Steiner, A. W. (2009). *Phys. Rev. C* **80**, 044606.

[90] Ono, A., Danielewicz, P, Friedman, W. A., Lynch, W. G., and Tsang, M. B. (2003). *Phys. Rev. C* **68**, 051601(R).

[91] Natowitz, J. B. *et al.* (2002). *Phys. Rev. C.* **65**, 034618.

[92] Bauer, W. and Botvina, A. S. (2002). *Phys. Rev. C* **52**, R1760.

[93] Elliott, J. B. *et al.* (2002). *Phys. Rev. Lett.* **88**, 4, 042701.

[94] Pienkowski, L. *et al.* (2002). *Phys. Rev. C* **65**, 6, 064606.

[95] Karnaukhov, V. A. (2004). *Phys. Rev. C* **70**, 041601.

[96] Karnaukhov, V. A. *et al.* (2003). *Phys. Rev. C* **67**, 011601(R).

[97] Rodionov V. K. *et al.* (2000). *Nucl. Phys. A* **700**, pp. 457.

[98] Buyukcizmeci, N. *et al.* (2013). *Nucl. Phys. A* **907**, pp. 13–54.

Chapter 3

Nuclear Liquid–Gas Phase Transition in Multifragmentation

Rohit Kumar[a] **and Arun Sharma**[b]

[a]*Facility for Rare Isotope Beams (FRIB), Michigan State University, East Lansing, Michigan 48824, USA*

[b]*Department of Physics, Govt. Degree College Billawar, Jammu 184204, India*

rohitksharma.pu@gmail.com

3.1 Introduction

One of the most important goals of the modern nuclear physics is the better understanding of the various phases of the dense and excited nuclear matter. Of special interest are two phase-transitions: (1) One occurs at high temperature (150 MeV) and high density (several times normal nuclear density), the hadrons to quark-gluon plasma phase. This phase transition is intensively investigated in the experiments at CERN and at the Relativistic Heavy-Ion Collider (RHIC) [Collins (1975)]. (2) The other phase transition occurs at low temperature (few tens of MeV) and density (near normal and subnormal density) nuclear Fermi-liquid to nucleons and fragments [Borderie (2008), Finn (1982), Campi (1988)]. The later one plays a vital role in supernova explosions.

Multifragmentation in Heavy-Ion Reactions: Theory and Experiments
Edited by Rajeev K. Puri, Yu-Gang Ma, and Arun Sharma

ISBN 978-981-4968-69-0 (Hardcover), 978-1-003-38513-4 (eBook)
www.jennystanford.com

The nucleus, if excited to energy greater than its binding energy, breaks into various different chunks (usually termed fragments). This phenomenon, termed multifragmengation, is found to depend on various different entrance channels such as incident energy of the projectile, system mass of the colliding nuclei, isospin asymmetry of the projectile/target etc. [Borderie (2008), Peaslee (1988), Tsang (1993), Schüttauf (1996)]. Also, since the first inclusive heavy-ion experiments, this phenomenon has been linked to the liquid-gas phase-transition or a critical phenomenon. This was considered due to the similarities between the nucleon–nucleon interactions in nuclear matter, and the van der Waals interactions in the gases. Both these interactions, differing by almost five orders of magnitude, have short-range repulsive core and long-range attractive part. To further support this link, the Fermilab-Purdue experiments of p+Kr and p+Xe observed the power-law behavior for the yields of fragments [Finn (1982)]. This behavior was in accordance with the predictions of Fisher's droplet model [Fisher (1967)]. In nuclear matter, the liquid–gas phase transition can be investigated using two different approaches. In the first approach, i.e., the kinetic approach, the phase transition is predicted based on the breaking of nuclei at subnormal density. In the second approach, the decay mechanism of the nucleus is studied as a function of the excitation energy. Among these, the former one is a purely theoretical concept. On the other hand, the advantage with the later one is that it can be studied both in theory and experiments.

In the last three decades, a fantastic progress has been achieved on the topic of liquid–gas phase transition in nuclear matter [Li (1994), Pochodzalla (1995), Ogilvie (1991), Cao (1996), Ma (1999), Raduta (1999), Chomaz (2001), Cole (2002), Borderie (2002), Natowitz (2002), Natowitz (2002b), Ma (2005), Pichon (2006), Gulminelli (2007), Lin (2018), Sood (2019), Sood (2021), Sood (2021b)]. For better insights into the liquid–gas phase transition, numerous order parameters have been introduced in the literature [Li (1994), Pochodzalla (1995), Ogilvie (1991), Cao (1996), Ma (1999), Raduta (1999), Chomaz (2001), Cole (2002), Borderie (2002), Natowitz (2002), Natowitz (2002b), Ma (2005), Pichon (2006), Gulminelli (2007)]. So in addition to the power-law behavior at the critical point, the characteristic signals such as the second moment of the charge distribution, variance of the charge distribution, fluctuation in the first largest fragment, the average charge of the second largest cluster, Zipf's law, and information entropy are also used. The simultaneous study of all these parameters helps to draw a consistent picture of liquid–gas phase transition, as it has been seen that each signal may have its own weak point and so is hardly a definitive proof if taken individually. In this chapter, we will review a few of the characteristic signals advocated to study the liquid–gas phase transition in nuclear matter.

3.2 Various Signals for Liquid–Gas Phase Transitions

Over the years, many different observables have been constructed to study the nuclear liquid–gas phase transition. To discuss all of the proposed characteristic signatures, one will require much time and space. Instead, here our focus will be on signatures that have attracted most attention on the topic.

3.2.1 The Power-Law Behavior

In 1982, the Purdue group in their experiments studied the mass yield of the fragments for the reactions of proton beam on Xe and Kr targets [Finn (1982)]. The fragment yield as a function of fragment mass (A_f) for Xenon target is shown in Fig. 3.1. They found that when the mass yield of the fragments is fitted as a function of the mass of the fragments, it obeys the power-law function ($Y(A_f) \propto A_f^{-\tau}$) over a broad range of yields. The particular value of the slope of the power-law was considered to be the signature of the mechanism of fragment formation. This behavior was conjectured by the Fisher droplet model in which the droplet yields obey a power-law function at the critical point [Fisher (1967)]. Now, this power-law has been well established in the heavy-ion induced reactions. For example, Ogilvie *et al.* studied the reactions of the Au projectile on the targets on C, Al, and Cu at the incident energy of 600 MeV/nucleon using the ALADiN forward spectrometer at GSI, Germany [Ogilvie (1991)]. They found that the fragments follow the power-law behavior and also found

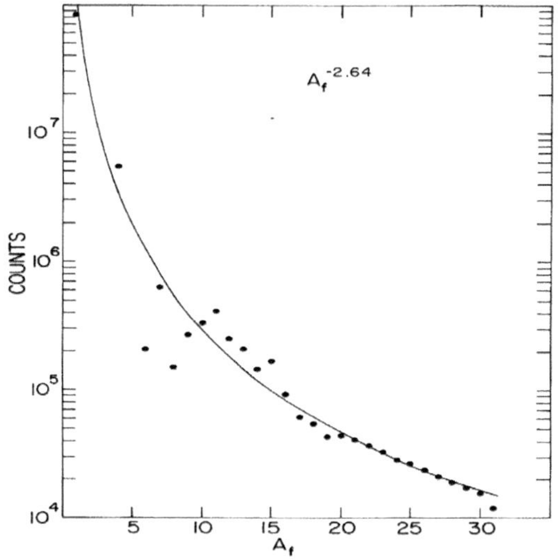

Figure 3.1 The mass yield of fragments as a function of fragment mass (A_f) for Xenon [Finn (1982)].

the minima in the power-law exponent, signal of liquid–gas phase transition, when plotted as a function of the deposited energy.

In another study, performed using the Michigan State University 4π Array, Li *et al.* performed an experiment of ^{40}Ar+^{45}Sc reactions at various projectile beam energies in the range of 15 to 115 MeV/nucleon [Li (1994)]. They fitted the charge spectra of the intermediate mass fragments (IMFs) [$3 \leq Z_f \leq 12$] to obtain the values of the power-law exponent τ. They observed that the slope parameter τ changes from $\simeq$ 1.2 at 25 MeV/nucleon to $\tau \simeq 4.72$ at 115 MeV/nucleon. The excitation energy value corresponding to these energies varies from 8 MeV/nucleon to 29 MeV/nucleon. They have observed the onset of multifragmentation/liquid–gas transition energy to be at 23.9$\pm$0.7 MeV/nucleon corresponding to the minima in τ with a value equal to 1.21. For these reactions, the percolation model calculations predicted the minima in τ at an energy of 28 MeV/nucleon with a value of 1.5. The values obtained for both the energy and slope parameter were overpredicted. In the recent studies, the QMD model was coupled with various clusterization algorithms such as the minimum spanning tree (MST) method, minimum spanning tree with momentum cut (MSTP), minimum spanning tree with binding energy cut (MSTB; thermal binding cut), and more sophisticated energy-based clusterization algorithm, i.e., simulated annealing clusterization algorithm (SACA) (please see chapter by Yogesh K. Vermani for details of the clusterization algorithms) [Sood (2019), Sood (2021b)]. These studies predicted the power-law fit for QMD+MST, MSTP, MSTB', and SACA at 18.03, 19.04, 18.03, and 20.1 MeV/nucleon, respectively. The values of τ were very small for the MST, MSTP, and MSTB' identified fragments. Whereas the QMD+SACA predicts the value of 1.39. It is also worth mentioning that the values of τ at the minima were lesser than the usually predicted value of 2.2. E. E. Zabrodin has shown that the values of τ depends on the fragment's mass range one is fitting. The smaller values of τ is due to the exclusion of the lighter charges from the fitting [Zabrodin (1995)]. In another study, Ma *et al.* have studied the variation of power-law exponent with incident energy for the $^{40}Ar+^{27}Al$ using the QMD model [Ma (1995)]. They obtained the minima in the values of τ at 65 MeV/nucleon. In some studies, exponential fits are used instead of power fits to search for liquid–gas phase-transition energy (see Fig. 3.5a). The value of λ also attains minimum values near the energies where τ attains minimum values. For example, in Fig. 3.5a, we have shown our results of λ as a function of the incident energy using the QMD+SACA model [Sood (2021b)]. We see a minimum value of λ at 23.1 MeV/nucleon, whereas τ shows minimum value at 20.1 MeV/nucleon.

It is worth mentioning that the observation of the power-law behavior is no longer taken as the "proof" of criticality because many other systems exhibit this kind of behavior, e.g., mass distributions of asteroids in the solar system, debris from the crushing of basalt pellets [Hufner (1986)], and the fragmentation of frozen potatoes [Oddershede (1993)]. It has also been seen that the lattice gas model, used extensively to study phase transition, shows a power-law behavior at the critical point, also at and away from the coexistence region [Pan (1995), Pan (1995b)]. It is worth mentioning that in the literature of heavy-ion collisions at intermediate energies, one assumes that seeing the phase transition means the existence of the criticality. One has to hit the right values of density at the right value of temperature, though the later can be controlled by fixing the incident energies that former cannot be controlled. Therefore, it is highly unlikely that the dissociation of the nuclear system occurs at the critical point.

3.2.2 Rise and Fall Behavior of IMFs

The multiplicity of the intermediate mass fragments (IMFs) — usually defined as the fragments heavier than the α clusters and on the higher mass end excludes the fission fragments — is often studied as a function of the incident energy or impact parameter. The copious production of IMFs is considered a consequence of the occurrence of phase transition in nuclear matter. If one studies the IMF production as a function of the incident energy, one sees that the multiplicity values are less at lower incident energies, a few fragments evaporate from the liquid. Also, the multiplicity values are smaller at higher incident energies because due to the complete breaking of the system into lighter charges, the liquid vaporizes into gas. In between these two limits, the maximum multiplicity of IMFs exists. Peaslee *et al.* observed such behavior of IMF multiplicity ($< N_{IMFs} >$) for the central reactions of ^{84}Kr + ^{197}Au in the incident energy range of 35 to 400 MeV/nucleon [Peaslee (1988)]. The results are shown in Fig. 3.2. They obtained a maximum value of IMF multiplicity at 100 MeV/nucleon. Whereas the charged particle multiplicity ($<N_c>$) keeps on increasing with the incident energy. The quantum molecular dynamics (QMD) and QMD + statistical multifragmentation model (SMM) also reported such rise and fall behavior of IMFs. Later on, Sisan *et al.* made a systematic study of the maximum IMF production, where they found the energy corresponding to the maximum production to vary with the system mass [Sisan (2001)]. Such a rise and fall behavior of IMFs with energy is also seen for the Au-induced reactions on Be, C, Al, Cu, and Au targets [Schüttauf (1996)]. The IQMD model is seen to well describe the physics related to the maximum production of IMFs for both symmetric and mass asymmetric reactions [Kaur (2013), Sharma (2021)]. Whereas the percolation model failed to describe the experimentally observed results. A similar rise and fall of IMFs is also seen as a function of the impact parameter of the reactions measured with the ALADiN forward spectrometer at SIS.

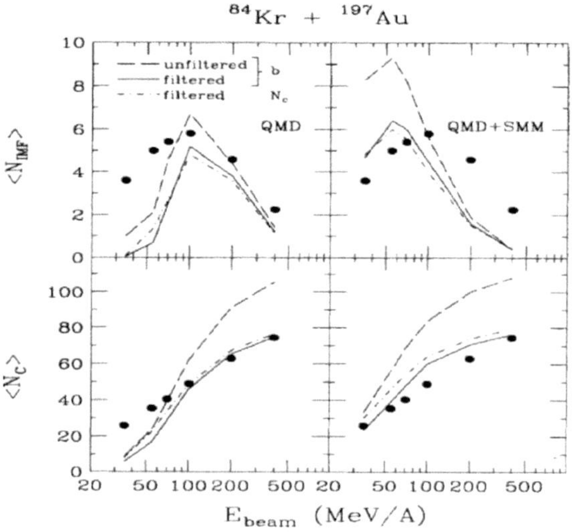

Figure 3.2 The multiplicity of IMFs (Z_f=3–20) (upper panels) and multiplicity of charged particles ($< N_c >$) (lower panels) are plotted as the function of projectile incident energy. The calculations using the QMD model without filters (dashed lines, left panels) and calculations filtered with the experimental acceptance (solid lines, left panels) are shown in the figure. The dashed-dotted lines represent the filtered QMD calculations that were analyzed as data to assess impact parameter fluctuations. The right-hand panels show a similar presentation after coupling the calculations with the secondary evaporation of the excited pre-fragments in the SMM step [Peaslee (1988)].

3.2.3 Flattening of the Caloric Curve

The ALADiN collaboration studied the reactions of the Au+Au at an incident energy of 600 MeV/nucleon and constructed the caloric curve [Pochodzalla (1995)] (results are displayed in Fig. 3.3). They extracted the temperature using the T_{iso} (He-Li) and observed almost constant temperature, i.e., plateau over a broad range of excitation energy range between 3 and 10 MeV/nucleon, followed by a strong increase in the behavior in accordance with the liquid–gas phase transition. This study triggered a lot of activities in the field. Various different thermometers were tested to predict the slope of the kinetic energies (kinetic temperature), the excited states population

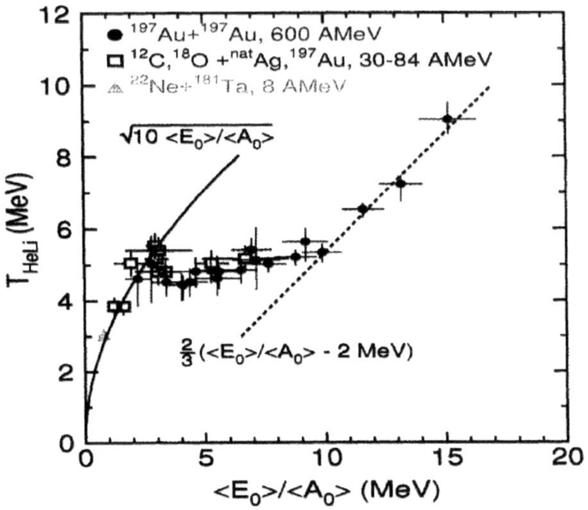

Figure 3.3 Caloric curve of nuclei determined by the dependence of the isotope temperature $T_{He,Li}$ on the excitation energy per nucleon [Pochodzalla (1995)].

(internal temperature), and the population of various isotopes (chemical temperature). Initially, there were some differences among the different thermometers, and different experiments were giving different results. Later, it was found that the disagreement was due to not including some important effects such as cooling of the fragments after freeze out (it can modify the populations of the clusters and their excited states), expansion effects (it can influence kinetic temperatures), and one should not treat the clusters as ideal gas (so its volume should be excluded).

Later, a systematic study was performed and it was found that the system mass of reacting partners can influence the results [Natowitz (2002), Natowitz (2002b)]. They observed that the temperature of the plateau decreases with the increase in the mass of the reacting systems or fragmenting systems. Finally, within the framework of microcanonical multifragmentation model (MMM), different isotopic temperatures were calibrated against pure

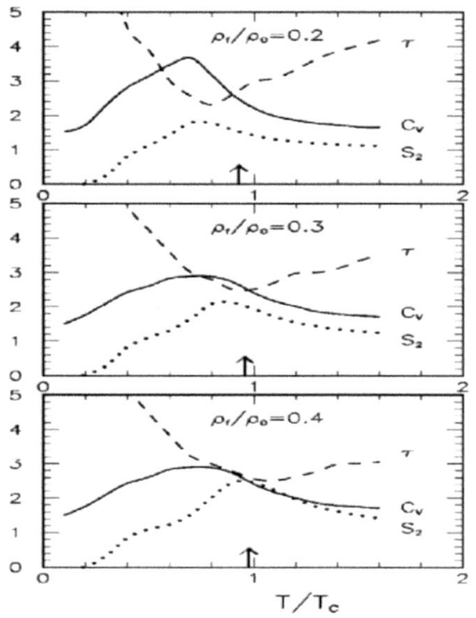

Figure 3.4 Here, values of τ, C_V (units of k_B), and S_2 at ρ_f/ρ_o= 0.2, 0.3, and 0.4 are plotted as a function of temperature normalized to critical temperature (T/T_c; where $T_c = 1.1275|\epsilon|$). At each point, 1000 events were taken [Pan (1998)].

microcanonical one [Raduta (1999)]. A universal relation was found for all masses, which they also used to re-evaluate the already published caloric curves. After corrections, the caloric curves exhibit three parts, the Fermi gas part, a plateau, followed by a linearly rising part, i.e., the classical gas region.

3.2.4 Maximal Fluctuations

Now, near the minimum value of the τ, the fluctuations in the system are maximum. Campi was the first to observe such fluctuations and introduced the other powerful methods based on the conditional moments of the charge distribution to characterize the critical behavior in fragmentation [Campi (1988)]. In general, the kth

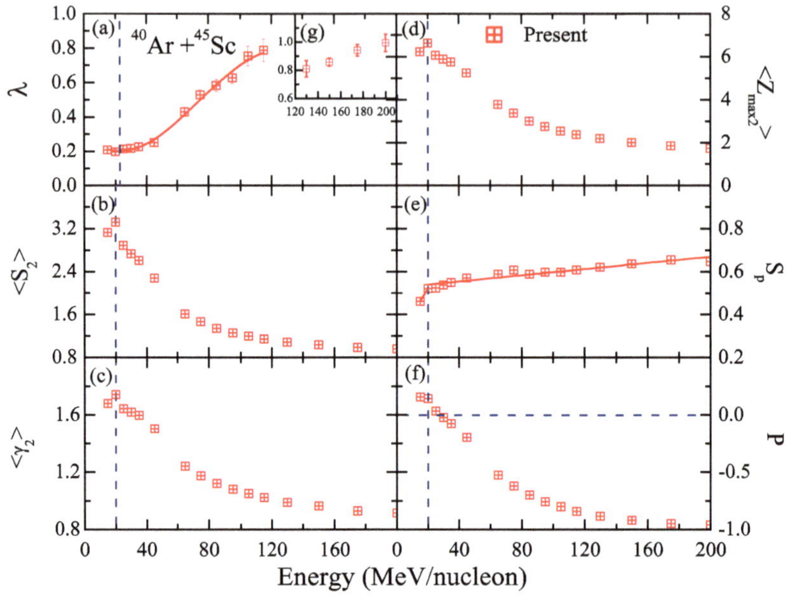

Figure 3.5 We display the values of various critical exponents, i.e., λ (a), S_2 (b), γ_2 (c), $<Z_{max2}>$ (d), S_p (e), and bimodal parameter P(f) for the fragments using the QMD+SACA model [Sood (2021b)].

moment of the cluster charges is defined as:

$$M_k = \sum_{Z_f \neq Z_{max}} Z_i^k n_i(Z_f), \tag{3.1}$$

where $n_i(Z_f)$ is the number of clusters with the charge value Z_f. Here, the sum runs over all the cluster charges except the largest cluster. The EOS collaboration has used these moments to analyze the data of Au+C reactions at 1 AGeV [Gilkes (1994)]. For these conditional moments, one expects an enhancement in the value of the moment M_k for k $> \tau - 1$, for a value of $\tau > 2$. Campi also proposed to use normalized moment ratios $S_k^i = M_k^i/M_1^i$. Of special interest is the S_2, which is expected to be proportional to the compressibility κ_T, if the charge of the first largest fragment is not included. Note that the Campi plots, i.e., the scattered plots between the ln(Z_{max1}) and

$\ln(S_2)$ have also proved to be very instructive to look for the critical behavior. Here one sees maximum fluctuations at the critical point. Within the lattice gas model, Pan *et al.* have observed a peak in the fluctuation for all freeze-out densities of $0.2\rho_o$, $0.3\rho_o$, and $0.4\rho_o$ at the temperature below the critical point but close to the first-order liquid–gas phase transition [Pan (1998)] (results are displayed in Fig. 3.4). Campi introduced another quantity to investigate the critical behavior based on moments, i.e., the relative variance γ_2, defined as:

$$\gamma_2 = \frac{M_2 M_0}{M_1^2} \tag{3.2}$$

The other quantity that attains peak values is the mean charge of the second largest cluster ($< Z_{max2} >$) to predict the critical point of the liquid–gas phase transition [Ma (1995)]. The values of γ_2 and Z_{max2} are expected to show peaks around the critical point. Within the QMD model coupled with various different cluster identifiers, we also observed a peak in fluctuations at the energy corresponding to the minima in τ. We also observed that there is no role of different fragment identifiers in the maximal fluctuations. Also, Ma *et al.* proposed to use the cross sections of different IMF multiplicities and plotted them against the incident energy to predict the critical point [Ma (1995)]. We found this proposed signature do not give a clear signature with the QMD model analysis using different clusterization techniques to obtain fragments [Sood (2021)].

3.2.5 Phase Separation Parameter

Cole introduced the phase separation parameter to separate the two phases in the percolation model calculations [Cole (2002)]. It is defined as the ratio of the masses/charges of the second largest and the first largest fragments ($S_p = \; < A_{max2} > / < A_{max1} >$ or $<Z_{max2}>/<Z_{max1}>$). This was based on the fact that for evaporation-like events, S_p remains almost always less than 0.4. Therefore, if one groups the produced fragments in groups with $S_p \; <$ or ≥ 0.4, the phases can be separated. Later, Ma *et al.*, in their study of reactions ^{40}Ar + ^{27}Al, ^{48}Ti, and ^{58}Ni at an incident energy of 47 MeV/nucleon, also utilized this parameter [Ma (2005)]. In Fig. 3.6, we display the

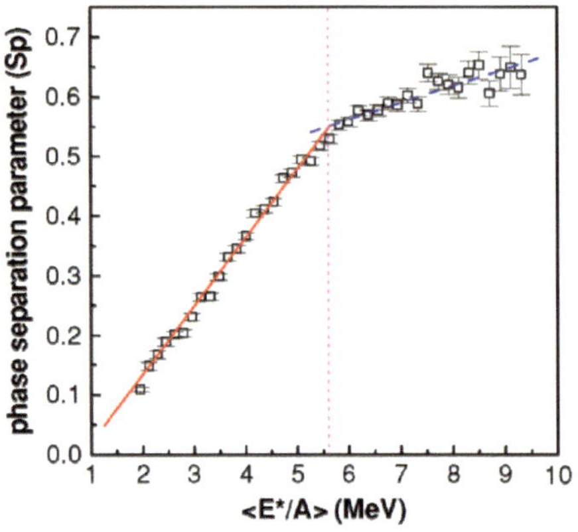

Figure 3.6 The phase separation parameter as a function of excitation energy for the QP formed in $^{40}Ar+^{58}Ni$ [Ma (2005)].

variation of S_p as a function of the excitation energy. We see that at an excitation energy of 5.2 MeV/nucleon, the S_p changes its slope at 0.5. The phase separation parameter (S_p) exhibits two different slopes below and above the transition energy. Within the dynamical model, a similar behavior was seen [Sood (2021b)]. The transition energy predicted by S_p is also consistent with the other characteristic signals and also helps to draw a consistent picture of the liquid–gas phase transition.

3.2.6 Bimodality

Chomaz *et al.* proposed bimodality as the test for the separation of the two phases [Chomaz (2001)]. They interpreted the bimodal behavior of the event distribution as coexistence, with each component supposed to represent the different phases. This leads to the definition of order parameter to isolate the two peaks of the distribution. Borderie *et al.* [Borderie (2002)] at INDRA took the Z_f=12 as the boundary between the two phases and used $\sum_{Z_i \geq 13} Z_i - \sum_{3 \geq Z_i \leq 12} Z_i / \sum_{Z_i \geq 3} Z_i$ as order parameter. This bimodal parameter

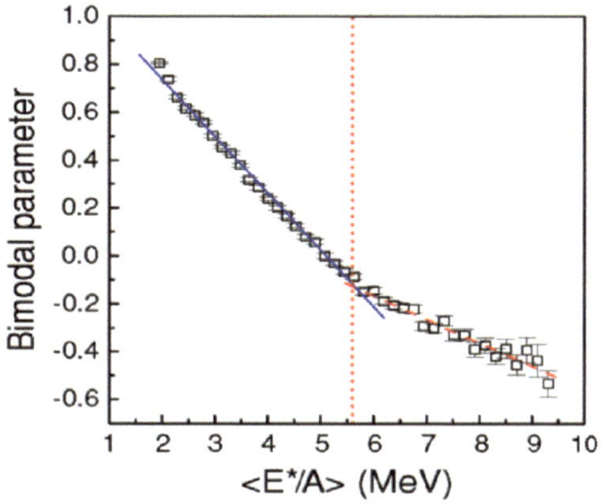

Figure 3.7 Same as Fig. 3.6, but here we display the results for the bimodal parameter (P) [Ma (2005)].

is also considered connected with the density difference between the liquid and gas phases. Ma *et al.* modified the definition for the study of light systems with $Z_f = 4$ as the limit between the two phases [Ma (2005)]. They defined $Z_f \geq 4$ as the liquid phase and $Z_f \leq 3$ as the gas phase. The bimodal parameter was defined as

$$P = \frac{\sum_{Z_i \geq 4} Z_i - \sum_{1 \geq Z_i \leq 3} Z_i}{\sum_{Z_i \geq 1} Z_i}. \tag{3.3}$$

where P shows two different slopes for liquid dominance and gas dominance regions and attains the value $P = 0$ at the point of equal distribution of charges in two phases. In Fig. 3.7, the results are displayed for the bimodal parameter P as a function of the excitation energy of the QP formed in the system of ^{40}Ar+^{58}Ni. Within the dynamical model, this behavior was also observed with the value of order parameter close to zero near the point of liquid–gas phase transition for the reactions of ^{40}Ar+^{45}Sc [Sood (2021b)]. The results are displayed in Fig. 3.5g. For the heavier system of Au+Au studied by the INDRA/ALADiN collaboration, the bimodal parameter was considered to be the asymmetry between the charges of the first two largest charges, $(Z_{max1} - Z_{max2})/(Z_{max1} + Z_{max2})$ [Pichon (2006)]. Later, they performed more elaborated analysis, where Z_{max1} was weighted by a probability given by the excitation energies [Gulminelli (2007)].

3.2.7 Information Entropy

In 1948, the concept of Shannon information entropy was introduced to measure the information contained in a message sent along the transmission line. This concept was introduced by C. E. Shannon [Shannon (1948)]. Later, this concept was used in other fields, e.g., nuclear/particle physics, astrophysics, life sciences, economics, engineering, etc. Cao and Hwa introduced this concept to the field of nuclear/particle physics and studied the copious production of particles in high-energy hadron collisions [Cao (1996)]. They defined information entropy in event space. Later, Y. G. Ma proposed to use the information entropy as a characteristic signal to predict the point of phase transition in fragmenting system [Ma (1999)]. The information entropy constructed in the multiplicity event space of fragments/particles attains a maximum value at the point of the liquid–gas phase transition. This shows that the fragmenting system has the maximum diorder/fluctuation/chaoticity in the event space at transition. The information entropy is constructed as:

$$H = -\sum_{i} p_i \, ln(p_i) \tag{3.4}$$

Here, $\sum_i p_i$=1, and p_i is the probability distribution of the total multiplicities with "I" particles/fragments produced in an event. One should keep this into mind that this quantity is constructed in the event space of fragments, not in phase space. One should make a careful distinction between the information entropy and thermodynamic entropy. The thermodynamic entropy illustrates the heat disorder in momentum space and always increases with temperature, whereas in the case of information entropy, one deals with the event space. In Fig. 3.8, we have shown the results for the isospin-dependent lattice gas model (LGM) with different freeze-out densities and molecular dynamics (MD) model with and without

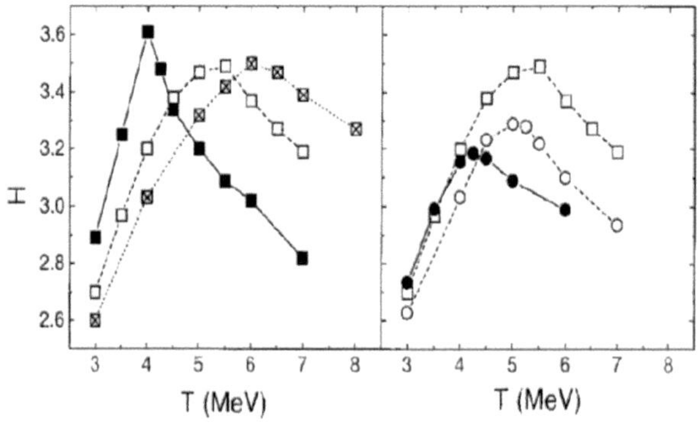

Figure 3.8 (Left panel) The isospin-dependent lattice gas model (LGM) results with different freeze-out densities of ρ_f=018ρ_o (filled), 0.38ρ_o (open), 0.60ρ_o (crossed-squares) and (right panel) the comparison of molecular dynamical (MD) with (filled) and without Coulomb forces (open circles) to LGM with freeze-out density of 0.38ρ_o [Ma (1999)].

Coulomb forces for the Xe nucleus [Ma (1999)]. It was observed that the information entropy attains a maximum value for all the cases, which shows the point of phase transition. This maximum occurs reflecting that the multiplicity values are fluctuating at the liquid–gas transition point. Also, the value of temperature where the phase transition occurs increases with an increase in the value of the freeze-out density for the LGM model. Whereas for the MD case, a slightly light temperature value corresponding to the liquid-gas phase transition is obtained when Coulomb forces are ignored. The value of temperature further decreases with the inclusion of Coulomb forces in the simulations. A recent prediction using the QMD+SACA model (labeled present) is shown in Fig. 3.9 (right panel) [Sood (2021b)]. The maximum of information entropy is seen at 20 MeV/nucleon with corresponding minima in τ at 20.1 MeV/nucleon for the reactions of ^{40}Ar+^{45}Sc. The results were also consistent with other characteristic signals.

3.2.8 Zipf's Law

Zipf's law was first found in linguistics by George Zipf. It states that if the words in a language are placed according to their rank, then the most frequent word appears twice as often as the second most frequent word, three times as often as the third most frequent word, and so on [Ma (1999)]. Later on, many examples were found to show such behavior outside of linguistics. Y. G. Ma [Ma (1999)] proposed to use Zipf's law as the characteristic signal for the liquid–gas phase transition in heavy-ion collisions. According to it, if the charges are placed according to their sizes (Z_r), they will follow Zipf's law at the point of liquid–gas phase transition. For example, the charge of the second largest cluster (Z_{max2}) with rank r=2 will be half the charge of the first largest cluster (Z_{max1}); the charge of the third largest cluster (Z_{max3}) will be one-third the charge of the largest cluster (Z_{max1}), and so on. Therefore, if the obtained values of the size of the clusters of rth rank, i.e., $< Z_r >$ is fitted with the power-law of the form $< Z_r > \propto r^{-\xi}$ (here, ξ is the order parameter). At the point of the liquid–gas phase transition, the ξ=1 was considered to be fulfilled [Ma (1999)]. Zipf's law was found to be fulfilled with the calculations using the LGM model and the MD model [Ma (1999)]. Later, Campi argued that the observation of Zipf's law is just the outcome of the power-law fit and will be fulfilled at energies where the power-law parameter has the value 2 [Campi (2005)]. Neindre also obtained similar results independently [Neindre (2007)]. Recently, we have used the QMD+SACA model to study the reactions of ^{40}Ar+^{45}Sc to check the consistency of Zipf's law with the minima of the power-law parameter and other characteristic signals [Sood (2021b)]. The results are displayed in the left panel of Fig. 3.9. We see that Zipf's law is fulfilled at 35 MeV/nucleon, which is much greater than the energy (23.9 MeV/nucleon) where a minima in τ was obtained. Also note that Bauer *et al.* extended Zipf's law to a more general Zipf–Mandelbrot distribution [Paech (2007)]. As Zipf's law is a direct observable that can be used to characterize the fragment hierarchy in nuclear disassembly, it is still considered a useful signal of phase transition.

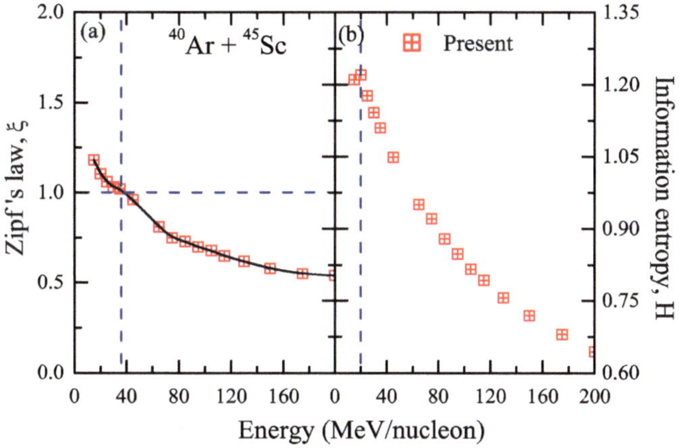

Figure 3.9 (Left) The ξ parameter (from the power-law fit, $<Z_r> \propto r^{-\xi}$), and (right) the information entropy (H) for the central reactions of ^{40}Ar+^{45}Sc is plotted as a function of the incident energy of the projectile. The dashed line corresponds to the value of energy where Zipf's law is followed when $\xi = 1$. The label present corresponds to the QMD+SACA results [Sood (2021b)].

3.3 Summary

In this chapter, we have presented a short review of the experimental as well as theoretical studies of different signals used to predict the liquid–gas phase transition in a multifragmentation heavy-ion reaction. Taken individually, each of the signals discussed in this chapter has its own drawback and weakness. However, taken as a whole, they start to draw a convicting picture of the actual observation of the liquid–gas phase transition in nuclei. Not only the reported signals are qualitative, they are becoming even quantitative, allowing to think about a real metrology of the nuclear phase diagram. However, the order and the nature of the transition are still subject to debate. For a large part, this debate is related to finite-size effects, which remain an important challenge.

References

Borderie, B. (2002). Dynamics and thermodynamics of the liquid–gas phase transition in hot nuclei studied with the INDRA array, *J. Phys. G* **28**, pp. 217.

Borderie, B. *et al.* (2008). Nuclear multifragmentation and phase transition for hot nuclei, *Prog. Part. Nucl. Phys.* **51**, pp. 551.

Campi, X. (1988). Signals of a phase transition in nuclear multifragmentation, *Phys. Lett. B* **208**, pp. 351; ibid., (1986). *J. Phys. A*: Math. Gen. **19**, pp. 917.

Campi, X. and Krivine, H. (2005). Zipf's law in multifragmentation, *Phys. Rev. C* **72**, pp. 057602.

Cao, Z. and Hwa, R. C. (1996). Chaotic behavior of particle production in branching processes, *Phys. Rev. D* **53**, pp. 6608.

Chomaz, Ph., Gulminelli, F., and Duflot, V. (2001). Topology of event distributions as a generalized definition of phase transitions in finite systems, *Phys. Rev. E* **64**, pp. 046114.

Cole, A. J. (2002). Separation and characterization of phases in bond percolation and implications for studies of nuclear multifragmentation, *Phys. Rev. C* **65**, pp. 031601.

Collins, J. C. and Perry, M. J. (1975). Superdense matter: Neutrons or asymptotically free quarks? *Phys. Rev. Lett.* **34**, pp. 1353.

Finn, J. E. *et al.* (1982). Nuclear fragment mass yields from high-energy proton-nucleus interactions, *Phys. Rev. Lett.* **49**, pp. 1321.

Fisher, M. E. (1967). The theory of condensation and the critical point, *Physics*, **3**, pp. 255.

Gilkes, M. L. *et al.* (1994). Determination of critical exponents from the multifragmentation of gold nuclei, *Phys. Rev. Lett.* **73**, pp. 1590.

Gulminelli, F. (2007). Looking for bimodal distributions in multi-fragmentation reactions, *Nucl. Phys. A* **791**, pp. 165–179.

Hufner, J. and Mukhopadhyay, D. (1986). Fragmentation of nuclei, stones and asteroids, *Phys. Lett. B* **173**, pp. 373.

Kaur, S. and Puri, R. K. (2013). Isospin effects on the energy of peak mass production, *Phys. Rev. C* **87**, pp. 014620.

Li, T., *et al.* (1994). Mass dependence of critical behavior in nucleus–nucleus collisions, *Phys. Rev. C* **49**, pp 1630–1634.

Lin, W. *et al.* (2018). Sensitivity study of experimentl measures for the nuclear liquid–gas phase transition in the statistical multifragmentation model, *Phys. Rev. C* **97**, pp. 054615.

Ma, Y. G. (1999). Application of information theory in nuclear liquid–gas phase transition, *Phys. Rev. Lett.* **83**, pp. 3617.

Ma, Y. G. *et al.* (1995). Onset of multifragmentation in intermediate energy light asymmetyric collisions, *Phys. Rev. C* **51**, pp. 710.

Ma, Y. G. *et al.* (2005). Critical behavior in light nuclear systems: Experimental aspects, *Phys. Rev. C* **71**, pp. 054606.

Natowitz, J. B. *et al.* (2002). Caloric curves and critical behavior in nuclei, *Phys. Rev. C* **65**, pp. 034618.

Natowitz, J. B. *et al.* (2002). Caloric curves and nuclear expansion, *Phys. Rev. C* **66**, pp. 031601.

Neindre, N. L. *et al.* (2007). Yield scaling, size hierarchy and fluctuations of observables in fragmentation of excited heavy nuclei, *Nucl. Phys. A* **795**, pp. 47–69.

Oddershede, L., Dimon, P., and Bohr, J. (1993). Self-organized criticality in fragmenting, *Phys. Rev. Lett.* **71**, pp. 3107.

Ogilvie, C. A. (1991). Rise and fall of multifragment emission, *Phys. Rev. Lett.* **67**, pp. 1214.

Paech, K., Bauer, W., and Pratt, S. (2007). Zipf's law in nuclear multifragmentation and percolation theory, *Phys. Rev. C* **76**, pp. 054603.

Pan, J. and Gupta, S. D. (1995). A schematic model for fragmentation and phase transition in nuclear collisions, *Phys. Lett. B* **344**, pp. 29.

Pan, J. and Gupta, S. D. (1995). Unified description for the nuclear equation of state and fragmentation in heavy-ion collisions, *Phys. Rev C* **51**, pp. 1384.

Pan, J., Gupta, S. D., and Grant, M. F. (1998). First-order phase transition in intermediate-energy heavy ion collisions, *Phys. Rev. Lett.* **80**, pp. 1182.

Peaslee, G. F. *et al.* (1994). Energy dependence of multifragmentation in ^{84}Kr + ^{197}Au collisions, *Phys. Rev. C* **49**, pp. 2271.

Pichon, M. *et al.* (2006). Bimodality: A possible experimental signature of the liquid–gas phase transition of nuclear matter, *Nucl. Phys. A* **779**, pp. 267.

Pochodzalla, J. *et al.* (1995). Probing the nuclear liquid–gas phase transition, *Phys. Rev. Lett.* **75**, pp. 1040.

Raduta, A. H. *et al.* (1999). Microcanonical calibration of isotopic thermometers, *Phys. Rev. C* **59**, pp. R1855.

A. Schüttauf *et al.* (1996). Universaility of spectator fragmentation at relativistic bombarding energies, *Nucl. Phys. A* **607**, pp. 457.

Shannon, C. E. (1948). A mathematical theory of communication, *Bell Syst. Tech. J.* **379**, pp. 623.

Sharma, S., Kumar, R., and Puri, R. K. (2021). Role of mass asymmetry on the peak energy of intermediate mass fragments production and its influence towards isospin effects, *Nucl. Phys. A* **1008**, pp. 122144.

Sisan, D. *et al.* (2001). Intermediate mass fragment emission in heavy-ion collisions: Energy and system mass dependence, *Phys. Rev. C* **63**, pp. 027602.

Sood, S., Kumar, R., Sharma, A., and Puri, R. K. (2019). Cluster formation and phase transition in nuclear disassembly using a variety of clusterization algorithms, *Phys. Rev. C* **99**, pp. 054612.

Sood, S., Kumar, R., Sharma, A., and Puri, R. K. (2021). On the fragment production and phase ransition using QMD + SACA model, *Adv. Phys.* pp. 65.

Sood, S., Kumar, R., Sharma, A., Gautam, S., and Puri, R. K. (2021). Fragment emission and critical behavior in light and heavy charged systems, *Chin. Phys. C* **45**, 014101.

Tsang, M. B. *et al.* (1993). Onset of nuclear vaporization in ^{197}Au+ ^{197}Au collision, *Phys. Rev. Lett.* **71**, pp. 1502.

Zabrodin, E. E. (1995). Vapor–liquid phase transition and multifragmentation of nuclei, *Phys. Rev. C* **52**, pp. 2608.

103.

Chapter 4

Nuclear Liquid–Gas Phase Transition: A Theoretical Overview

S. Mallik and G. Chaudhuri

Physics Group, Variable Energy Cyclotron Centre, 1/AF Bidhan Nagar, Kolkata 700064, India

swagato@vecc.gov.in, gargi@vecc.gov.in

4.1 Introduction

Phase transition is a thermodynamic process where a system changes from one phase or state to another by transfer of energy [1]. The study of phase transition is a very exciting topic in different fields of physics like statistical mechanics, atomic and molecular physics, superconductivity, magnetism, etc. For the last four decades, one of the most important motivations of the nuclear physics community is to probe the liquid–gas coexistence region in the phase diagram of nuclear matter [2–8]. The nuclear liquid–gas phase transition plays an important role in estimating the nuclear equation of state (EoS) at finite temperatures. A detailed knowledge of the EoS (i.e., the dependence of the pressure or, alternatively, of the energy per nucleon on the temperature and the density) is essential to describe different aspects of nuclear physics as well as nuclear astrophysics.

Multifragmentation in Heavy-Ion Reactions: Theory and Experiments
Edited by Rajeev K. Puri, Yu-Gang Ma, and Arun Sharma

ISBN 978-981-4968-69-0 (Hardcover), 978-1-003-38513-4 (eBook)
www.jennystanford.com

The most common example of phase transition is water-to-vapor transition [9]. The Lenard–Jones potential for water molecules is repulsive at a very short range due to the overlapping of the electron cloud, and then at a comparatively higher intermolecular separation, it becomes attractive (upper left panel of Fig. 4.1). If one heats water, then initially the supplied heat energy is converted into kinetic energy and the temperature increases. But when the temperature (T) reaches 100^0C, the supplied energy (latent heat) is wholly used to overcome the attractive potential between water molecules; therefore, the temperature remains constant and the water is converted into vapor. After the completion of the conversion

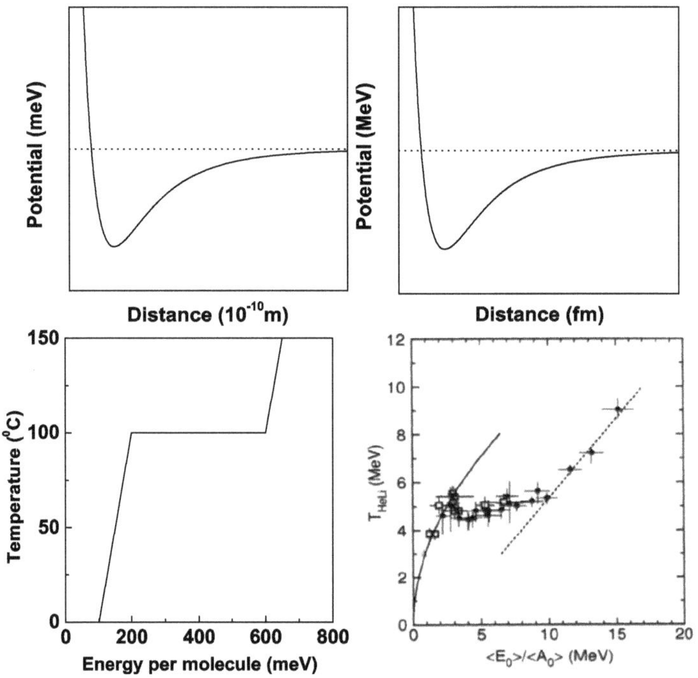

Figure 4.1 Upper panels: Schematic view of radial dependence of molecular (left) and nucleon–nucleon interaction (right) potential. Lower panels: Caloric curves for water to vapor phase transition (left) and nuclear liquid–gas phase transition (right). The diagram of lower right panel is taken from Ref. [10].

from water to vapor, again the temperature of vapor starts to increase. Turning to nuclear physics, the nuclear EoS provides a way to describe the bulk properties of a nuclear many-body system in thermodynamical equilibrium, governed at the microscopic level by the two-body nucleon–nucleon interaction. If one studies the nucleon–nucleon interaction potential (upper right panel of Fig. 4.1), it is observed that its variation with separation distance is similar to that of the Lenard–Jones potential (though the scales are completely different). This occurs at sub-saturation nuclear densities and at a temperature of the order of a few MeVs (1 MeV$= 1.2 \times 10^{10}$ Kelvin).

In the laboratory, the only possible way to achieve such high temperatures is through collisions between atomic nuclei (which can be considered "chunks" of nuclear matter) at intermediate energies (which correspond to projectile beam energy between 25 AMeV and 1 AMeV). Also there are no direct probes to measure this high temperature in experiments. Indirect methods based on models are used to measure it. The collisions between the nuclei are over in 10^{-22} seconds, so one cannot keep the matter in an exotic state long enough to study its properties. The detectors measure only the final products of these collisions, which are in normal states. Hence, one needs to extrapolate from the end products to what happened during disassembly. Traditionally, phase transition is studied in the thermodynamic limit, and for normal liquid or normal gas, the number of particles is very high (10^{23}). But in laboratories, one can get a system containing at most a few hundreds of nucleons, which is far away from the thermodynamic limit. Also, the signals of phase transition are affected due to the presence of the Coulomb force between the protons. Hence, both from theoretical and experimental point of view, nuclear liquid–gas phase transition research is very interesting and highly challenging. Due to the different limitations mentioned above, the conventional definition of normal liquid and gas is not applicable here directly. Generally, a combination of a large nucleus (size almost same as that of the fragmenting system) and a few nucleons is termed nuclear liquid; in addition, there may be a few smaller clusters. On the other side, a large number of free nucleons and a few very light fragments are referred to as a nuclear gas. Therefore, nuclear multifragmentation reaction at intermediate energies is the unique reaction mechanism to study the nuclear liquid–gas phase transition as it covers the entire fragment spectrum.

In order to study the nuclear liquid–gas phase transition and nuclear multifragmentation reactions, different experimental facilities have been developed in different parts of the world like the National Superconducting Cyclotron Laboratory (NSCL) at Michigan State University (MSU, USA), Superconducting Cyclotron at Texas A&M University (USA), Grand Accelerateur National D'ions Lourds (GANIL, France), Heavy-Ion Synchrotron SIS accelerator at Gesellschaft fur Schwerionenforschung mbH (GSI, Germany), Superconducting Cyclotron at Laboratori Nazionali del Sud in INFN, Catania (Italy), etc. In India, the experimental facility is expected to be operational soon at the Variable Energy Cyclotron Centre, Kolkata. Different theoretical models have been developed for understanding the reaction mechanism and explaining the relevant experimental data. These models differ from one another by the respective physical pictures and mathematical foundations adopted by the authors. The theoretical models can be classified into two main categories: (1) dynamical models (Boltzmann–Uehling–Uhlenbeck models [11] and quantum molecular dynamics models [12, 13]) and (2) statistical models (microcanonical statistical multifragmentation model [14], statistical multifragmentation model [15], canonical thermodynamic model [16], grand-canonical model [17], etc.). In addition to the statistical and dynamical models, percolation model [18, 19] and lattice gas model [20] calculations have very remarkable contribution in explaining the nuclear liquid–gas phase transition.

In addition to the nuclear liquid–gas phase transition, at very high energy and/or high baryon density, the nucleons themselves undergo phase transition and produce quark–gluon plasma (QGP), i.e., the transition between the hadronic phase and the QGP phase [21]. A detailed knowledge of this phase transition is important for studying the dynamics of the early universe (deconfined nuclear matter) as well as the inner core of the neutron stars. This is a separate detailed topic, and the discussions of this chapter will be restricted to liquid–gas phase transition only.

After introducing the nuclear phase transition, the questions to be addressed are (1) what is the order of this phase transition and (2) what are the accessible signatures from theoretical calculations as well as from experiments? These will be discussed in the next section.

4.2 Searching for the Signatures and Order of Nuclear Phase Transition

Phase transition is usually characterized by the specific behaviour of state variables like pressure, density, energy, entropy, etc. [22, 23]. The order of phase transition, according to Ehrenfest, is determined by the lowest-order derivative of free energy that shows a discontinuity. In heavy-ion collisions, there is no direct way of accessing these state variables and hence unambiguous detection of phase transition becomes difficult. This shortcoming motivated one to look for signatures of phase transition that can be extracted from the observables that are easily accessible in experiments. Ideally, phase transition exists in the thermodynamic limit and for a first-order one, entropy should have a finite discontinuity and specific heat a divergence at the phase transition temperature. In finite nuclei, the discontinuity or divergence is replaced by a sudden jump or maxima. A lot of effort has been made in order to search for the signatures of nuclear liquid–gas phase transition in heavy-ion collisions around the Fermi energy domain both from theoretical and experimental point of view. The different signals that have been explored so far are the caloric curve [10], the negative heat capacity [24, 25], bimodality in order parameters [26–30], Landau free energy approach [31, 32], fluctuation properties of the largest cluster [33, 34], the moment of the charge distributions [18], Fisher's power-law exponent of fragments [35], and Zipf's law [36].

It has been observed that the variation of total multiplicity or size of the largest cluster (Z_{max}) with temperature is very much similar to that of entropy or excitation energy (caloric curve) with temperature, and this can be seen from Fig. 4.2. Hence, the first-order derivative of these observables with respect to temperature is expected to behave in a similar way as those of entropy or energy. This observation led to the investigation of the nature of the derivatives of these multifragmentation observables, which can be easily measured in experiments. Encouraging results have been obtained from this study, and it has been observed that the first-order derivative of the order parameters related to the total multiplicity and largest cluster size (produced in heavy-ion collisions) exhibits similar behavior as that of the variation of specific heat at constant volume C_V, which is an established signature of the first-order phase transition. This motivates us to propose these derivatives of total multiplicity [37], largest cluster size [38] as confirmatory signals of liquid–gas phase transition. Another observable being studied is related to the difference (normalized) between the sizes of the first and second largest clusters, which also serve as order parameter for phase transition in nuclear fragmentation and has been studied experimentally [39, 40] too. The derivatives of all these peak at the same temperature as specific heat and hence confirm the occurrence of phase transition in the fragmentation process. The measurement of these signals is easily feasible in most experiments as compared to the other signatures like specific heat, caloric curve, or bimodality. In the next section, we will use the statistical model to describe the results using these newly proposed signatures, which involve the derivatives of multiplicity or largest cluster size.

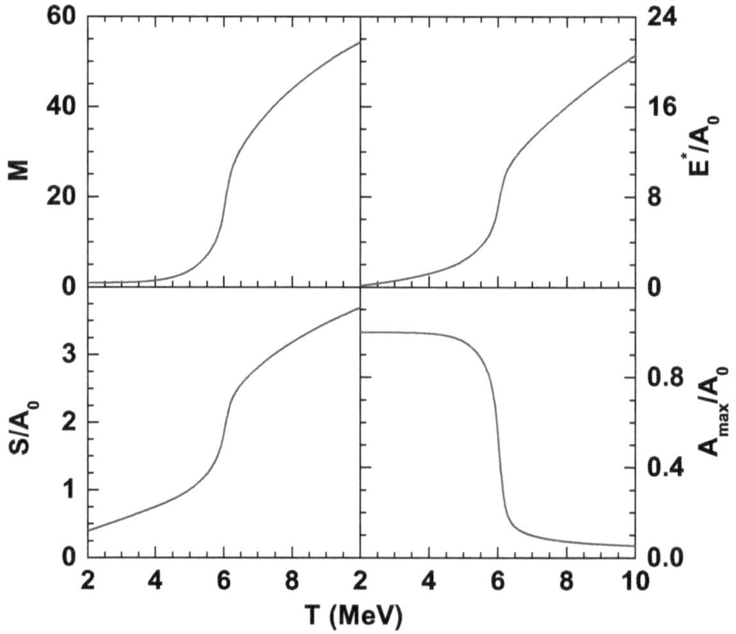

Figure 4.2 Variation of total multiplicity M (upper left panel), excitation energy per nucleon E^*/A_0 (upper right panel), entropy per particle S/A_0 (lower left panel), and average size of the largest cluster A_{max}/A_0 with temperature for the fragmentation of a system of 200 nucleons.

4.3 Statistical Model and Phase Transition Signatures

The statistical models for nuclear multifragmentation are based on the assumption that because of multiple nucleon–nucleon collisions, a statistical equilibrium is reached and disintegration pattern is solely decided by the statistical weights in the available phase space. The temperature rises, and the system expands from normal density and composites are formed on the way to disassembly as a result of density fluctuation. As the system reaches between three and six times the normal volume, the interactions between composites become unimportant (except for the long-range Coulomb interaction), and one can do a statistical equilibrium calculation to obtain the yields of composites at a volume called the freeze-out volume. This model can be implemented in different statistical ensembles (microcanonical, canonical, and grand canonical) [14–16]. For nuclear systems, where the number of particles is finite (as it would be in experiments), the partitioning into the available channels is solved in the canonical ensemble, and the standard version, the Canonical Thermodynamical Model (CTM) [16], is essentially used for the purpose. The study is done for different nuclear sizes, freeze-out volumes, and temperatures. Since Coulomb interaction is long range and suppresses the signatures of phase transition, it has been switched off in some part of the study in order to have a better idea of the signatures. In such cases, symmetric nuclear matter is considered and no distinction is made between neutron and proton.

One can measure the total multiplicity $M = \sum_A M_A$ (A being the mass number of the composites) with 4π detectors in the laboratory. In CTM, the derivative of M with T is seen to have a maximum. Figure 4.3 (left panel) shows the derivative of the total multiplicity dM/dT for fragmenting system having the proton number (Z_0)=82 and the neutron number (N_0)=126 (top) and that having Z_0=28 and N_0=30 (bottom), respectively. Results for both real nuclei (left panel) and the one for one kind of particles with no Coulomb (right panel) have been displayed in order to emphasize the effects of Coulomb interaction. The rise and the peak are much sharper in the absence of Coulomb interaction clearly indicating the role of the long-range

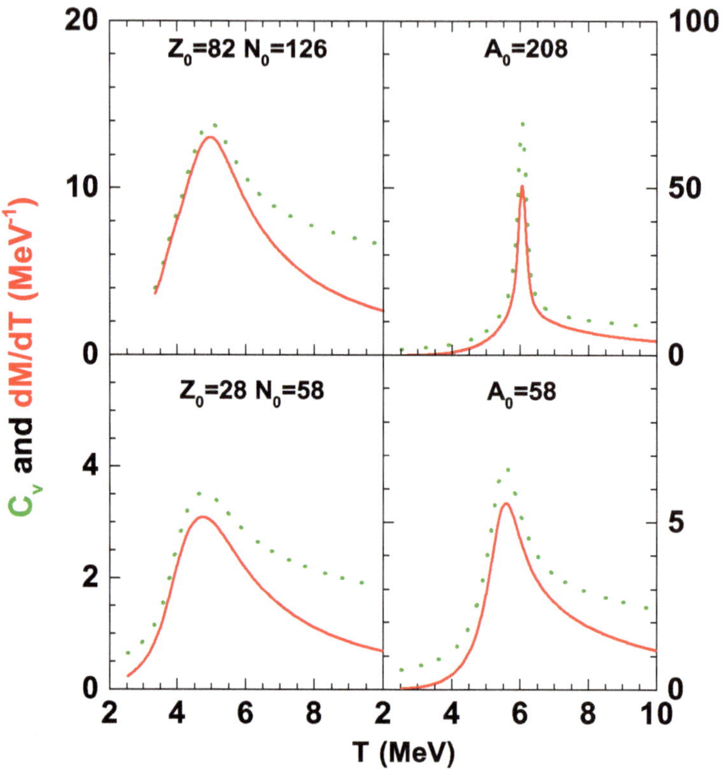

Figure 4.3 Variation of $\mathrm{d}M/\mathrm{d}T$ (red solid lines) and C_v (green dashed lines) with temperature from CTM for fragmenting systems having Z_0=82 and N_0=126 (upper left panel), Z_0=28 and N_0=58 (lower left panel), and for a hypothetical system of one kind of particle with no coulomb interaction of mass number 208 (upper right panel) and 58 (lower right panel). To draw $\mathrm{d}M/\mathrm{d}T$ and C_v in the same scale, C_v is normalized by a factor of 1/50.

interaction in suppressing the signatures of phase transition. The features become less sharp in Z_0=28 and N_0=30, as the system size decreases and can be attributed to the finite size effect of phase transition. We have also plotted C_v for all these systems, and it is seen that the peak in dM/dT coincides with the maximum of the specific heat at constant volume C_v as a function of temperature for all the cases. It is an established fact that the specific heat at constant volume peaks at the transition temperature, and this is a signature of the first-order phase transition. Hence, based on our results as presented in Fig. 4.3, we conclude that dM/dT can be a signature of phase transition and the advantage is that it gives an exact value of the transition temperature where the maximum of dM/dT occurs.

In laboratory experiments, the excited fragments produced after the multifragmentation stage decay to their stable ground states before reaching the detectors. Therefore, it is important to study how secondary decay affects the multiplicity derivative signal. In order to do that, the secondary decay of excited fragments is studied from the evaporation model [41] based on Weisskopf formalism and the multiplicities of the primary and the secondary fragments and their derivatives with respect to temperatures are plotted in Fig. 4.4. It is apparent that the effect of secondary decay does not alter our previous observation. Moreover, it enhances the signals, the total multiplicity changes more rapidly, and the peak in dM/dT is sharper in the case of the secondary fragments. Thus, the maxima of multiplicity derivative can be extracted successfully through experiments with an unaltered transition temperature.

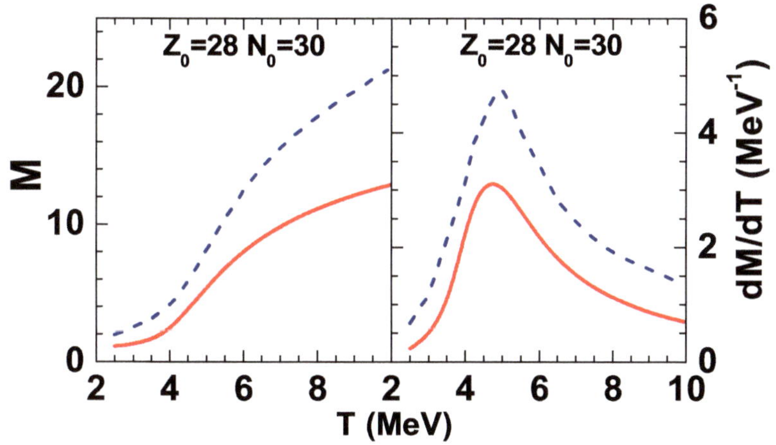

Figure 4.4 Effect of secondary decay on M (left panel) and dM/dT (right panel) for fragmenting systems having Z_0=28 and N_0=30. Red solid lines show the results after the multifragmentation stage (calculated from CTM), whereas blue dashed lines represent the results after secondary decay of the excited fragments.

The proposed signal of multiplicity derivative dM/dT was tested and verified in different statistical and dynamical models like the statistical multifragmentation model (SMM) [42, 43], quantum molecular dynamics (QMD) model [44], and nuclear statistical equilibrium (NSE) model [45]. The theoretical proposition of this signal got further support when it was experimentally verified by

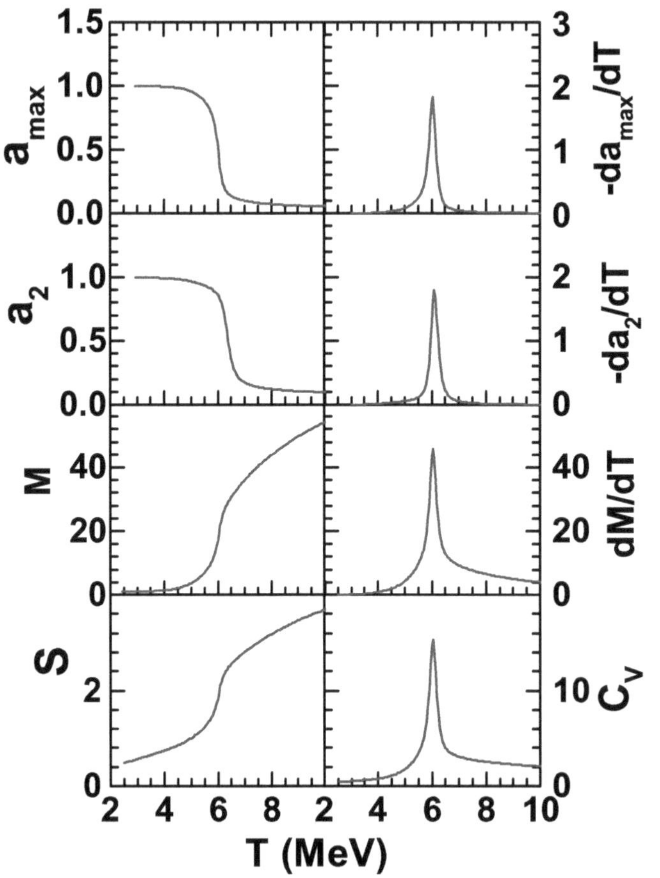

Figure 4.5 Variation of a_{max} (upper left panel), $-da_{max}/dT$ (upper right panel), da_2 (middle upper left panel), $-da_2/dT$ (middle upper right panel), M (middle lower left panel), dM/dT (middle lower right panel) and S (lower left panel) and C_v (lower right panel) with temperature for fragmenting system of mass $A = 200$.

measuring the quasi-projectile reconstructed from the reactions $^{40}Ar + ^{58}Ni$, $^{40}Ar + ^{27}Al$, and $^{40}Ar + ^{48}Ti$ at 47 AMeV performed at Texas A&M university K=500 superconducting cyclotron [46].

The average size of the largest cluster A_{max} formed in the fragmentation of the excited nuclei acts as an order parameter for the first-order phase transition. The variable a_2, which is a measure of the difference between the average size of the largest (A_{max}) and the second largest cluster size (A_{max-1}) divided by the sum of these two ($a_2 = (A_{max} - A_{max-1})/(A_{max} + A_{max-1})$), also has similar behaviour as that of A_{max}. So this observable, which is measured in some experiments, can also act as an order parameter. The analytical expressions leading to the calculation of the average size of the first and second largest clusters can be found in Ref. [38]. We consider an ideal system of A=200 identical nucleons with no Coulomb force acting between them in order to have a better idea of these proposed signatures. The left panels of Fig. 4.5 display the variations of the four variables, the normalized size of the average largest cluster ($a_{max} = A_{max}/A_0$), a_2, total multiplicity M, and entropy per particle (S/A_0) with temperature. a_{max} and a_2 are almost constant and assume a value ≈ 1 up to approximately 5 MeV, in the temperature scale. This implies that in this temperature range, the size of the largest fragment produced is almost the same as the size of the fragmenting source. Around $T = 6$ MeV, both of them fall sharply to a very low value near zero, which indicates the entire system fragments into the light mass nuclei. After that, they remain almost unchanged. These observables, clearly, indicate a sharp transition near $T = 6$ MeV and therefore behave as an order parameter of the nuclear phase transition. a_{max} and a_2 display similar behaviour as that of the multiplicity and the entropy; the sudden jump (or fall) of these four variables occurs almost at the same temperature around 6 MeV. This similarity motivates us to investigate the behaviour of the derivatives of a_{max} and a_2. In the right panel of Fig. 4.5, the temperature derivatives of all the four quantities are plotted as a function of temperature. In the right bottom panel of Fig. 4.5, C_v is plotted, which is related to the temperature derivative of the entropy (S). The negatives of the derivatives of a_{max} and a_2 exhibit maxima just like the total multiplicity and specific heat, and almost at the same temperature, which we call the transition temperature. This establishes these two variables as signatures of phase transition. This signature is much

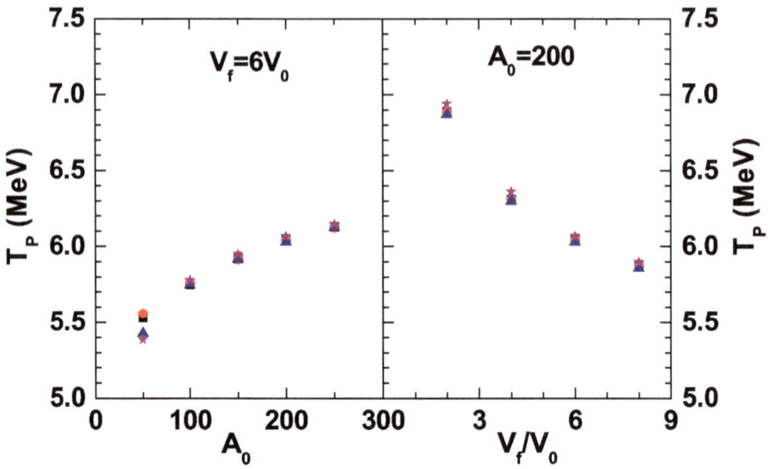

Figure 4.6 Dependence of the peak position of $-da_{max}/dT$ (blue triangles), $-da_2/dT$ (magenta stars), dM/dT (red circles) and C_v (black squares) on fragmenting system size (left panel) and freeze-out volume (right panel).

easier to access both theoretically and experimentally as compared to the bimodality in the probability distribution of the largest cluster. The later has been used so far in order to detect the existence of phase transition in nuclear multifragmentation, but detecting two peaks (bimodality) of equal height in a distribution at a particular temperature (or excitation energy) is far more a difficult job than to simply calculate the derivative in its size with temperature or excitation energy. This new proposed signature related to the largest cluster size will definitely provide a great impetus to the study of liquid–gas phase transition in heavy-ion collisions.

In Fig. 4.6, the variation of the transition temperatures as a function of system size A_0=50, 100, 200 at a fixed freeze-out volume $V_f = 6V_0$ is shown (left panel), and as a function of freeze-out volume V_f=3V_0, 4V_0, 8V_0 with fixed source A_0=200 (right panel). In each panel, four different sets of transition temperatures are plotted. Those sets are obtained from the position of the maxima in $-da_{max}/dT$, $-da_2/dT$, dM/dT, and C_v. The transition temperatures obtained from all the four observables give consistent results. Small differences between them can be attributed to the finiteness of the fragmenting system. The variation implies that the smaller system

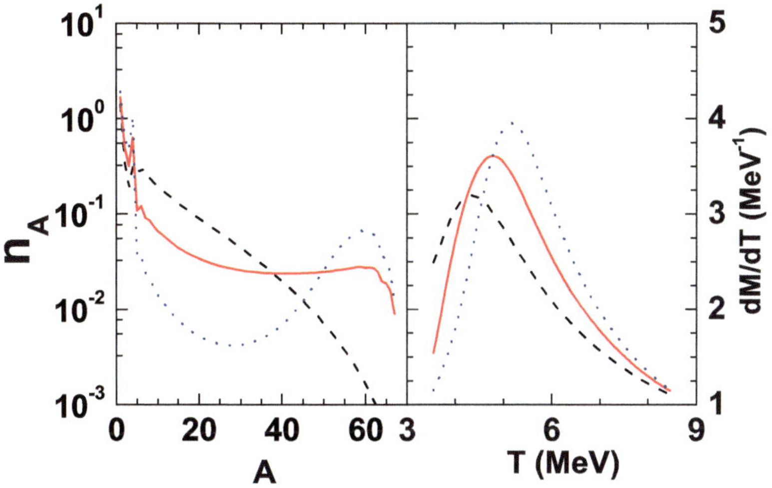

Figure 4.7 Mass distribution at temperature T=5 MeV (left panel) and dM/dT variation with temperature for three different surface energy coefficients (a_{s0}) 15 MeV (blue dotted lines), 18 MeV (red solid lines), and 21 MeV (black dashed lines).

fragments more easily at a lower transition temperature as compared to its bigger counterparts. The peak also becomes sharper for bigger sources, which once again proves that phase transition signals are enhanced in larger systems. For freeze-out volume, the result that we have obtained is expected, since higher freeze-out volume (lower density) will favour the disintegration of the nucleus, resulting in lower transition temperature.

The effect of the different parameters used in the liquid drop model of the nuclear binding energy, which is the main ingredient of the canonical thermodynamical model (CTM), has been examined on the phase transition temperature, and it is clearly seen that the surface energy part has a significant influence on the process of liquid–gas phase transition [47]. The change in surface energy coefficient a_{s0} (the temperature-dependent surface energy $[a_s(T) = a_{s0}\{(T_c^2 - T^2)/(T_c^2 + T^2)\}^{5/4}$ with $T_c = 18.0$ MeV]) shifts the transition temperature considerably (right panel of Fig. 4.7), and this is also evident when one calculates the mass distribution of a fragmenting nucleus of mass 67 and charge 32 for three different values of a_{s0} at a fixed temperature (left panel of Fig. 4.7). This variation is pretty similar to the variation in mass distribution at different temperatures, and one can conclude that surface energy plays the equivalent role of temperature in the nuclear liquid–gas phase transition. The effects of other parameters in the liquid drop model as well as those in the Fermi gas model were also examined, and they were found not to be significant [47]. Hence, the surface energy coefficient needs to be determined with better precision from microscopic calculation in order to have a better understanding of the phase transition process.

4.4 Dynamical Model and Phase Transition Signatures

The standard statistical model studies on the liquid–gas phase transition at intermediate energy collisions assume that because of two-body collisions, nucleons reach thermodynamical equilibrium at a constant volume (for example CTM studies described in Section 4.3) or constant pressure, and then multifragmentation occurs. This section will focus the phase transition signatures obtained from the dynamical model based on the Boltzmann–Uehling–Uhlenbeck (BUU) equation [4, 11], which bypasses all such assumptions.

The BUU transport model calculation starts with two nuclei in their respective ground states approaching each other with specified velocities and impact parameters. The ground-state energies and densities of the projectile (mass number A_p) and target (mass number A_t) nuclei are constructed using the Thomas–Fermi approximation (Appendix E of Ref. [4]). The Thomas–Fermi phase space distribution is then sampled using the Monte Carlo technique by choosing test particles (we use $N_{test} = 100$ for each nucleon) with appropriate positions and momenta. As the projectile and target nuclei propagate in time, the test particles move in a mean field and occasionally suffer two-body collisions, with probability determined by the nucleon–nucleon scattering cross section, provided the final state of the collision is not blocked by the Pauli principle. The standard BUU model describes the properties of the average of all events. But to study signatures like mass distribution, the largest cluster probability distribution, etc. from the transport model calculation, one needs an event-by-event description, not just the average of all events. In order to do that, we have followed the recently developed computationally efficient prescription described in Ref. [48], which lies midway between the original BUU calculation [11] and the original fluctuation added model [49]. According to this prescription, the nucleon–nucleon collisions are computed at each time step with the physical cross section σ_{nn} only among the $A_p + A_t$ test particles belonging to the same event. For each event, if a collision between two test particles i and j is allowed, the method proposed in Ref. [49] is followed: The $N_{test} - 1$ test particles closest to i are picked and the same momentum change $\Delta\vec{p}$ as ascribed to i is given to all of them. Similarly, the $N_{test} - 1$ test particles closest to j are selected, and these are ascribed the same momentum change $-\Delta\vec{p}$ suffered by j. As a function of time, this is continued till the event is over. For the mean-field propagation, the Vlasov technique is employed: All test particles are used, and the Lattice Hamiltonian method [50] is used for calculating the mean field potential. This procedure is repeated for as many events as one needs to build up enough statistics. The details of the BUU transport model calculation can be found in [4, 48, 51]. This method allows the calculation of fluctuations in systems much larger than what was considered feasible in a well-known and already existing model [49], which is required for studying phase transition (as finite number effects often hide the bulk effects). As we are interested in phase transition under the influence of nuclear force, throughout this section the Coulomb effects are switched off. The BUU model with Coulomb and isospin effect can be found in Ref. [52, 53] for different applications related to symmetry energy. Comparison of different observables from different BUU and QMD models can be found in Ref. [54].

In order to study the nuclear phase transition, central collision reactions between projectile of mass $A_p = 120$ on target of mass $A_t = 120$ are simulated at four different beam energies (E_p). Figure 4.8 shows the mass distribution obtained from BUU calculation. For each energy, 1000 events are taken and time evolution is stopped at 200 fm/c. The results of averages for groups of five consecutive mass numbers are shown. At low beam energy (50 MeV/nucleon), the multiplicity first falls with mass number a, reaches a minimum, then rises again, and finally reaches a maximum before disappearing. As the beam energy increases, the height of the

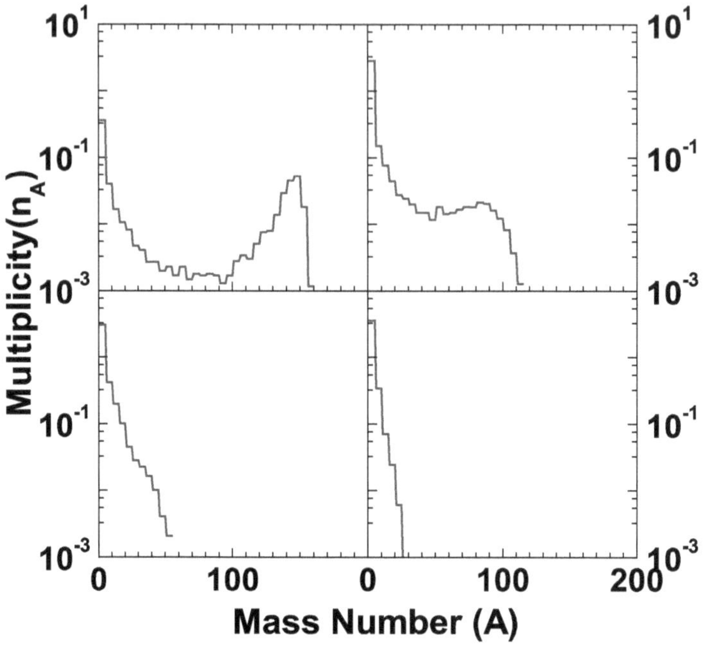

Figure 4.8 Mass distribution from BUU calculation for $A_p = 120$ on $A_t = 120$ reaction at beam energies (a) 50 AMeV, (b) 75 AMeV, (c) 100 AMeV, and (d) 150 AMeV.

second maximum decreases. At E_p =75 MeV/nucleon, the second maximum is still there but barely visible. At higher energies, the multiplicity is monotonically decreasing, the slope becoming steeper as the beam energy increases. The disappearance of the second maximum indicates phase transition, which was obtained earlier from CTM calculation.

The double humped distribution (hence the name bimodality) of the largest cluster probability in nuclear multifragmentation is also a measurable signature of the nuclear liquid–gas phase transition [3, 7, 55]. Some theoretical calculations [39, 40, 51, 56] as well as experimental measurements [57–59] conclude that it is due to the presence of memory effect of entrance channel where thermal equilibrium is not achieved in this energy domain. The signal was interpreted in these studies as a dynamical bifurcation of reaction mechanism, induced by the fluctuation of collision rate,

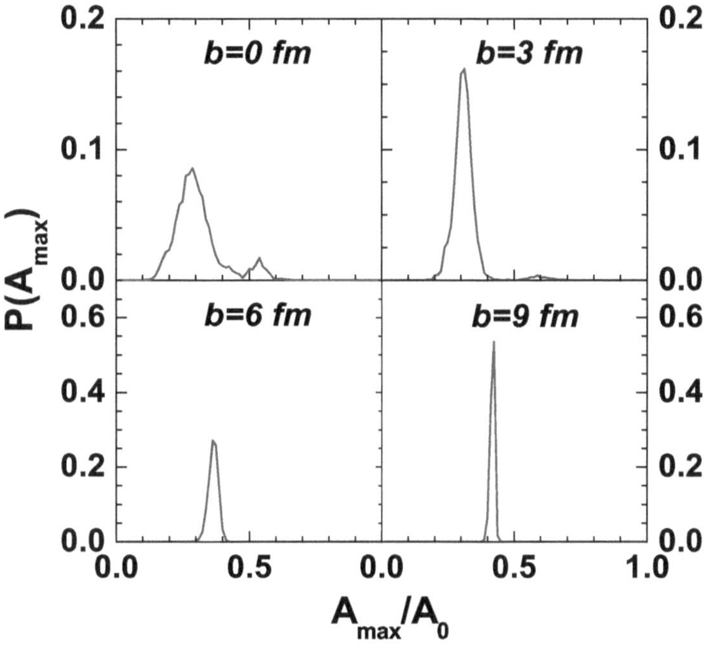

Figure 4.9 Largest cluster probability distribution $P(A_{max})$ at freeze-out stage (t=175 fm/c) for constant projectile beam energy 100 AMeV but four different impact parameters (a) b =0 fm, (b) b =3 fm, (c) b =6 fm, (d) b =9 fm calculated from the BUU model.

which leads to fluctuations of collective momentum distribution. Other successive theoretical studies (from lattice gas model and statistical models) [26, 60–65] as well as experimental observations [66–68] establish the equilibrium scenario of bimodality, which would rather point toward a thermal phase transition. In order to study the combined effect of entrance channel and exit channel on bimodality, a single symmetric system $^{40}Ca + ^{40}Ca$ is considered with projectile beam energy 100 MeV/nucleon at different impact parameters by switching off the Coulomb interaction. The dynamical stage is simulated by the BUU transport model. The largest cluster probability distribution is shown in Fig. 4.9 for four different impact parameters at freeze-out time where we have decided to stop the dynamical calculation. For central collision (b=0 fm), two peaks are seen, which can be interpreted as dynamical bimodality very similar to the phenomenon described in Ref. [39]. Fluctuations in the collision rates lead to fluctuations in the momentum distribution, that is, in the degree of stopping of the reaction. A higher mass peak represents stopped events having nearly a zero z-component (beam direction) of momentum and scattered isotropically in the center of mass frame, whereas fragments in the lower mass peak represent crossed events having a high z-component of momentum and scattered either in the forward direction (projectile-like fragments) or in the backward direction (target-like fragments) [69]. For non-central cases, only liquid phase is present (crossed events). Freeze-out condition is identified (175 fm/c for 100 AMeV reaction) in transport simulation from the isotropy of momentum distribution and maxima of the average size of second largest cluster [69]. The distribution can still evolve in subsequent time because of the secondary decay, which has been calculated by switching over to

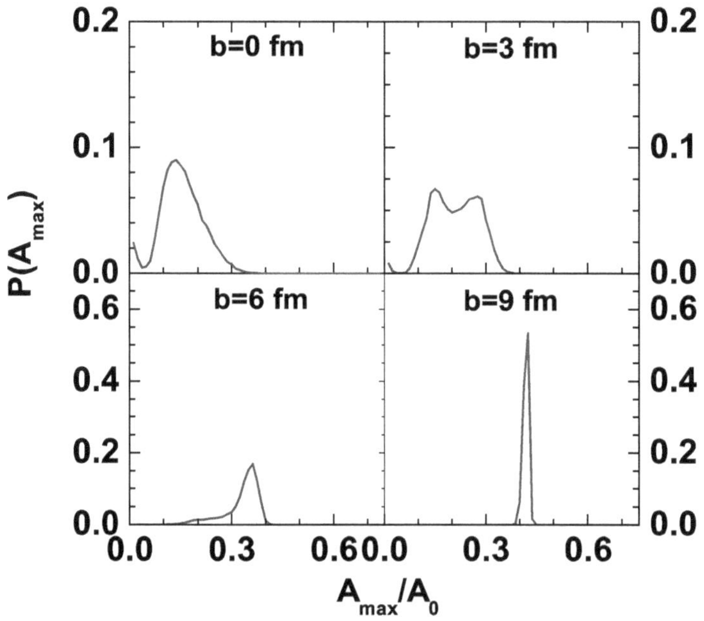

Figure 4.10 Largest cluster probability distribution $P(A_{max})$ after secondary decay for constant projectile beam energy 100 AMeV but four different impact parameters (a) $b = 0$ fm, (b) $b = 3$ fm, (c) $b = 6$ fm, (d) $b = 9$ fm calculated from the BUU model.

the CTM from the transport one and the final result is shown in Fig. 4.10. The behavior at $b = 0$ fm is structures less and typical of multifragmentation reactions: The average excitation energy is so high in this case that both fully stopped and incompletely stopped events undergo multiple decay. As a consequence, the bimodality signal observed in Fig. 4.9 disappears. At mid-central collision, the situation is reversed. The probability distribution of the largest cluster now shows a bimodal behaviour, which is indicative of the existence of two phases simultaneously. This, however, strongly depends on the entrance channel conditions. In particular, central collisions at lower bombarding energy (40 AMeV) lead to a situation where the freeze-out distribution is not distorted by secondary decay and bimodal behaviour can be observed both after transport calculation, and after the statistical model calculation [69]. Hence, one can conclude that depending on the incident energy and the impact parameter of the reaction, both entrance channel and exit channel effects can be at the origin of the observed bimodal behavior. Specifically, fluctuations in the reaction mechanism induced by fluctuations in the collision rate, as well as thermal bimodality, are directly linked to the nuclear liquid–gas phase transition.

Apart from statistical and dynamical model studies, remarkable progress on theoretical studies of nuclear phase transition has been made by percolation and lattice gas model calculations. Brief descriptions of these two models and some recent application of them on phase transition study will be described in the next section.

4.5 Phase Transition Signatures from Lattice Gas Model and Percolation Model

The bond percolation [70] is a model of continuous phase transition, which has been extensively used in the past to establish a link with experimental data [18, 49]. A nucleus of N^3 nucleons is considered, each of which are in N^3 boxes with no distinction being made between neutrons and protons. Nearest neighbours (these have a common wall) can bind together with a probability p_s, which is the sole parameter in the model. If p_s is 1, there is just one nucleus with N^3 nucleons and multiplicity M=1. If p_s is 0, there are N^3 monomers and M=N^3. For intermediate values of p_s, nucleons can group into several composites, and one needs to do Monte Carlo sampling to generate clusters. If the average number of clusters of A nucleons is n_A, then M=$\sum_A n_A$. Instead of plotting M against p_s, we plot M against p_b=$1 - p_s$, which is the bond-breaking probability. The left panel of Fig. 4.11 plots M and $\frac{dM}{dp_b}$ in the range of p_b 0 to 1. Both multiplicity and its derivative differ from those of CTM results (see Section 4.3). This confirms that the percolation model has no first-order phase transition and hence there is no maximum in the M derivative.

The M derivative is next examined in the lattice gas model [8, 71, 72], which uses similar geometry as that of percolation but

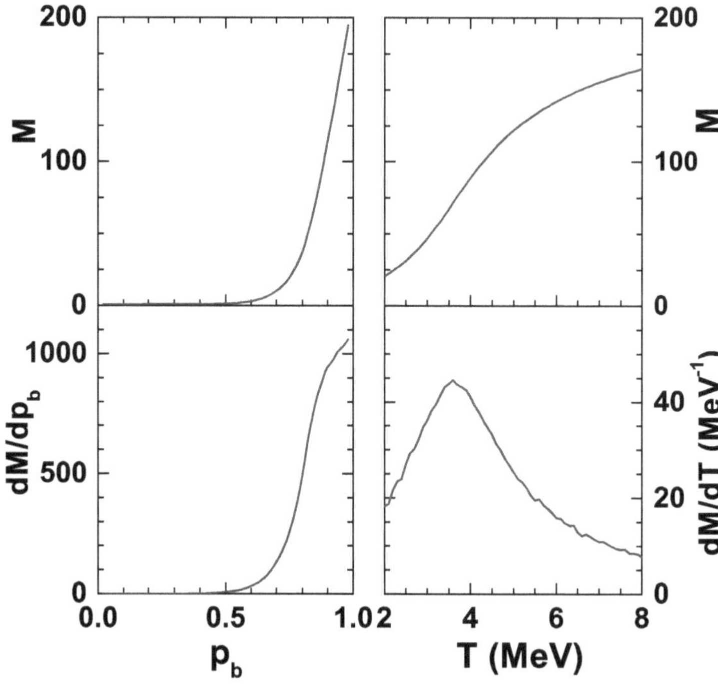

Figure 4.11 Left Panels: Variation of M (upper left panel) and $\frac{dM}{dp_b}$ (lower left panel) with p_b obtained from bond percolation model for a system of 6^3 nucleons. Right Panels: Variation of M (upper right panel) and $\frac{dM}{dT}$ (lower right panel) with T obtained from lattice gas model for fragmenting system having Z=82 and N=126.

has a Hamiltonian. Let $A_0 = N_0 + Z_0$ be the number of nucleons in the system that dissociates. We consider D^3 cubic boxes where each cubic box has volume $(1.0/0.16)fm^3$. D^3 is larger than A_0 (they have the same value in the bond percolation model). Here $D^3/A_0 = V_f/V_0$ where V_0 is the normal volume of a nucleus with A_0 nucleons and V_f is the freeze-out volume where partitioning of nucleons into clusters is calculated. One adopts nearest neighbour interactions for nuclear forces. We use neutron–proton interactions v_{np}=-5.33 MeV and set $v_{nn} = v_{pp}$=0.0 following the standard practice. The coulomb interaction between protons is included. Each cube can contain 1 or 0 nucleon. A very large number of configurations are possible (a configuration designates which cubes are occupied by neutrons, which by protons, and which are empty; we sometimes call a configuration an event). Each configuration has an energy. If a temperature is specified, the occupation probability of each

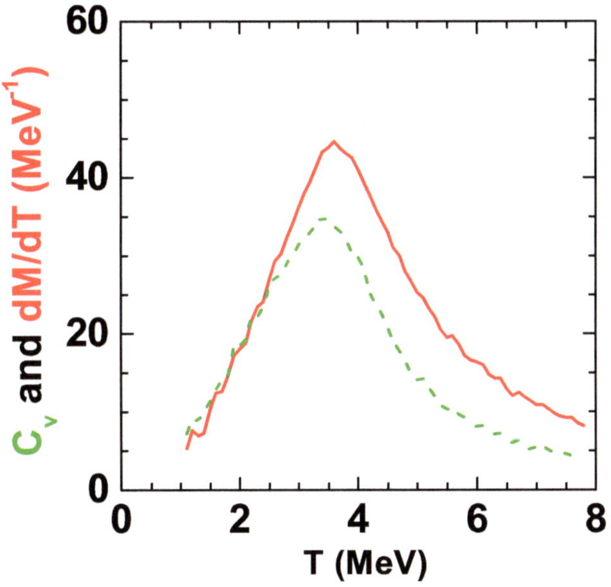

Figure 4.12 Variation of $\mathrm{d}M/\mathrm{d}T$ (red solid lines) and C_v (green dashed lines) with temperature from lattice gas model at D=8 (see text) for fragmenting system having Z_0=82 and N_0=126. To draw $\mathrm{d}M/\mathrm{d}T$ and C_v in the same scale, C_v is normalized by a factor of 1/10.

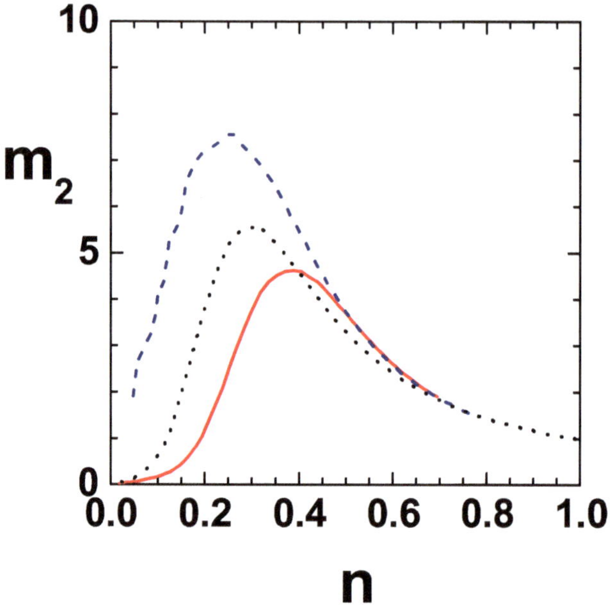

Figure 4.13 Variation of m_2 with n calculated from the lattice gas model at D=7 (red solid line) and D=8 (blue dashed line) and the percolation model (black dotted line) for the fragmenting system having Z=82 and N=126.

configuration is proportional to its energy: P $\propto$ exp(-E/T). This is achieved by the Monte Carlo sampling using Metropolis algorithm. The details of the model can be found in [72, 73].

The variation of M and dM/dT with the temperature obtained from the lattice gas model is shown in the right panel of 4.11. The plots of dM/dT and $d < E > /dT$ are shown in Fig. 4.12. Note that C_v goes through a maximum at some temperature, which is a hallmark of the first-order phase transition, and this occurs at the same temperature where dM/dT maximizes [73]. This is remarkably different from the percolation model results but very similar to the CTM results of Section 4.3 corroborating the evidence that the appearance of a maximum in dM/dT is indicative of a first-order phase transition.

One can define a reduced multiplicity $n = M/A_0$ where $A_0 = N^3$ is the number of nucleons. Also define $m_2 = [\sum A^2 n_A - A^2_{max}]/A_0$ where

A_{max} is the largest cluster in an event. For the percolation model, we pick a p_b and get n, n_A, and A_{max}^2 by averaging over 50,000 events. Thus, we can plot m_2 against n as seen in Fig. 4.13. m_2 displays a maximum, which becomes sharper as particle number is increased. The experimental data of Au on emulsion [18, 74] display the same behaviour of m_2. This is regarded as a manifestation of the second-order transition. The results from the lattice gas model also display the same behaviour, but it has a first-order phase transition. Hence, one can conclude that m_2 is not a confirmatory signal.

4.6 Hypernuclear Phase Transition

The study of hypernuclei formed in heavy-ion collisions (when the strange hadrons are captured by the nuclei) is a major area of research in the domain of high-energy nuclear physics. The role of the strange particles on the disintegration of the excited system is the subject of investigation here. The simultaneous existence of liquid- and gas-like fragments over a temperature interval is linked to the liquid–gas phase transition, and the motivation here is to investigate if this coexistence survives in the presence of hyperons too. For heavy-ion peripheral collisions in the 3–10 GeV range, the hyperons are mainly produced in the participant zone and then some of them get attached to the projectile-like fragment (PLF) and target-like fragment (TLF). In the present study, we will concentrate on the PLF only. Because of the excitation energy (usually characterized by a temperature, T), the PLF will break up into many fragments [75–77] and the velocities of the fragments are centered around the velocity of the projectile. In the participating region, apart from original neutrons and protons, other particles (pions, hyperons, etc.) are also produced. The produced Λ's have an extended rapidity range. Those produced in the rapidity range close to that of the projectile and having total momenta in the PLF frame up to the Fermi momenta can be trapped in the PLF and form hypernuclei [78–80]. The Λ-nucleon interactions are well studied, and the potential depth of Λ hyperons is such that bound Λ hypernuclear states exist. At higher energies, multiple hyperons can get attached to the PLF. Here we consider up to eight hyperons being attached to the PLF. The fragmentation of the excited hypernuclear system has been studied using canonical thermodynamical model being extended to three component systems, namely the neutrons, protons, and hyperons.

The details of the three-component canonical model can be found in Ref. [81–84]. The intrinsic partition function of the fragments is modified by using the appropriate term for the hyperons in the liquid drop formula. The study of the liquid-drop model formula reveals that by adding hyperons, the stability of the fragments increases for mass numbers $a > 8$. Hence, this implies that the hyperon–nucleon interaction is attractive for this mass range. For $a \leq 8$, this liquid-drop formula is not suitable, and so we have used experimental binding energies for these lower mass nuclei or hypernuclei. It is known that ^{4}H or ^{5}He are not stable, but when one Λ is added, the corresponding nuclei ($^4_\Lambda$H or $^5_\Lambda$He) become stable.

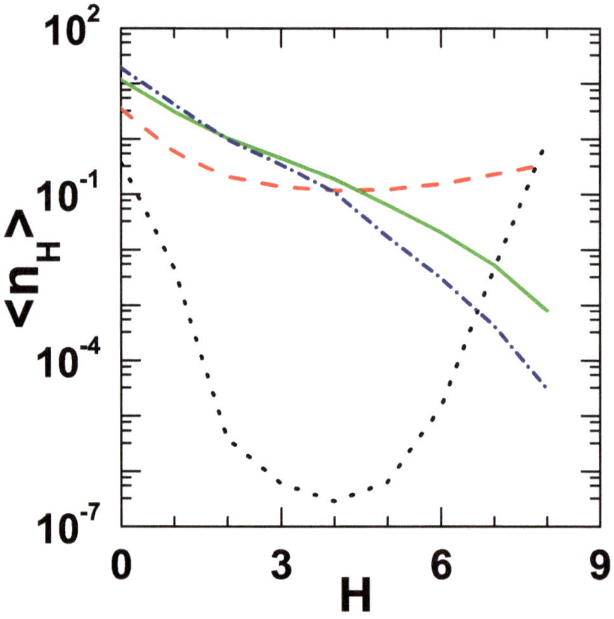

Figure 4.14 Distribution of hyperfragments produced from the fragmentation of $A_0 = 128$, $Z_0 = 50$, $H_0 = 8$ at $T = 3$ MeV (black dotted line), 5 MeV (red dashed line), 7 MeV (green solid line), and 10 MeV (blue dash-dotted line).

Hence, this establishes the attractive nature of the hyperon–nucleon interaction.

In addition to the liquid-drop formula, we have also included the contribution to $z_{int}(a, z, h)$ due from the excited states. This gives a multiplicative factor $exp(r(T)Ta/\epsilon_0)$ where we have introduced a correction term $r(T) = \frac{12}{12+T}$ to the expression used in Ref. [15]. This slows down the increase in $z_{int}(a, z, h)$ due to excited states as T increases. Figure 4.14 shows the variation of $\langle n_A \rangle$ with the mass number A(mass distribution) for the fragmentation of nuclei with H_0=0 and H_0 = 8 at T = 3 MeV. The mass distribution of ordinary nuclei displays an 'U'-shaped variation, which is expected at lower temperature (3 MeV) and which gradually disappears as the temperature is increased. This feature indicates the liquid–gas phase transition or phase coexistence, i.e., the existence of "liquid-like"(heavier) and "gas-like" (lighter) fragments described in Section 4.4. It is quite amazing that the nature of mass distribution is similar to that with the strange nucleons, and the two curves are pretty close to each other. This establishes the fact that the first-order phase transition (coexistence) still persists in the presence of hyperfragments. This feature is independent of the strangeness

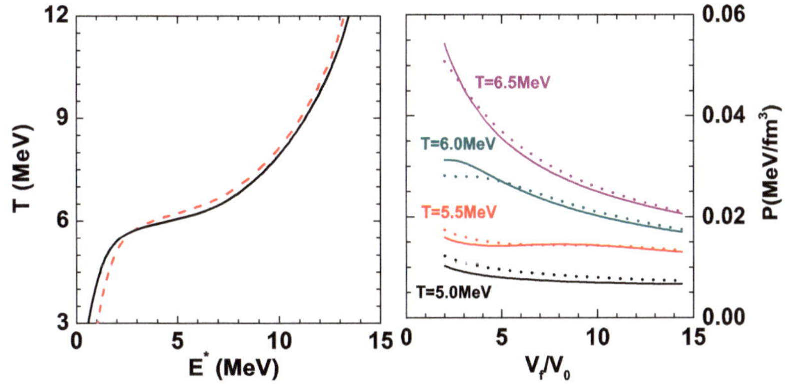

Figure 4.15 Left panel: Variation of temperature (T) with excitation energy (E^*) for two fragmenting systems of same $A_0 = 128$, $Z_0 = 50$ but different $H_0 = 8$ (solid line) and $H_0 = 0$ (dashed line). Right panel: Variation of pressure with volume at four different temperatures $T = 5.0$, 5.5, 6.0, and 6.5 MeV.

content of the fragments. This motivated us to study the caloric curve (left panel of Fig. 4.15) for both normal and strange nuclei by switching off the Coulomb interaction, which also confirms the first-order phase transition in the presence of hyperons. The right panel of Fig. 4.15 displays the variation of pressure with volume for four different temperatures (isotherms) both for normal nuclei and those with eight hyperons. Initially, the pressure decreases with volume after which it remains more or less constant, irrespective of the change in volume. This is another strong signature of liquid–gas coexistence or first-order phase transition. It is also observed from this figure that for a particular temperature, the strange system disintegrates at a smaller volume as compared to the normal one.

References

[1] Stanley, H. E. (1971). *Introduction to Phase Transitions and Critical Phenomena*, Oxford University Press.

[2] Siemens, P. J. (1983). Liquid–gas phase transition in nuclear matter, *Nature*, **305**, pp. 410.

[3] Gupta, S. D., Mekjian, A. Z., and Tsang, M. B. (2001). *Advances in Nuclear Physics*, pp. 89, edited by J. W. Negele and E. Vogt, Plenum Publishers, New York.

[4] Gupta, S. D., Mallik, S., and Chaudhuri, G. (2019). *Heavy Ion Reaction at Intermediate Energies: Theoretical Models*, World Scientific Publishers.

[5] Borderie, B. and Frankland, J. D. (2019). Liquid–gas phase transition in nuclei, *Prog. Part. Nucl. Phys.* **105**, pp. 82.

[6] Gross, D. H. E. (1993). Nuclear rings and gas–liquid phase transitions, *Prog. Part. Nucl. Phys.* **30**, pp. 155.

[7] Gross, D. H. E. (2001). *Microcanonical Thermodynamics: Phase Transitions in Small Systems, World Scientific Lecture Notes in Physics*, vol. 66, World Scientific, Singapore.

[8] Chomaz, Ph., Collona, M., and Randrup, J. (2004). *Phys. Rep.* **389**, pp. 263.

[9] Huang, K. (1987). *Statistical Mechanics*, Wiley & Sons, New York.

[10] Pochodzalla, J. *et al.* (1995). Probing the nuclear liquid-gas phase transition, *Phys. Rev. Lett.* **75**, pp. 1040.

[11] Bertsch, G. F. and Gupta, S. Das. (1988). A guide to microscopic models for intermediate energy heavy ion collisions, *Phys. Rep* **160**, pp. 189.

[12] Aichelin, J. (1991). Quantum molecular dynamics: A dynamical microscopic n-body approach to investigate fragment formation and the nuclear equation of state in heavy ion collisions, *Phys. Rep.* **202**, pp. 233.

[13] Hartnack, Ch. *et al.* (1998). Modelling the many-body dynamics of heavy ion collisions: Present status and future perspective, *Eur. Phys. Journ. A* **1**, pp. 151.

[14] Gross, D. H. E. (1997). Microcanonical thermodynamics and statistical fragmentation of dissipative systems. The topological structure of the N-body phase space, *Phys. Rep* **279**, pp. 119.

[15] Bondorf, J. P., Botvina, A. S., Iljinov, A. S., Mishustin, I. N., and Sneppen, K. (1995). Statistical multifragmentation of nuclei, *Phys. Rep.* **257**, pp. 133.

[16] Das, C. B., Gupta, S. D., Lynch, W. G., Mekjian, A. Z., and Tsang, M. B. (2005). The thermodynamic model for nuclear multifragmentation, *Phys. Rep.* **406**, pp. 1.

[17] Gupta, S. D. and Mekjian, A. Z. (1981). *Phys. Rep.* **72**, pp. 131.

[18] Campi, X. (1988). Signals of a phase transition in nuclear multifragmentation, *Phys. Lett B* **208**, pp. 351.

[19] Bauer, W. (1988). Extraction of signals of a phase transition from nuclear multifragmentation, *Phys. Rev. C* **38**, pp. 1297.

[20] Pan, J. and Gupta, S. D. (1995). A schematic model for fragmentation and phase transition in nuclear collisions, *Phys. Lett B* **344**, pp. 29.

[21] Shuryak, E. V. (1978). Quark-gluon plasma and hadronic production of leptons, photons and psions, *Phys. Lett. B* **78**, pp. 150.

[22] Reif, F. (1965). *Fundamentals of Statistical and Thermal Physics*, McGraw-Hill, New York.

[23] Pathria, R. K. (1972). *Statistical Mechanics*, Pergamon Press.
[24] Agostino, M. *et al.* (2000). Negative heat capacity in the critical region of nuclear fragmentation: An experimental evidence of the liquid–gas phase transition, *Phys. Lett. B* **473**, pp. 219.
[25] Agostino, M. *et al.* (2002). On the reliability of negative heat capacity measurements, *Nucl. Phys. A* **699**, pp. 795.
[26] Chomaz, P., Duflot, V., and Gulminelli, F. (2000). Caloric curves and energy fluctuations in the microcanonical liquid–gas phase transition, *Phys. Rev. Lett.* **85**, pp. 3587.
[27] Gulminelli, F. and Chomaz, Ph. (2005). Distribution of the largest fragment in the lattice gas model, *Phys. Rev. C* **71**, pp. 054607.
[28] Krishnamachari, B., McLean, J., Cooper, B., and Sethna, J. (1996). Gibbs–Thomson formula for small island sizes: Corrections for high vapor densities, *Phys. Rev. B* **54**, pp. 8899.
[29] Pleimling, M. and Selke, W. (2000). Droplets in the coexistence region of the two-dimensional Ising model, *J. Phys. A* **33**, pp. 199.
[30] Chaudhuri, G. and Gupta, S. D. (2007). Properties of the largest fragment in multifragmentation: A canonical thermodynamic calculation, *Phys. Rev C* **75**, pp. 034603.
[31] Huang, M. *et al.* (2010). Power law behavior of the isotope yield distributions in the multifragmentation regime of heavy ion reactions, *Phys. Rev. C* **82**, pp. 054602.
[32] Bonasera, A. *et al.* (2008). Quantum nature of a nuclear phase transition, *Phys. Rev. Lett.* **101**, pp. 122702.
[33] Borderie, B. (2002). Nuclear multifragmentation and phase transition for hot nuclei, *Jour. Phys. G* **28**, pp. 217.
[34] Dorso, C. O., Latora, V. C., and Bonasera, A. (1999). Signals of critical behavior in fragmenting finite systems, *Phys. Rev. C* **60**, pp. 034606.
[35] Elliott, J. B. *et al.* (2002). Liquid to vapor phase transition in excited nuclei, *Phys. Rev. Lett.* **88**, pp. 042701.
[36] Ma, Y. G. (1999). Application of information theory in nuclear liquid gas phase transition, *Phys. Rev. Lett.* **83**, pp. 3617.

[37] Mallik, S., Chaudhuri, G., Das, P. and Gupta, S. D. (2017). Multiplicity derivative: A new signature of a first-order phase transition in intermediate-energy heavy-ion collisions, *Phys. Rev. C* **95**, pp. 061601.

[38] Das, P., Mallik, S., and Chaudhuri, G. (2018). First-order derivative of cluster size as a new signature of phase transition in heavy ion collisions at intermediate energies, *Phys. Lett B* **783**, pp. 364.

[39] Fevre, A. Le *et al.* (2008). Bimodality: A sign of critical behavior in nuclear reactions, *Phys. Rev. Lett.* **100**, pp. 042701.

[40] Fevre, A. Le *et al.* (2009). Bimodality: A general feature of heavy ion reactions, *Phys. Rev. C* **80**, pp. 044615.

[41] Chaudhuri, G. and Mallik, S. (2011). Effect of secondary decay on isoscaling: Results from the canonical thermodynamical model, *Nucl. Phys. A* **849**, pp. 190.

[42] Lin, W., Ren, P., Zheng, H., Liu, X., Huang, M., Wada, R., and Qu, G. (2018). Sensitivity study of experimental measures for the nuclear liquid–gas phase transition in the statistical multifragmentation model, *Phys. Rev. C* **97**, pp. 054615.

[43] Lin, W., Ren, P., Zheng, H., Liu, X., Huang, M., Yang, K., Qu, G., and Wada, R. (2019). Solidarity of signal of measures for the liquid–gas phase transition in the statistical multifragmentation model, *Phys. Rev. C* **99**, pp. 054616.

[44] Liu, H. L., Ma, Y. G., and Fang, D. Q. (2019). Finite-size scaling phenomenon of nuclear liquid-gas phase transition probes, *Phys. Rev. C* **99**, pp. 054614.

[45] Bakeer, R., Awad, W., and Jaqam, H. R. (2002). *Jour. Phys. G* **28**, pp. 217.

[46] Wada, R. *et al.* (2019). Experimental liquid–gas phase transition signals and reaction dynamics, *Phys. Rev. C* **99**, pp. 024616.

[47] Chaudhuri, G. and Mallik, S. (2019). Effect of liquid drop model parameters on nuclear liquid–gas phase transition, *Phys. Rev. C* **99**, pp. 054602.

[48] Mallik, S., Gupta, S. D., and Chaudhuri, G. (2015). Event simulations in a transport model for intermediate energy

heavy ion collisions: Applications to multiplicity distributions, *Phys. Rev. C* **91**, pp. 034616.

[49] Bauer, W., Bertsch, G. F., and Gupta, S. D. (1987). Fluctuations and clustering in heavy-ion collisions, *Phys. Rev. Lett.* **58**, pp. 863.

[50] Lenk, R. J. and Pandharipande, V. R. (1989). Nuclear mean field dynamics in the lattice Hamiltonian Vlasov method, *Phys. Rev. C* **39**, pp. 2242.

[51] Mallik, S., Gupta, S. D., and Chaudhuri, G. (2016). Bimodality emerges from transport model calculations of heavy ion collisions at intermediate energy, *Phys. Rev. C* **93**, pp. 041603.

[52] Mallik, S., Chaudhuri, G., and Gulminelli, F. (2019). Sensitivity of the evaporation residue observables to the symmetry energy, *Phys. Rev. C* **100**, pp. 024611.

[53] Mallik, S. and Chaudhuri, G. (2020). Isospin dependent hybrid model for studying isoscaling in heavy ion collisions around the Fermi energy domain, *Nucl. Phys. A* **1002**, 121948.

[54] Zhang, Y. X. *et al.* (2018). Nuclear mean field dynamics in the lattice Hamiltonian Vlasov method, *Phys. Rev. C* **97**, pp. 034625.

[55] Gulminelli, F. (2004). Phase coexistence in nuclei, *Ann. Phys. Fr.* **29**, pp. 6.

[56] Zbiri, K. *et al.* (2007). Transition from participant to spectator fragmentation in Au+Au reactions between 60 A and 150 A MeV, *Phys. Rev. C* **75**, pp. 034612.

[57] Pichon, M. *et al.* (2006). Bimodality: A possible experimental signature of the liquid–gas phase transition of nuclear matter, *Nucl. Phys. A* **779**, pp. 267.

[58] Lautesse, P. *et al.* (2005). Experimental overview of Ni+Ni collisions at 32 MeV/nucleon: Discriminant analysis and duality in the decay modes of a fusionlike system, *Phys. Rev. C* **71**, pp. 034602.

[59] Lopez, O. and Rivet, M. F. (2006). Bimodalities: A survey of experimental data and models, *Eur. Phys. Journ. A* **30**, pp. 263.

[60] Chomaz, Ph. and Gulminelli, F. (2003). First-order phase transitions: Equivalence between bimodalities and the Yang-Lee theorem, *Physica A* **330**, pp. 451.

[61] Gulminelli, F. *et al.* (2003). Transient backbending behavior in the Ising model with fixed magnetization, *Phys. Rev. E* **68**, pp. 026119.

[62] Gulminelli, F. and Chomaz, Ph. (2005). Distribution of the largest fragment in the lattice gas model *Phys. Rev. C* **71**, pp. 054607.

[63] Kastner, M. and Pleimling, M. (2009). Microcanonical phase diagrams of short-range ferromagnets, *Phys. Rev. Lett* **102**, pp. 240604.

[64] Lehaut, G. *et al.* (2010). Phase diagram of the charged lattice-gas model with two types of particles, *Phys. Rev. E* **81**, pp. 051104.

[65] Chaudhuri, G. and Gupta, S. D. (2007). Properties of the largest fragment in multifragmentation: A canonical thermodynamic calculation, *Phys. Rev. C* **75**, pp. 034603.

[66] Bruno, M. *et al.* (2008). Signals of bimodality in the fragmentation of Au quasi-projectiles, *Nucl. Phys. A* **807**, pp. 48.

[67] Bonnet, E. *et al.* (2009). Bimodal behavior of the heaviest fragment distribution in projectile fragmentation, *Phys. Rev. Lett.* **103**, pp. 072701.

[68] Borderie, B. *et al.* (2010). The prominent role of the heaviest fragment in multifragmentation and phase transition for hot nuclei, *Int. Journ. Mod. Phys. E* **19**, pp. 1523.

[69] Mallik, S., Chaudhuri, G. and Gulminelli, F. (2018). Dynamical and statistical bimodality in nuclcar fragmentation, *Phys. Rev. C* **97**, pp. 024606.

[70] Stauffer, D. and Aharony, A. (1992). *Introduction to Percolation Theory*, Taylor and Francis, London.

[71] Pan, J., Gupta, S. D., and Grant, M. (1998). First-order phase transition in intermediate-energy heavy ion collisions, *Phys. Rev. Lett* **80**, pp. 1182.

[72] Samaddar, S. K. and Gupta, S. D. (2000). Nuclear fragmentation characteristics from isotopic spin dependent lattice-gas model, *Phys. Rev. C* **61**, pp. 034610.

[73] Gupta, S. D., Mallik, S., and Chaudhuri, G. (2018). Further studies of the multiplicity derivative in models of heavy ion collision at intermediate energies as a probe for phase transitions, *Phys. Rev. C* **97**, pp. 044605.

[74] Waddington, C. J. and Freier, P. S. (1985). Interactions of energetic gold nuclei in nuclear emulsions, *Phys. Rev. C* **31**, pp. 888.

[75] Mallik, S., Chaudhuri, G., and Gupta, S. D. (2011). Model for projectile fragmentation: Case study for Ni on Ta and Be, and Xe on Al, *Phys. Rev. C* **83**, pp. 044612.

[76] Mallik, S., Chaudhuri, G., and Gupta, S. D. (2011). Improvements to a model of projectile fragmentation, *Phys. Rev. C* **84**, pp. 054612.

[77] Mallik, S., Gupta, S. D., and Chaudhuri, G. (2014). Estimates for temperature in projectile-like fragments in geometric and transport models, *Phys. Rev. C* **89**, pp. 044614.

[78] Wakai, M. *et al.* (1988). Hypernucleus formation in high-energy nuclear collisions, *Phys. Rev. C* **38**, pp. 748.

[79] Satio, T. R. *et al.* (2012). Production of hypernuclei in peripheral HI collisions: The HypHI project at GSI, *Nucl. Phys. A* **881**, pp. 218.

[80] Gaitanos, Th. *et al.* (2009). Formation of hypernuclei in high energy reactions within a covariant transport model, *Phys. Lett. B* **675**, pp. 297.

[81] Gupta, S. D. (2009). Extending the canonical thermodynamic model: Inclusion of hypernuclei, *Nucl. Phys. A* **822**, pp. 41.

[82] Topor, P. V. and Gupta, S. D. (2010). Model for hypernucleus production in heavy ion collisions, *Phys. Rev. C* **81**, pp. 054911.

[83] Mallik, S. and Chaudhuri, G. (2015). Liquid–gas phase transition in hypernuclei, *Phys. Rev. C* **91** pp. 054603.

[84] Das, P., Mallik, S., and Chaudhuri, G. (2017). Effect of hyperons on phase coexistence in strange matter, *Phys. Rev. C* **95** pp. 014603.

Chapter 5

A 3D Calorimetry of Hot Nuclei

E. Vient

LPC Caen, ENSICAEN, Université de Caen, CNRS/IN2P3, F-14050 Caen cedex, France

vient@lpccaen.in2p3.fr

5.1 Introduction

The quality of the reproduction of the binding energies of the nuclei by the Bethe–Weizsäcker formula, which is based on the liquid drop model introduced by G. Gamow, has led nuclear physicists to define the concept of nuclear matter. The formal analogy between the nuclear potential of Skyrme and the molecular potential of van der Waals [Jaqaman *et al.* (1983)] has confirmed the interest of this notion. This has obviously led nuclear physicists to carry out a thermodynamics of nuclear matter, the previous analogy suggesting the possibility of observing a first-order phase transition. The knowledge of nuclear matter appears important in nuclear astrophysics for modelling the gravitational collapse of type II supernovae and for the physical study of neutron stars [Durand *et al.* (2001)].

Multifragmentation in Heavy-Ion Reactions: Theory and Experiments
Edited by Rajeev K. Puri, Yu-Gang Ma, and Arun Sharma

ISBN 978-981-4968-69-0 (Hardcover), 978-1-003-38513-4 (eBook)
www.jennystanford.com

In the laboratory, a way to undergo nuclear thermodynamics studies is to send a projectile nucleus onto a fixed target nucleus using a heavy-ion accelerator. Doing this way, we hope to heat, compress, and expand nuclei, i.e., a piece of nuclear matter, sufficiently to observe an eventual phase transition. The order of magnitude of the incident energy per nucleon to be used to heat the nucleus must often be from tens to a few hundred MeV per nucleon (thus avoiding exciting the sub-nucleonic degrees of freedom). It should be kept in mind that this way of heating matter is not at all the usual way used in thermodynamics; here we use a nuclear reaction to deposit energy in the nuclei. This implies understanding and "controlling" the progress of the nuclear collision. We will see that this is a very difficult experimental task. In this energy range, mainly in a domain close to the Fermi energy, the collisions are extremely violent and complex. There is an influence competition between the nuclear mean field and the nucleon–nucleon interaction. For symmetrical collisions, allowing the best energy deposition in the nuclei, there is a strong memory of the input channel. These collisions are mainly binary and correspond to the process of deeply inelastic diffusion [Steckmeyer *et al.* (1996)]. During some of these collisions, it can be observed that an important production of intermediate mass fragments (IMFs) sets in the interface of the two colliding nuclei, often called "Neck Emission" [Casini *et al.* (1993), Montoya *et al.* (1994), Łukasik *et al.* (1997)]. In the first moments of the collision, at the time of contact, there are collisions between nucleons, giving rise to the emission of light particles called pre-equilibrium particles [Doré *et al.* (2001), Germain *et al.* (2000), Lefort *et al.* (2000)]. This production is located at the velocity of the nucleon–nucleon center of mass, identical to the collision center of mass for symmetric reactions. It is often referred to as "mid-velocity production." At the same time, there may be compression of nuclear matter followed eventually by an expansion. The fusion of the two nuclei still exists, but its cross section decreases drastically with the incident energy as of 25 AMeV (MeV per nucleon), until it disappears at higher energies [Eudes *et al.* (2014)]. The respective probabilities of these processes are obviously dependent on the incident energy, the nuclear systems in collision, and the impact parameter.

A typical nuclear collision lasts approximately from 50 to 300 fm/c depending on the nuclear system, the incident energy, and the impact parameter. In the first moments, pre-equilibrium particles are produced in the overlap region of the nuclei from 10 to 30 fm/c after the beginning of the collision. At the same time, there are compression of the nuclear matter and then expansion during 100 fm/c. The mechanical energy dissipated during the collision will be thermalized in the nuclei. Hot nuclei are thus produced. This thermalization time can last for a long time from 15 to 100 fm/c approximately according to the theoretical models. These excited nuclei will obviously de-excite. Depending on the stored excitation energy, there will be particle evaporation, fission, multifragmentation, and even apparent vaporization of the two initial nuclei [Borderie *et al.* (1999)]. The time required to evacuate this energy can take from 15 fm/c to several thousand fm/c depending on the physical process involved.

On a human scale, such a collision lasts a very short time (for example 5000 fm/c $\approx 1.7 \times 10^{-20} s$) and is detected experimentally in about a hundred nanoseconds. There is, therefore, no direct measurement of the thermodynamic characteristics of the hot nuclei formed, but an after-the-fact reconstruction from the debris of the collision. This is, therefore, an extremely difficult experimental task.

Obviously, we must try to have the most exhaustive detectors possible. They must cover a solid angle as close as possible to 4π around the collision site to try to detect all the particles produced. These ones can be neutral (gammas or neutrons) or charged (light-charged particles or heavy ions). In reality, there is no perfect detector to detect all these particles. The detectors are specialized, for the detection of charged nuclei (DELF, MUR, TONNEAU, INDRA, FOPI, MULTICS, CHIMERA, NIMROD, FAZIA), for neutrons (ORION, DEMON, NEBULA), and for gammas (PARIS, HIRA, EXOGAM, SeGA, GRETINA, AGATA). In general, we first use a charged particle detection system, to which we may add neutral particle detectors.

Based on the information provided by a multidetector of charged particles and our knowledge of reaction mechanisms, we try to do the best possible calorimetry of hot nuclei, i.e., reconstruct their static characteristics (charge and mass), kinematics (velocity), and dynamics (kinetic energy of translation and rotation eventually, excitation energy therefore internal energy). A large number of classical calorimetries have been developed so far [Viola and Bougault (2006)]. The calorimetry presented in this chapter can be considered a step forward. It was developed as part of the INDRA detector collaboration. This multidetector is described in detail in Ref. [Pouthas *et al.* (1995)]. It has a high angular coverage in the laboratory between 2° and 176°, covering 90% of 4π. It has excellent granularity provided by 336 telescopes. It is capable of identifying all detected nuclei in charge with low energy thresholds of 0.8 MeV/nucleon for $Z \leqslant 12$ and 1.3 MeV/nucleon for the others. It identifies in mass up to the Boron element.

The prominent binary nature of the collisions studied and the detection characteristics of INDRA, allowing a better detection of the quasi-projectile and its products, lead us to want to do essentially the calorimetry of the quasi-projectile (QP). In the following sections, we will discover and study how to trace back to the average properties of the QP as a function of a global experimental variable to estimate the average impact parameter associated with the selection of events studied.

5.2 Necessary Selections

We know that the detector used, whatever its qualities, is imperfect. We must, therefore, define experimental criteria to judge the quality of the detection of a nuclear reaction. We will take into account the conservation laws that govern nuclear reactions. There must be conservation of the charge, the number of nucleons, the linear momentum, the angular momentum, and the energy. But we must also take into account the experimental capabilities of the detector used. This led us to define two global variables as quality criteria for INDRA detection: the total sum of the detected charges and the sum, not of the total parallel momentum of all the nuclei but the total parallel pseudo-momentum (product of the charge by the parallel velocity of the nucleus) of the latter. The latter global variable is, therefore, in reality only approximately conserved. To simplify the use of these quantities, we have normalized them to the initial total charge of the studied system for one and to the initial pseudo-parallel momentum (projectile charge times its velocity) for the other, which gives the following mathematical expressions:

$$Z_{tot}\,Norm = \frac{\sum_{i=1}^{Mul} Z_i}{Z_{Proj} + Z_{Targ}}$$

$$(ZV_{//})_{tot}\,Norm = \frac{\sum_{i=1}^{Mul} Z_i \times (V_{//})_i}{Z_{Proj} \times (V_{//})_{Proj}}$$

With Z_i: charge of the ith detected particle, *Mul*: detected multiplicity, and $(V_{//})_i$: parallel velocity in the laboratory of the ith detected particle, Z_{Proj} and Z_{Targ}: projectile charge and target charge, $(V_{//})_{Proj}$: projectile parallel velocity in the laboratory frame.

At the top of Fig. 5.1, we present $Z_{tot}\,Norm$ *vs.* $(ZV_{//})_{tot}\,Norm$ for ^{124}Xe+^{124}Sn at 32 MeV/nucleon studied with INDRA for 21 million events. In this figure, the red dot located at the point (1,1) corresponds to the events for which all the charged particles produced were detected. We clearly see three main areas appearing in the figure. Zone 3 corresponds to poorly measured events for which both QP and quasi-target (QT) residues are not detected. For zone 2, only the QT residue is missing, as indicated by the values of the parallel pseudo-momentum. Zone 1 is in the vicinity of the red point. As we want to rebuild the QP, we will select as events of interest those in Zone 1 and Zone 2. To do so, we define an additional selection based on the global variable $Z_{Tot\,CM\,F}\,Norm$. We determine the total charge detected at the front of the center of mass of the reaction (i.e., the front of the collision) and normalize it to the projectile charge. We keep only the events such as:

$0.7 \leqslant Z_{Tot\,CM\,F}\,Norm \leqslant 1.1$ and $0.7 \leqslant (ZV_{//})_{tot}\,Norm \leqslant 1.1$.

We can observe the consequence of this selection at the bottom of Fig. 5.1. We only keep about one third of the events, but events for which the detection of products of hot QP de-excitation must be correct.

For a possible comparison with models or to use models to validate the experimental method used, it seems interesting to have a global variable that is very strongly correlated to the collision impact parameter. Several studies [Marie (1995), Péter *et al.* (1990)] have shown that the transverse kinetic energy of light-charged particles (LCPs, with Z = 1 or 2) is an excellent variable for this. It is defined as

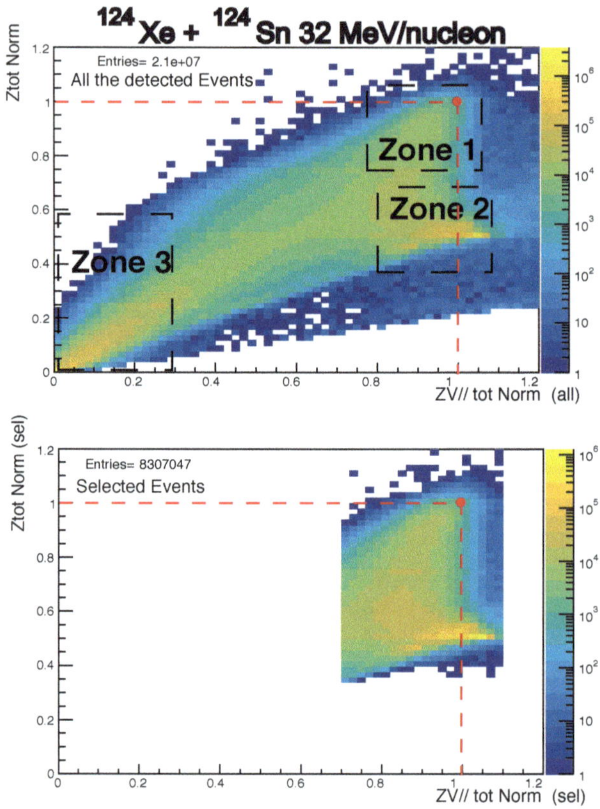

Figure 5.1 At the top, Z_{tot} *Norm vs.* $(ZV_{//})_{tot}$ *Norm* for all detected events, and at the bottom, Z_{tot} *Norm vs.* $(ZV_{//})_{tot}$ *Norm* for all the selected and studied events.

follows:

$$E_{t12} = \sum_{i=1}^{Mul_{LCP}} T_i \times \sin^2(\theta_i)$$

With T_i: kinetic energy of the ith detected LCP (with Z = 1 or 2), Mul_{LCP}: Multiplicity of detected LCP, θ_i: polar angle associated with the ith detected LCP, in the laboratory frame.

In these references, it is indicated how to experimentally determine the average correlation between the impact parameter and the transverse energy of the LCPs. Again, we use a specific normalization to the transverse energy available in the center of mass (i.e., two thirds of the total energy available in the center of mass). We know that if there is a complete thermalization of the incident energy, then the normalized transverse energy must be equal to 1, because the energy available in the center of mass is in this case distributed equitably throughout the 3D space. We also know that the violence of the collision increases with this variable. We are, therefore, certain to obtain hotter nuclei when this variable increases. Three-dimensional (3D) calorimetry makes it possible to determine only average values of the physical quantities characterizing the hot nuclei over a statistical ensemble of events. To reconstruct the QP, we chose to make a selection based on the transverse energy of LCP normalized in 10 slices representing the same number of events among the ensemble of events well measured. We can see effectively this selection in Fig. 5.2. An indication of the impact parameters involved is given in this figure.

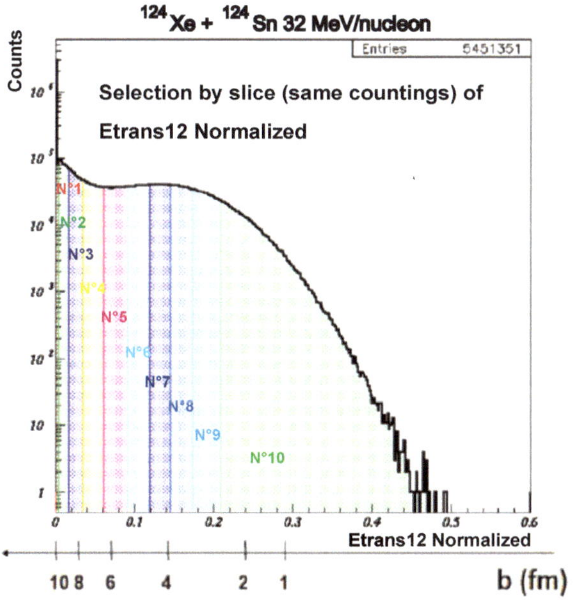

Figure 5.2 Distribution of E_{t12}*Normalized*, with the different slices numbered and colored corresponding to the different selections (the vertical scale is logarithmic).

5.3 Reconstruction of the Quasi-Projectile Velocity and Associated Reference Frame

Now it is necessary to recover event by event the initial velocity of the QP. We will then be able to define the QP reference frame and the experimental reaction plane event by event. This will be important to define whether or not a particle is emitted by the QP.

But we already know that this task is actually impossible. The QP debris were eventually more or less accelerated during the reaction by the Coulomb fields of the other collision partners. Since we do not know the temporal sequence of production of all these pieces of the QP and the others, it is impossible to correct this eventual Coulomb effect. We chose here to neglect this effect knowing that this approximation will be all the less true as the collision is central.

We also obviously assume that the two hot nuclei reached thermodynamic equilibrium during the collision. For this reconstruction, we will not use light-charged particles. This is mainly to avoid the possible influence of light pre-equilibrium particles, not emitted by the hot nuclei. We will, therefore, take into consideration only the two heaviest fragments and the so-called IMFs (such as $Z \geq 3$). With all these different nuclei, we will calculate the momentum tensor of the event in the frame of the center of mass [Cugnon and L'hote (1983)]. After diagonalization, this tensor can be described by an ellipsoid that will give information in the velocity space on the overall shape of the studied event. We then divide the velocity space into two halves at the velocity of the center of mass perpendicularly to the main axis of the momentum ellipsoid. Each fragment located in the forward part of the ellipsoid (forward of the center of mass (FCM)), thus defined, is then considered to reconstruct the total momentum of the QP and its velocity in the reference frame of the center of mass by the following relation:

$$\overrightarrow{V_{QP}} = \frac{\sum_{i=1}^{Mul_{IMF(FCM)}} \overrightarrow{V_i} \times A_i}{\sum_{i=1}^{Mul_{IMF(FCM)}} A_i}$$

We thus take into account any rotation or other of the binary system in collision to consider only the forward part of the collision. To define the reaction plane, we assume that it is the plane that contains both the reconstructed QP vector and the vector associated with the initial projectile in the reference frame of the center of mass. In fact, we make a cross product of the two unit vectors that can be associated with these two velocities. The vector product is made in such a way that the QP velocity is always to the left of the beam velocity in this reaction plane. We thus orientate the velocity space event by event.

5.4 Selection Criteria and Characterization of the Evaporation Component of Quasi-Projectile

We will now characterize the kinematics of each particle or fragment produced during a collision in the reconstructed QP frame as described in the previous section. For this, we will define two angles: the polar angle θ_{spin} and the azimuthal angle φ of a particle in the QP frame and reaction plane. These two angles are presented in Fig. 5.3. θ_{spin} is the angle between the velocity of the particle in the QP frame and the unit vector normal to the reaction plane. φ is the angle

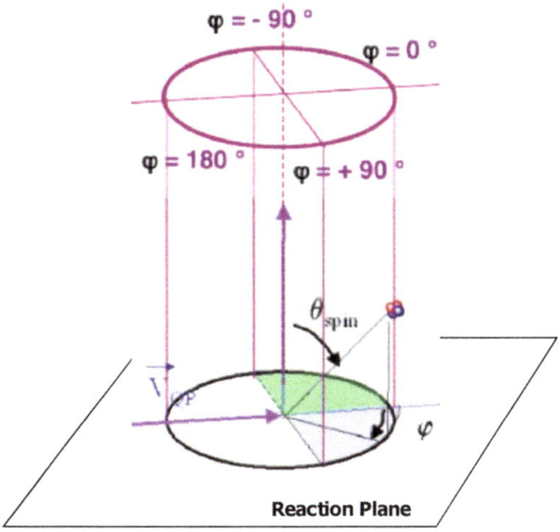

Figure 5.3 Definitions of the polar angle θ_{spin} and the azimuthal angle φ of a particle in the QP frame and in relation to the reaction plane (the azimuthal angle φ presented in the figure is positive).

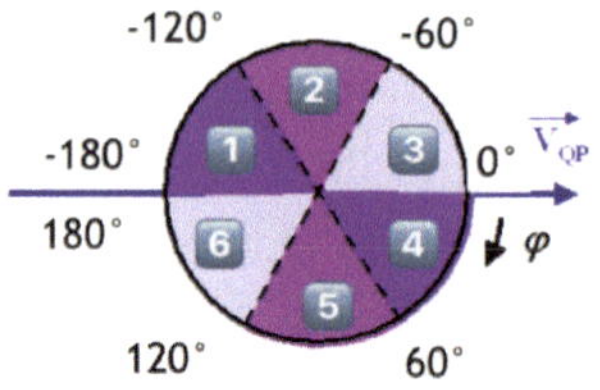

Figure 5.4 Presentation of the six zones of selection in the QP frame with top view of the reaction plane. Each selection zone is numbered.

between the orthogonal projection of the velocity of the particle in the QP frame on the reaction plane and the velocity of the QP.

To study all the nuclei produced during the collision, we divide the entire velocity space into six spatial zones corresponding to the same solid angle in the QP frame. This means cutting the space into six slices of 60 degrees in the azimuthal angle φ from -180° to 180° as shown in Fig. 5.4. These zones are numbered.

We will now construct for each type of particle the kinetic energy distribution defined in the QP frame and the $cos(\theta_{spin})$ distribution for each φ zone defined previously and for each selection in $E_{t12}Normalized$ (impact parameter selection). This is, of course, done for all well-measured events, and we take all the particles and fragments of the event to build these distributions. We present for illustrative purposes what these distributions are for protons in Fig. 5.5. Normally, if the protons were all emitted by thermally equilibrated QPs, we would have for both physical quantities the same spectra of a given color for all selections regardless of the numbering of the spatial zone. $cos(\theta_{spin})$ distributions should be flat, and kinetic energy distributions should be Maxwellian distributions according to Weisskopf's theory of evaporation for a hot nucleus [Weisskopf (1937), Vient *et al.* (2018b)]. This type of curve is only shown on the left with the graph numbered 3 in Fig. 5.5. If there is an angular momentum, the $cos(\theta_{spin})$ distribution is no longer flat and has a hump at the value 0 while remaining symmetrical with respect to 0 [Steckmeyer *et al.* (2001)].

It is normal that the distributions are different. We know that not all protons are emitted by the hot QP. We will see it clearly afterward. The collisions being mainly binary, there will be also

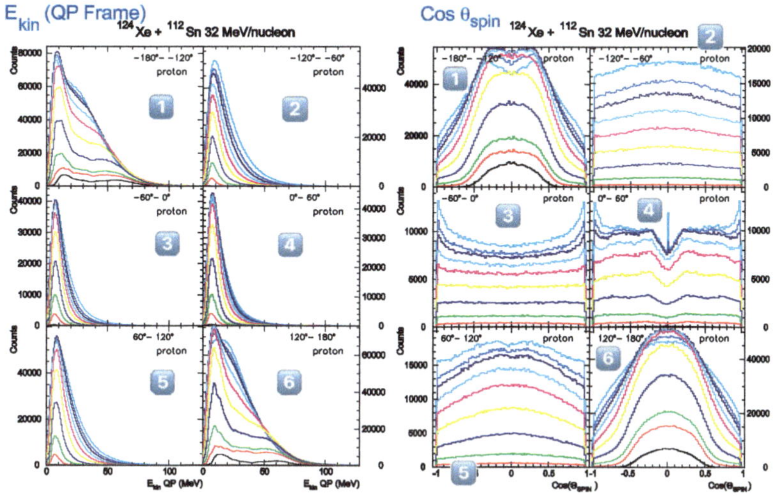

Figure 5.5 On the left, kinetic energy distributions of protons in the reconstructed frame of the QP, for the different selections in φ indicated by one of the numbers defined in Fig. 5.4, obtained by the INDRA collaboration for the system ^{124}Xe + ^{112}Sn at 32 MeV/nucleon. On the right, $cos(\theta_{spin})$ distributions of protons in the reconstructed frame of the QP, for the different selections in φ indicated by one of the numbers defined in Fig. 5.4, obtained by the INDRA collaboration for the system ^{124}Xe + ^{112}Sn at 32 MeV/nucleon. The graphs of different colors correspond to the different selections of impact parameters (the integral increases when the impact parameter diminishes). The studied events are only the selected events as defined previously.

protons emitted by the QT. There is possibly between the two partners a neck of matter that can produce protons. In addition, there are, of course, also potentially other protons produced during the first contacts between the two nuclei: the pre-equilibrium protons. This was shown for example in Ref. [Vient *et al.* (2018a)] using the HIPSE event generator in Fig. 9 of this paper.

To fully understand what the different selection areas in φ correspond to, we will present four different figures concerning the protons. They will also allow a better understanding of the diversity of origin of the particles produced during a nuclear collision. They are also representative of what happens with most other light-charged particles. We will, in fact, present proton velocities in different ways. First of all, it should be noted that the reference frame chosen to

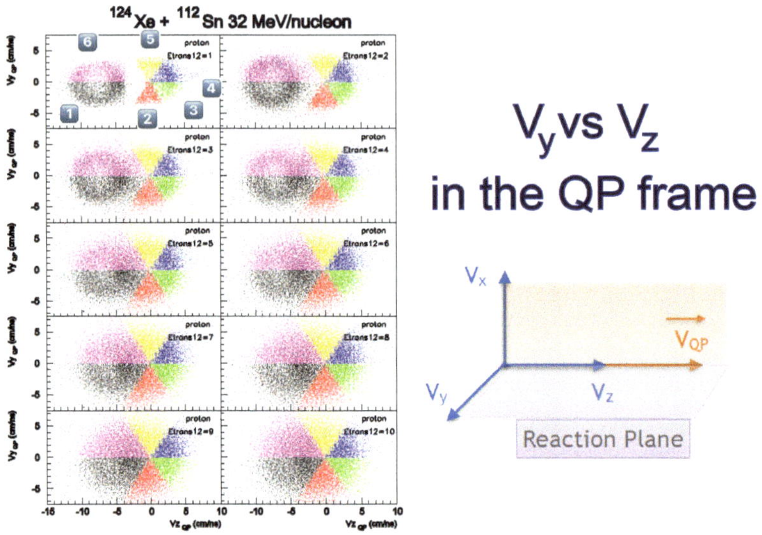

Figure 5.6 Two-dimensional graphs of the y-components versus z-components of the proton velocities in the reconstructed frame of the QP defined in the reconstructed reaction plane, for the different selections in φ indicated by one of the numbers defined in Fig. 5.4 and a different color. These data are obtained by the INDRA collaboration for the system ^{124}Xe + ^{112}Sn at 32 MeV/nucleon. Each two-dimensional graph corresponds to a selection in E_{t12}*Normalized* (impact parameter selection). Collisions are more and more central when going from left to right and from top to bottom. The studied events are only the selected events as defined previously.

define these velocities is the reference frame associated with the reconstructed QP.

In Fig. 5.6, we, therefore, present the V_y component of the proton velocity as a function of the V_z component in the QP reference frame. The direction along the z-axis corresponds event by event to the direction of the QP. The V_y component is the component of the velocity in the reaction plane. In this figure, we observe, therefore, the orthogonal projection of proton vectors in the reaction plane, as if we were observing the collisions in the velocity space seen from above. The positive V_y component corresponds to a positive azimuthal angle φ. The presented data have been obtained by the INDRA collaboration for the system ^{124}Xe + ^{112}Sn at 32 MeV/nucleon. Each two-dimensional graph was constructed by superimposing

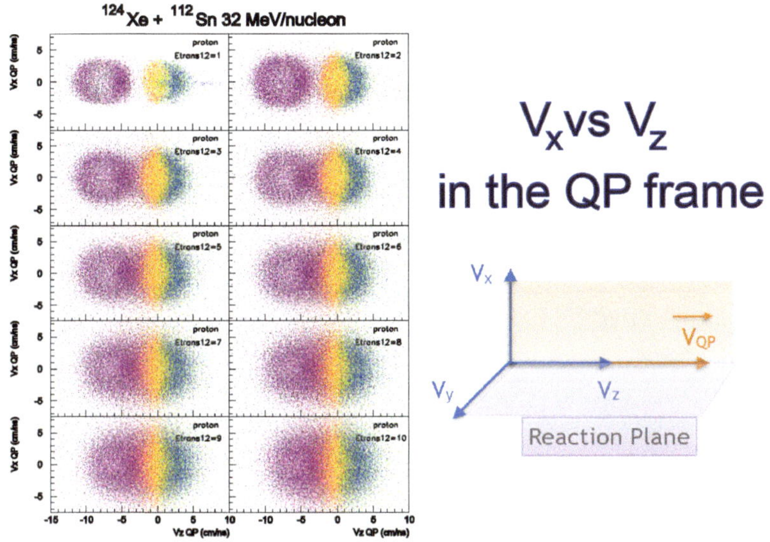

Figure 5.7 Two-dimensional graphs of the x-components versus z-components of the proton velocities in the reconstructed frame of the QP defined in a plane perpendicular to the reconstructed reaction plane, for the different selections in φ, indicated by a different color. These data are obtained by the INDRA collaboration for the system ^{124}Xe + ^{112}Sn at 32 MeV/nucleon. Each two-dimensional graph corresponds to a selection in E_{t12}*Normalized* (impact parameter selection). Collisions are more and more central when going from left to right and from top to bottom. The studied events are only the selected events as defined previously.

the different graphs obtained for each of the selection in φ. To differentiate them, these graphs have different colors. In Fig. 5.6 on the graph at the top left, we can see which numbered selective area is associated with each color. Figure 5.6 shows 10 two-dimensional graphs associated to the 10 selections in E_{t12}*Normalized* (impact parameter selection). Collisions are more and more central when going from left to right and from top to bottom. The studied events are only the selected events as defined in Section 14.2.

In Fig. 5.6, we see that zones labeled 1 and 6 (pink and black) are clearly associated with the emission sphere of the QT for the most peripheral collisions. Then for the following graphs, there is overlap for these two zones at the front, the rarest contribution of the QP and a mid-velocity contribution. As the centrality of the

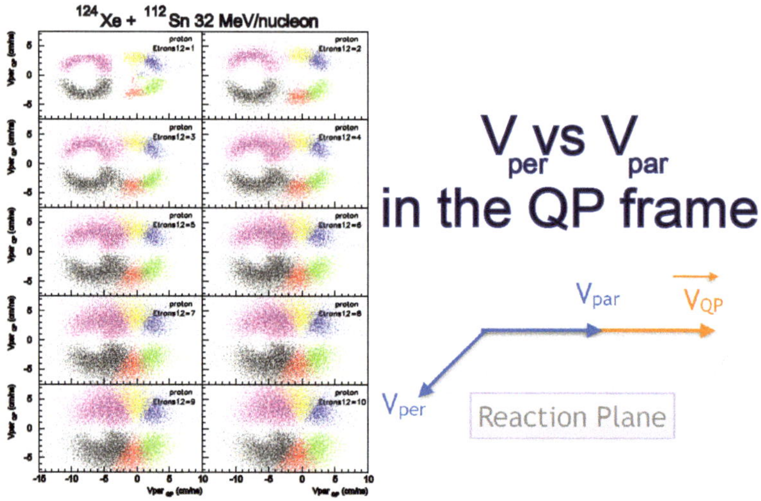

Figure 5.8 Two-dimensional graphs of the perpendicular components versus parallel components of the proton velocities in the reconstructed frame of the QP defined in the reconstructed reaction plane, for the different selections in φ, indicated by a different color. These data are obtained by the INDRA collaboration for the system ^{124}Xe + ^{112}Sn at 32 MeV/nucleon. Each two-dimensional graph corresponds to a selection in E_{t12}*Normalized* (impact parameter selection). Collisions are more and more central when going from left to right and from top to bottom. The studied events are only the selected events as defined previously.

collision increases, the two emission spheres of the two hot nuclei seem to overlap more and more. This will be confirmed and more clearly visible in Fig. 5.8. For the two graphs corresponding to the most peripheral collisions, we can notice that the QP emission sphere seems truncated at the rear of the QP. This is an effect due to the detection and our completeness criterion at the front of the center of mass, which selects some particular topologies of the QP contribution in the velocity space. There are a few particles emitted by the QP, and they must remove the QP residue from the detection hole between 0° and 2° of the INDRA multidetector.

Figure 5.7 is complementary to Fig. 5.6 as it provides a side view in velocity space of protons produced during collisions. The V_x component of the proton velocity is presented as a function of the V_z component in the QP reference frame. Again, we can very clearly observe the emission spheres of the two collision participants. We also see the increasing overlap between these two contributions as the dissipation increases with the centrality of the collisions. We also note a non-equilibrated component at the center of mass of the collision.

To better understand and confirm the image of the collisions that the velocity space of protons gives us, we present in Fig. 5.8 the perpendicular components versus parallel components of the proton velocities in the reconstructed frame of the QP brought back in the reconstructed reaction plane. Presented in this way, the emission spheres become thick circles and are very clearly visible. Their thicknesses increase with the centrality of the collisions and, therefore, the violence of the collision. We also find the progressive overlap of the two spheres. The different colored selection areas defined in Fig. 5.4 appear very clearly on this representation. We can, therefore, unambiguously associate the green zone numbered 3 with the QP emission. The others, which could also correspond to a contribution from the QP, always have an additional contribution, even if it is small, as we will see later in Fig. 5.10.

It should be kept in mind that the graphs of previous figures show the occupancy of the velocity space but not in an absolute way from a quantitative point of view, especially since there is an overlap of the graphs. For this reason, Fig. 5.9 is also presented. In this figure, we have the two-dimensional graphs in invariant cross section of the perpendicular components versus parallel components of the proton velocities in the reconstructed frame of the QP defined in the reconstructed reaction plane. There is no longer any selection according to φ. We find again the fact that for the most peripheral collisions, there are a few collisions detected with protons coming from the QP allowing to have completeness in front of the center of mass of the reaction. We also observe an effect related to the choice to reconstruct the reaction plane by always placing the QP to the left of the beam in it. It is the right–left effect, which makes more particles on the emission circle on the left than on the right with respect to the direction of the QP. This effect is described in Refs. [Steckmeyer *et al.* (2001), Vient *et al.* (2002), Vient *et al.* (2018a)].

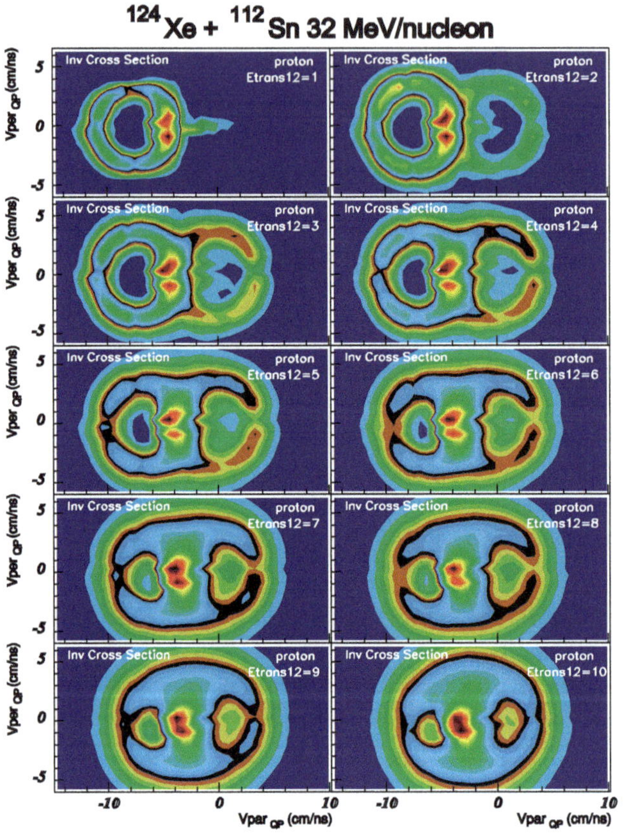

Figure 5.9 Two-dimensional graphs in invariant cross section of the perpendicular components versus parallel components of the proton velocities in the reconstructed frame of the QP defined in the reconstructed reaction plane. These data are obtained by the INDRA collaboration for the system ^{124}Xe + ^{112}Sn at 32 MeV/nucleon. Each two-dimensional graph corresponds to a selection in E_{t12}*Normalized* (impact parameter selection). Collisions are more and more central when going from left to right and from top to bottom. The studied events are only the selected events as defined previously.

5.5 Calculation of the Emission Probabilities by the QP

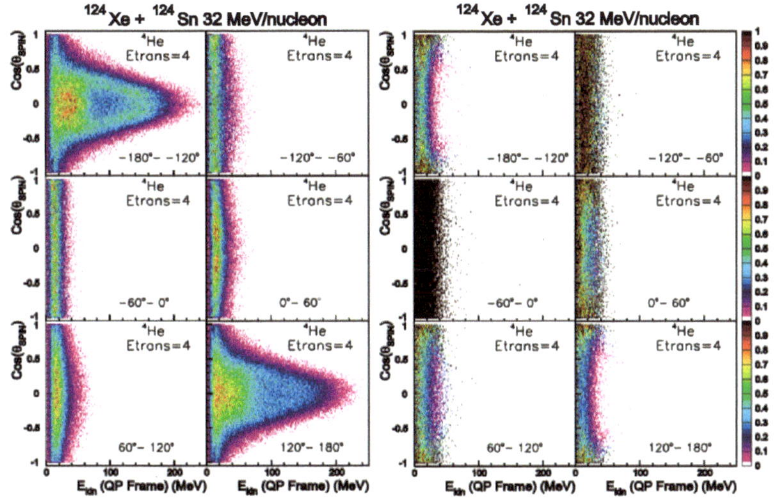

Figure 5.10 On the left side, two-dimensional graphs of the kinetic energy in the reconstructed frame of the QP versus the cosinus of the angle θ_{spin} for the alphas. Each graph corresponds to a different selection in φ as indicated. These data are obtained by the INDRA collaboration for the system ^{124}Xe + ^{124}Sn at 32 MeV/nucleon. On the right side, two-dimensional graphs of the kinetic energy in the reconstructed frame of the QP versus the cosinus of the angle θ_{spin} for the alphas. Each graph corresponds to a different selection in φ as indicated. The third dimension of these graphs corresponds to the probability for an alpha to be evaporated by the QP when it has a given kinetic energy and a given $cos(\theta_{spin})$. The selection number 4 in $E_{t12}Normalized$ (impact parameter selection) corresponds to rather peripheral collisions. The studied events are only the selected events as defined previously.

As explained in detail in Refs. [Vient (2006), Vient *et al.* (2018a)], 3D calorimetry is based on the assumption that the emission of the hot QP has a symmetry of revolution around the direction perpendicular to the reaction plane. For this reason, we defined the six selection areas in φ, all corresponding to the same solid angle in the velocity space in the QP reference frame. The extensive studies made in Refs. [Vient (2006), Vient *et al.* (2018a)] as well as the careful study of two-dimensional graphs such as those presented in Figs. 5.6, 5.7, and 5.8 concerning not only protons but also all charged light particles, led us to infer that only the contributions associated with the green zone numbered 3 can be considered as exclusively evaporated by hot QPs and hence representative of this evaporation. This is confirmed in Refs. [Vient (2006), Vient *et al.* (2018a)] using the HIPSE event generator, created by Durand and Lacroix [Lacroix *et al.* (2004)]. The comparison of this zone with the others allows to deduce a probability of emission by the QP for each particle or fragment produced during a nuclear reaction.

In Ref. [Vient (2006)], for each zone in φ, for each selection in $E_{t12}Normalized$, for all the types of detected particles, the polar angular distributions as well as the kinetic energy distributions, defined in the frame of the reconstructed QP (as in Fig. 5.5), have been built with all the detected nuclei, except for the two heaviest fragments emitted at the front of the center of mass, both considered QP residues. From these distributions, for a type of given particle, a selection in $E_{t12}Normalized$ and for a selection area, therefore, an angular domain between φ_1 and φ_2, an experimental probability of emission by the QP for each type of particle, can be determined: $Prob(E_k, \theta_{spin}, E_{t12}Normalized, \varphi_1 - \varphi_2))$. In this study, for any particle, the probability to be emitted in a polar angle θ_{spin} is assumed independent from the kinetic energy and reciprocally. Consequently, the probability is calculated from the following relation:

$$\begin{aligned} &Prob(E_k, \theta_{spin}, E_{t12}Normalized, \varphi_1 - \varphi_2) \\ &\quad = Prob(E_k, E_{t12}Normalized, \varphi_1 - \varphi_2) \\ &\quad \times Prob(\theta_{spin}, E_{t12}Normalized, \varphi_1 - \varphi_2) \end{aligned} \tag{5.1}$$

Given our different hypotheses about the QP emission, for a type of particle, the experimental probability is determined using the experimental histograms by the following relation:

$$Prob(E_k, \theta_{spin}, \varphi_1 - \varphi_2) = \frac{\frac{dN(E_k, 0^\circ - (-60^\circ))}{dE_k}}{\frac{dN(E_k, \varphi_1 - \varphi_2)}{dE_k}} \times \frac{\frac{dN(\theta, 0^\circ - (-60^\circ))}{d\cos\theta_{spin}}}{\frac{dN(\theta, \varphi_1 - \varphi_2)}{d\cos\theta_{spin}}} \tag{5.2}$$

The reference spectrum of zone 3 is divided by those of the other zones to obtain the probability distribution associated with each of

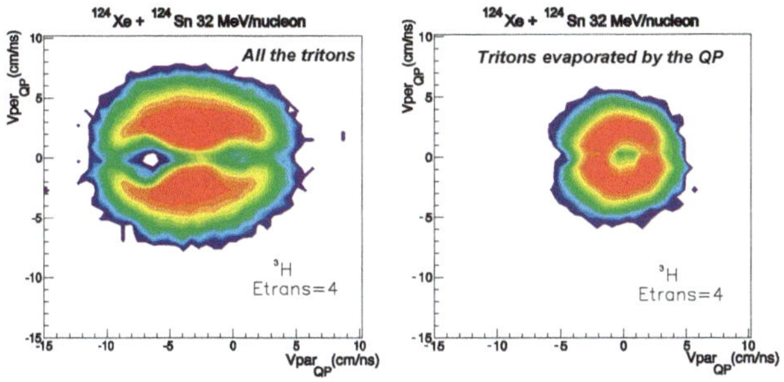

Figure 5.11 Two-dimensional graphs of the perpendicular components versus parallel components of the triton velocities in the reconstructed frame of the QP defined in the reconstructed reaction plane. On the left for all detected tritons, on the right for the contribution associated with the tritons considered evaporated by the QP. These data are obtained by the INDRA collaboration for the system ^{124}Xe + ^{124}Sn at 32 MeV/nucleon. The selection Number 4 in E_{t12}*Normalized* (impact parameter selection) corresponds to rather peripheral collisions (approximately between 6.5 and 7.5 fermis). The studied events are only the selected events as defined previously.

them. There is also a supplementary hypothesis that the probability of detection is independent of E_k, θ_{spin}, and φ.

It appears that the independence hypothesis between the probabilities associated with kinetic energy and the polar angle is perhaps a little too simplistic with respect to the component not evaporated by the QP, even if it allows more accurate measurements of the excitation energy per nucleon of the QP than usual techniques as shown in Refs. [Vient *et al.* (2018a), Vient *et al.* (2018c)]. This is why it seems more interesting to obtain experimental probabilities apart from two-dimensional graphs of the kinetic energy in the reconstructed frame of the QP versus the cosinus of the angle θ_{spin} as in Fig. 5.10 on the left side. This preserves the possible correlations existing between these two variables for the non-evaporated contribution, which could consequently influence the evaporation probabilities by the QP determined experimentally. The graph on the left of Fig. 5.10 shows the alphas produced in semi-peripheral collisions (selection Number 4 in E_{t12}*Normalized*) for the system ^{124}Xe + ^{124}Sn at 32 MeV/nucleon. Each graph corresponds to a different selection in φ. The graph corresponding to zone number 3 in φ $(0°, -60°)$ completely confirms that it corresponds only to the

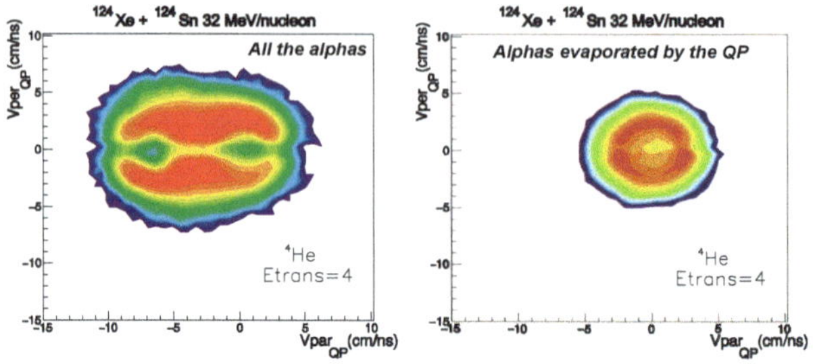

Figure 5.12 Two-dimensional graphs of the perpendicular components versus parallel components of the alpha velocities in the reconstructed frame of the QP defined in the reconstructed reaction plane. On the left for all detected alphas; on the right for the contribution associated with the alphas considered evaporated by the QP. These data are obtained by the INDRA collaboration for the system ^{124}Xe + ^{124}Sn at 32 MeV/nucleon. The selection Number 4 in E_{t12}*Normalized* (impact parameter selection) corresponds to rather peripheral collisions (approximately between 6.5 and 7.5 fermis). The studied events are only the selected events as defined previously.

evaporation of the QP. There is a uniform band from bottom to top indicating the same energy spectrum regardless of the $cosine(\theta_{spin})$. It is the only graph exhibiting such behavior. All the others have a non-uniformity. For the zones in φ numbered 1 $(-180°, -120°)$ and 6 $(120°, 180°)$, we can even observe the evaporation contribution of the QT. For the others, there is the mid-velocity contribution at different levels.

To obtain the experimental probabilities, it is then sufficient to divide the yield of the graph of zone 3 by each yield of the other graphs. We then obtain two-dimensional graphs of the experimental probabilities for being evaporated by the QP. These graphs are presented on the right of Fig. 5.10 for each of the zones in φ. Of course, it is necessary to have a large statistics of events and particles to limit statistical fluctuations and also to optimize the dimensioning of the graph. To a couple of values for kinetic energy and θ_{spin} corresponds in this graph a probability. We can, therefore, determine for a particle from its type, its kinetic energy, its polar angle, its azimuthal angle, and finally from the selection in normalized transverse energy its probability for being evaporated by the QP. We see in Figs. 5.11 and 5.12 what the use of these probabilities produces for tritons and alphas for rather peripheral collisions ^{124}Xe + ^{124}Sn at 32

MeV/nucleon. We show in these figures two-dimensional graphs of the perpendicular components versus parallel components of the velocities of alphas or tritons in the reconstructed frame of the QP. On the left, it is for all detected alphas or tritons; on the right, it is for the contribution associated with the alphas or tritons considered evaporated by the QP. To obtain this, we weight each particle by its probability for being evaporated by the QP in the graph on the left. This gives us the average aspect of the contribution evaporated by the QP in the velocity space for a kind of particle, for a certain range of impact parameters.

5.6 Hot QP Reconstruction

Knowing for each nucleus produced during a collision its probability $Prob_n$ of being emitted by the QP, it is possible, event by event, to reconstruct the initial hot QP. The QP charge is obtained as follows:

$$Z_{QP} = \sum_{n=1}^{mul_{tot}} Prob_n \times Z_n \tag{5.3}$$

where mul_{tot} is the number of detected charged particles in the event.

For the mass, we assume that the QP keeps the initial isotopic ratio of the projectile and that the produced cold nuclei follow the valley of stability. The number of neutrons produced by the QP is deduced from mass conservation as indicated by the following equation:

$$A_{QP} = Z_{QP} \times A_{Proj}/Z_{Proj} = \sum_{n=1}^{mul_{tot}} Prob_n \times A_n + N_{neutron} \tag{5.4}$$

We can also deduce the reaction Q-value.

$$Q = \delta(A_{QP}, Z_{QP}) - \sum_{n=1}^{mul_{tot}} Prob_n \times \delta(A_n, Z_n) - N_{neutron} \times \delta(1, 0) \tag{5.5}$$

$\delta(A_n, Z_n)$ is the mass excess of the nucleus ${}^{A_n}_{Z_n}X$.

The QP velocity is determined in the reference frame of the center of mass of the reaction only from charged particles by using this

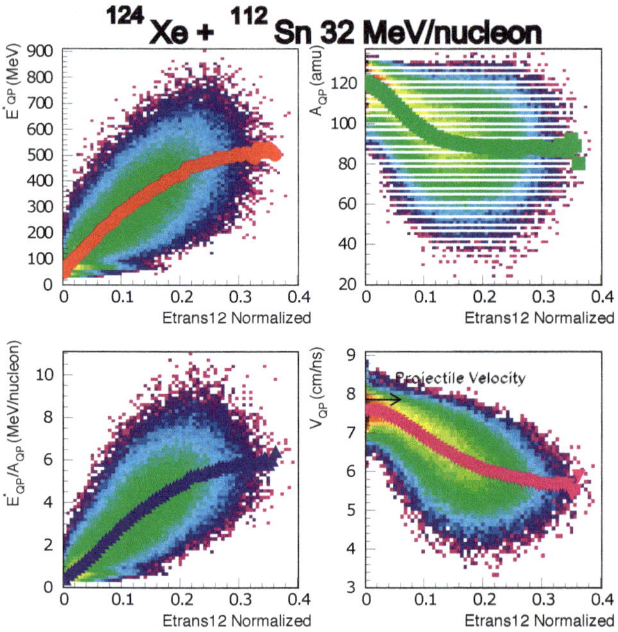

Figure 5.13 Top left, two-dimensional graph of $E_{t12}Normalized$ versus E^*_{QP} and superimposed the average correlation between these two physical quantities. Top right, two-dimensional graph of $E_{t12}Normalized$ versus A_{QP} and superimposed their average correlation. Bottom left, two-dimensional graph of $E_{t12}Normalized$ versus E^*_{QP}/A_{QP} and superimposed their average correlation and bottom right two-dimensional graph of $E_{t12}Normalized$ versus V_{QP} (velocity norm in the laboratory frame) and superimposed their average correlation. These data are obtained by the INDRA collaboration for the system ^{124}Xe + ^{112}Sn at 32 MeV/nucleon. The studied events are only the selected events as defined previously.

equation:

$$\overrightarrow{V}_{QP}^{CM} = \frac{\sum_{n=1}^{mul_{tot}} Prob_n \times \overrightarrow{P}_n^{CM}}{(A_{QP} - N_{neutron})} \tag{5.6}$$

with $\overrightarrow{P}_n^{CM}$ the linear momentum of the nth particle in the reference frame of the center of mass.

We can then compute the QP excitation energy:

$$E_{QP}^{*} = \sum_{n=1}^{mul_{tot}} Prob_n \times E_{k_n} + N_{neutron} \times \langle E_k \rangle_{p+\alpha} - Q - E_{k_{QP}} \qquad (5.7)$$

E_{k_n} being the kinetic energy of the nth particle in the center of mass frame, $\langle E_k \rangle_{p+\alpha}$ the mean kinetic energy of the neutrons deduced from that of the protons and alphas (Coulomb energy is subtracted) and finally with $E_{k_{QP}}$ the QP kinetic energy in the center of mass frame.

We now present as an example the result of the application of these different equations to the system ^{124}Xe + ^{112}Sn at 32 MeV/nucleon in Fig. 5.12. This kind of calorimetry only makes sense on average. This is why we present in this figure not only the two-dimensional graphs of the mass, the excitation energy, the norm of the velocity in the laboratory frame and excitation energy per nucleon of the QP as a function of the normalized transverse energy of LCP (of $Z = 1$ and 2) but especially the average correlation between these quantities.

5.7 Conclusion

The main purpose of this chapter on 3D calorimetry was to explain the principles of this experimental method for characterizing a hot nucleus and how it can be applied. We hope that we also demonstrated that a mastery and understanding of nuclear reactions in the domain of Fermi energy, given the complexity of physical processes, can only be achieved through a thorough study of the velocity space. This must be done in the reference frame of the hot nucleus. It shows that only a limited domain in this space corresponds to a "pure" emission of the hot nucleus of interest here for $\varphi \in (0°, -60°)$. There is indeed a strong mix of contributions from various sources in much of the velocity space. This shows that calorimetric methods based on fits with three sources (QP, QT, and mid-velocity), mainly when it presupposes the characteristics of the source at mid-velocity, are dangerous. It also implies that all methods of doubling the contribution forward of the QP to rebuild the QP are incorrect. This intrinsic difficulty has separated the fragments and particles according to their origin and, therefore, implies a knowledge of the characteristics of hot nuclei, which can only be valid on average. It should be kept in mind that this calorimetry, as it is, is only correct and can only be used for binary collisions. This is a problem for some central collisions in the range of incident energies studied and obviously when there is a complete or not complete fusion process. A section lacks in this chapter that would present the application of this calorimetry to data provided by an event generator and passed through a program simulating the functioning of the experimental device. It would make it possible to verify and validate the experimental method used and also to better understand the role of the experimental device and of all the selections made for this analysis. This kind of study is done in Refs. [Vient (2006), Vient *et al.* (2018a), Vient *et al.* (2018c)]. This is an absolute necessity if we want to do quantitative science in this field of physics.

References

Borderie, B. *et al.* (1999). Thermal and chemical equilibrium for vaporizing sources, *The European Physical Journal A - Hadrons and Nuclei* **6**, 2, pp. 197–202.

Casini, G. *et al.* (1993). Fission time scales from anisotropic in-plane distributions in $^{100}Mo +^{100} Mo$ and $^{120}Sn +^{120} Sn$ collisions around 20 A.MeV, *Phys. Rev. Lett.* **71**, 16, pp. 2567–2570.

Cugnon, J. and L'hote, D. (1983). Global variables and the dynamics of relativistic nucleus-nucleus collisions, *Nuclear Physics A* **397**, pp. 519–543.

Doré, D. *et al.* (2001). Properties of light particles produced in Ar+Ni collisions at 95 A.MeV: Prompt emission and evaporation, *Phys. Rev. C* **63**, 3, p. 034612.

Durand, D., Suraud, E. and Tamain, B. (2001). *Nuclear dynamics in the nucleonic regime* (Bristol, USA: IOP (2001)).

Eudes, P., Basrak, Z., de la Mota, V. and Royer, G. (2014). Is there incomplete fusion mechanism beyond 100 A MeV? *Nucl. Phys.* **A930**, pp. 131–138, doi: 10.1016/j.nuclphysa.2014.07.035.

Germain, M. *et al.* (2000). Evidence for dynamical proton emission in peripheral Xe+Sn collisions at 50 MeV/u, *Physics Letters B* **488**, 3-4, pp. 211–217.

Jaqaman, H., Mekjian, A. Z. and Zamick, L. (1983). Nuclear condensation, *Phys. Rev. C* **27**, pp. 2782–2791.

Lacroix, D., Van Lauwe, A. and Durand, D. (2004). Event generator for nuclear collisions at intermediate energies, *Phys. Rev. C* **69**, 5, p. 054604.

Lefort, T. *et al.* (2000). Study of intermediate velocity products in the Ar+Ni collisions between 52 and 95 A.MeV, *Nuclear Physics A* **662**, 3-4, pp. 397–422.

Łukasik, J. *et al.* (1997). Dynamical effects and intermediate mass fragment production in peripheral and semicentral collisions of Xe+Sn at 50 MeV/nucleon, *Phys. Rev. C* **55**, 4, pp. 1906–1916.

Marie, N. (1995). *Mouvement collectif et multifragmentation dans les collisions centrales du système Xe + Sn à 50 MeV par nucléon*, Ph.D. thesis, Université de Caen.

Montoya, C. P. *et al.* (1994). Fragmentation of necklike structures, *Phys. Rev. Lett.* **73**, 23, pp. 3070–3073.

Péter, J. *et al.* (1990). Strong impact parameter dependence of pre-equilibrium particle emission in nucleus-nucleus reactions at intermediate energies, *Physics Letters B* **237**, 2, pp. 187–191.

Pouthas, J. *et al.* (1995). Indra, a 4π charged product detection array at ganil, *Nuclear Instruments and Methods in Physics Research Section A: Accelerators, Spectrometers, Detectors and Associated Equipment* **357**, 2-3, pp. 418–442.

Steckmeyer, J. C. *et al.* (1996). Properties of very hot nuclei formed in ^{64}Zn+^{nat}Ti Collisions at Intermediate Energies, *Phys. Rev. Lett.* **76**, pp. 4895–4898.

Steckmeyer, J. C. *et al.* (2001). Excitation energy and angular momentum of quasiprojectiles produced in the Xe+Sn collisions at incident energies between 25 and 50 MeV/nucleon, *Nuclear Physics A* **686**, 1-4, pp. 537–567.

Vient, E. (2006). *Méthodologie de la calorimétrie et de la thermométrie des noyaux chauds formés lors de collisions aux énergies de Fermi.*, Mémoire d'habilitation à diriger des recherches, Université de Caen.

Vient, E. *et al.* (2002). Can we really measure the internal energy of hot nuclei with a 4π detection array? *Nuclear Physics A* **700**, 1-2, pp. 555–576.

Vient, E. *et al.* (2018a). New "3D calorimetry" of hot nuclei, *Phys. Rev. C* **98**, p. 044611.

Vient, E. *et al.* (2018b). Understanding the thermometry of hot nuclei from the energy spectra of light charged particles, *Eur. Phys. J.* **A54**, 6, p. 96.

Vient, E. *et al.* (2018c). Validation of a new "3D calorimetry" of hot nuclei with the HIPSE event generato, *Phys. Rev. C* **98**, p. 044612.

Viola, V. and Bougault, R. (2006). Calorimetry, *The European Physical Journal A - Hadrons and Nuclei* **30**, 1, pp. 215–226.

Weisskopf, V. (1937). Statistics and nuclear reactions, *Phys. Rev.* **52**, pp. 295–303.

Chapter 6

Early Recognition of Fragment Configuration in Intermediate Energy Heavy-Ion Collisions

Yogesh K. Vermani

Department of Physics, Sri Guru Granth Sahib World University, Fatehgarh Sahib 140407, India

yugs80@gmail.com

6.1 Cluster Production in Heavy-Ion Collisions: An Overview

The phenomenon of multifragment production in heavy-ion (HI) collisions is one of the fast-growing research themes at low and intermediate energy regimes. While the low-energy domain ($E_{lab} \lesssim 20AMeV$) is marked by fusion–fission events [Dutt (2010), Naderi (2013), Aguilera (2013), Wang (2016)], beyond $E_{lab} \approx$ 35 AMeV, one expects the heavier colliding nuclei to break into many pieces (fragments). This process is contemporary to other important phenomena occurring in the intermediate energy domain such as collective flow, elliptic flow, squeeze-out, stopping, and sub-threshold particle production [Hubbele (1991), Blaich (1993),

Multifragmentation in Heavy-Ion Reactions: Theory and Experiments
Edited by Rajeev K. Puri, Yu-Gang Ma, and Arun Sharma

ISBN 978-981-4968-69-0 (Hardcover), 978-1-003-38513-4 (eBook)
www.jennystanford.com

Schüttauf (1996), Aichelin (1986), Bertsch (1988), Aichelin (1991), CM Ko (1996)]. For the last two to three decades, these physical observables have been largely confined to collisions of stable nuclei with the choice of symmetric size of target and projectiles.

The dynamics of mutlifragmentation is characterized by interesting features such as power law yields for mass spectra [Hubbele (1991)], linear dependence of peak fragment yields on the total mass of the system [Sisan (2001), Vermani (2009)], universality behavior in the production of intermediate mass fragments as observed on ALADiN experimental setup [Blaich (1993), Schüttauf (1996)]. The data recorded on the ALADiN experimental setup have been particularly established as detailed systematics of multifragmentation as a function of mass and incident energy in the intermediate energy domain ranging from 100 AMeV to 1000 AMeV [Ogilvie (1991), Blaich (1993), Lynch (1995), Schüttauf (1996)].

The data related to fragmentation observables have been extensively used to test microscopic transport models in order to propose and explain the underlying mechanism behind the multifragmentation process [Aichelin (1987), Cugnon (1989), Tsang (1993), Bowman (1991), Bauer (1987), Peilert (1989)]. Many scenarios have been proposed for the underlying mechanism behind the multifragmentation channel such as density fluctuations [Bauer (1987), Peilert (1989)], and statistical model of fragmentation [Friedman (1983)]. Interestingly, an isoentropic expansion of hot nucleus to densities $\rho \leq 0.4\rho_\circ$ under the *rapid mass cluster formation* (RMCF) model [Friedman (1990)] is found to give a significant yield of intermediate mass fragments in the head-on $^{129}Xe + ^{197}Au$ collisions at $E_{lab} = 50$ AMeV [Bowman (1991)]. Predicted yields using RMCF calculations reproduce the observed charged particle multiplicities assuming decay from an expanding nuclear source characterized by a *soft* equation of state.

Statistical models have been extensively used in the past to compare their predictions with experimental data. For example, two statistical approaches "COPENHAGEN" and "GEMINI" were used with the same excitation energy and size input parameters to compare with multifragmentation data on the collisions of Au projectile with C, Al, Cu, and Pb targets at an incident energy of 600 AMeV [Kreutz (1993)]. Experimental charge distributions were found to

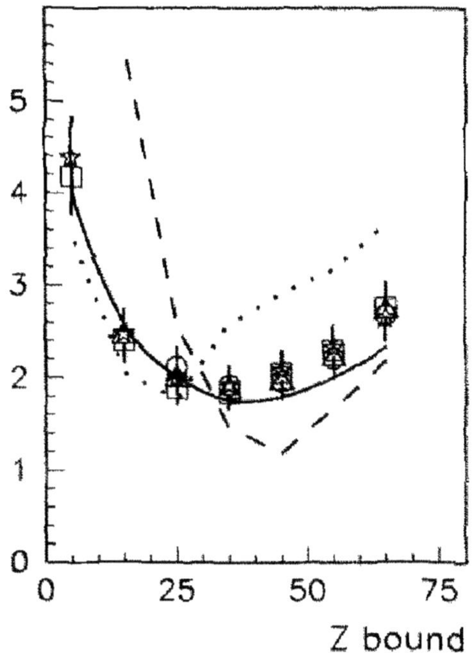

Figure 6.1 The extracted τ parameter as a function of Z_{bound} for collisions of Au projectiles with C (circles), Al (triangles), and Cu (squares) targets at an incident energy of 600 AMeV. Predictions are shown for COPENHAGEN (dashed line), GEMINI (dotted line), and percolations (solid line). The figure has been taken from Ref. [Kreutz (1993)].

exhibit power law parametrization:

$$\sigma(Z) = Z^{-\tau} \tag{6.1}$$

The "GEMINI" model overpredicted the extracted τ parameter for the bound charge $Z_{bound} > 25$ as shown in Fig. 6.1.

The "COPENHAGEN" model, on the other hand, predicts too many heavier intermediate mass fragments and underpredicted maximum charge $\langle Z_{max} \rangle$ as shown in Fig. 6.1. One can see that results from statistical calculations differ significantly from experimental data as well as from the percolation calculations. These results gave an insight into the multifragmentation phenomenon that how these fragment observables are sensitive to phase space dynamics. This led to a need for rigorously discriminating among model predictions. The disagreement of experimental data with the statistical model calculations necessitated full dynamical description for multifragment production.

6.2 Molecular Dynamics Approach to Multifragmentation: Role of Secondary Clusterization Algorithms

Fragmentation data for $^{197}Au +^{197} Au$ have been extensively studied by combining the ALADiN spectrometer and MSU-Miniball/Miniwall array [Kunde (1995)]. The experimental measurements taken at incident energies of 100 and 1000 AMeV revealed the importance of peripheral collisions for multifragmentation channel particularly at higher incident energies. With an increase in beam energy from 100 AMeV to 1000 AMeV for symmetric collisions of *Au* on *Au*, the peak multiplicity of intermediate-mass fragments was found to shift from central to peripheral collisions. In other words, the source of fragments changes from interaction zone (*fireball*) to two spectator nuclei in peripheral geometries.

Later on, experimental measurements were extended for systematic investigation of the impact parameter and incident energy dependence of the multifragmentation process [Tsang (1993), Lynch (1995)]. At the incident energy $E_{lab} = 100$ AMeV, there was observed enhancement in the production of intermediate mass fragments (IMFs) for central geometry. At higher incident energies, however, there was decline in the yields of IMFs. This decline is accompanied by a shift in the multifragmentation channel to peripheral collision geometries as shown in Fig. 6.2. Transport models based upon the molecular dynamics picture, namely the *quantum molecular dynamics* (QMD) model [Peilert (1989), Aichelin (1991)] and the *quasiparticle dynamics* (QPD) approach [Boal (1988)], successfully modelled the multifragmentation phenomenon in $^{197}Au +^{197} Au$ collisions at $E_{lab} = 100$ AMeV as illustrated in Fig. 6.2. Model calculations significantly underestimate the multiplicity of IMFs $\langle N_{IMF} \rangle$ at larger impact parameters as well as at incident energies > 100 AMeV. This statistical suppression of fragment emission in QMD and QPD calculations is not fully understood. It may be due to

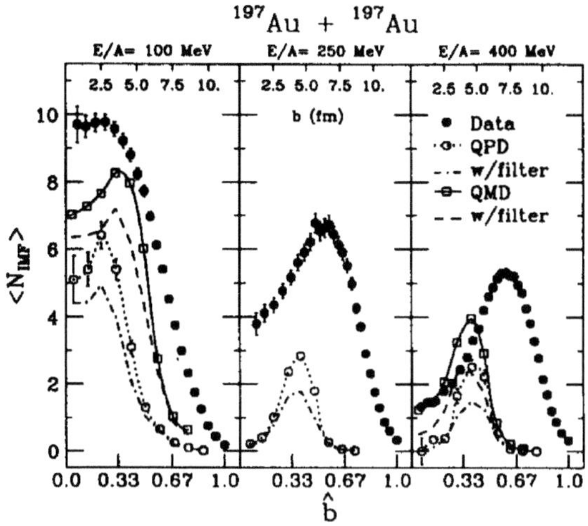

Figure 6.2 The IMF multiplicity as a function of impact parameter $\widehat{b}$ ($= b/b_{max}$) for $^{197}Au + ^{197}Au$ collisions. The figure has been taken from Ref. [Lynch (1995)].

the inadequate treatment of classical heat capacities [Boal (1989)], neglect of Fermi motion [Peilert (1992)], and neglect of quantum fluctuations in hot residual nuclei.

The comparison of fragment observables obtained using molecular dynamics approaches with experimental data still kept the puzzle unsolved that whether these dynamical approaches would ever explain the multifragmentation process. Particularly, the questions concerning entrance channel characteristics, i.e., the beam energy dependence and the role of collision geometry at higher incident energies were not satisfactorily explained by these dynamical calculations.

As far as the dynamical picture of the fragmentation process is concerned, the average behavior of fragment observables is shown to vary as a function of entrance channel properties, i.e., impact parameter, beam energy, and mass of projectile and target nuclei. Taking into account the entrance channel characteristics, namely, the projectile energy and impact parameter, Das *et al.* [Das (1996), Das (1997)] proposed a statistical simultaneous multifragmentation model for intermediate energies. In this model, Coulomb fragment–fragment interactions and nuclear interactions were incorporated through their respective mean fields using a statistical procedure [Satpathy (1990), Das (1993)]. This model depicted a tripartition picture of the total nuclear system consisting of participant *fireball*, two projectile-like and target-like *spectator* zones. Model-calculated IMF multiplicities as a function of the impact parameter $\widehat{b}(= b/b_{max})$ are then compared with the experimental data [Tsang (1993)] on $^{197}Au + ^{197}Au$ collisions at 100 and 400 AMeV up to midcentral impact parameters. Model predictions show significant contribution from spectator matter region. The overall IMF yields and peak IMF multiplicities were, however, underestimated, and these failed to reproduce the experimental trend for impact parameter dependence at an incident energy of 400 AMeV [Das (1997)]. This study based upon the *dynamical statistical multifragmentation* (DSM) model indicates that spectator matter, in spite of being relatively cold, contributes effectively toward intermediate and heavy mass fragment formation. The low energy fragmentation in peripheral HI collisions is, however, dominated by fireball.

It may be mentioned that cluster yields predicted by molecular dynamics approaches were obtained within the phase-space coalescence picture. The secondary algorithm recognizing the cluster yield based upon coalescence principle is known as the *minimum spanning tree* (MST) method [Aichelin (1991)]. In Section 6.3, we shall discuss details of the MST procedure along with its relevant extensions commonly used in cluster identification.

6.3 Minimum Spanning Tree Clusterization Algorithm and Its Extensions

The minimum spanning tree (MST) is the simplest cluster recognition algorithm based on the phase space coalescence principle. The MST procedure is characterized by clusterization radius parameter $R_\circ$, which corresponds to the range of attractive part of nuclear potential. In the MST approach, a cluster is defined in the following way:

Let *i* and *j* be two members of a given cluster C, then

$$\begin{aligned} &\forall i \in C, \exists j \in C, \ and \\ &|\mathbf{r}_i - \mathbf{r}_j| \leq R_\circ. \end{aligned} \tag{6.2}$$

Here, $\mathbf{r}_i$ and $\mathbf{r}_j$ are the spatial coordinates of the two nucleons. The clusterization radius $R_\circ$ is used as a free parameter, which may lie between 2 and 4 fm. An improvisation over the MST approach has also been tried, which looks for a bound two-nucleon structure in the same cluster. This algorithm labeled as MSTE [Pratt (1995), Pan (1995), Strachan (1997)] recognizes two nucleons i and j bound in a cluster, if

$$e_{ij} \leq 0, \tag{6.3}$$

with $e_{ij} = V_{ij} + \frac{(p_i - p_j)^2}{4\mu}$; μ being the reduced mass.

6.3.1 Minimum Spanning Tree with Momentum Cut (MSTM) Method

In this extension of MST algorithm, in addition to the spatial cut of Eq. (6.2), we introduce a cut on relative momentum of two nucleons, i.e.,

$$|\mathbf{r}_i - \mathbf{r}_j| \leq R_\circ,$$
$$|\mathbf{p}_i - \mathbf{p}_j| \leq P_F. \tag{6.4}$$

Here, P_F is the average Fermi momentum of nucleons bound in a nucleus ($\approx$150 MeV/c) in its ground state [Kumar (1998), Singh (2001), Dhawan (2007)]. This improvisation checks the formation of artificial and unbound fragments by excluding those nucleons having relative momenta larger than P_F. This algorithm identifies the largest fragment $\langle A^{max} \rangle$ as early as 50–60 fm/c [Kumar (1998), Singh (2001), Dhawan (2007)]. As a result, fragment emission starts earlier with MSTM, when the MST method just detects a single biggest cluster.

6.3.2 Minimum Spanning Tree with Binding Energy Check (MSTB) Method

In this modified version, pre-clusters obtained with the conventional MST approach are subjected to the binding energy check:

$$\zeta = \frac{1}{N_f} \sum_{i=1}^{N_f} \left[\frac{\left(\mathbf{p}_i - \mathbf{P}_{N_f}^{c.m.}\right)^2}{2m_i} + \frac{1}{2} \sum_{j \neq i}^{N_f} V_{ij}\left(\mathbf{r}_i, \mathbf{r}_j\right) \right] < -E_{bind}. \tag{6.5}$$

We take E_{bind} = 4.0 MeV if $N_f \geq 3$, else $E_{bind} = 0$. N_f is the number of nucleons in a fragment, and $\mathbf{P}_{N_f}^{cm}$ is the center-of-mass momentum of a fragment. This criterion forbids the formation of loosely bound clusters. The role of the MSTB method is quite important in central symmetric reactions [Kumar (1998), Vermani (2009)]. The MSTB method does not recognize the largest fragment A^{max} during the early violent phase. A properly bound A^{max} is identified only around 120 fm/c [Singh (2001)]. The time evolution of the fragments' binding energy using MST and MSTB approaches suggests that the total binding energy of all fragments will be $\leq$ 4.0 AMeV around 150 fm/c [Singh (2001)]. In other words, the binding energy check helps to recognize a properly bound and stable fragment structure quite early and reduces the computation time for multifragmentation.

In the present chapter, we shall not go much into the details of these extensions of the MST algorithm as all these extended versions, in first principle, ride on the spatial constraint criterion and the fragment structure is realized only at the end of the reaction.

6.4 Early Cluster Recognition Algorithm (ECRA)

C. O. Dorso and collaborators put forth an idea and showed that for the final fragment spectra, one does not need to wait for a very long time when the colliding system is a dilute mixture of free nucleons and fragments of different sizes. It was argued in Refs. [Dorso (1994), Dorso (1995)] that fragments are already formed in the phase space before they can be recognized in the configuration space using the MST procedure. This can be achieved via analyzing the evolution of the fragmenting system using a secondary clusterization algorithm, namely the *Early Cluster Recognition Algorithm* (ECRA) [Dorso (1993)]. In fact, it was illustrated that the populations of different fragment species reached their asymptotic values at times much earlier than the final end of the HI reaction. For this study, head-on collisions of the symmetric nuclei of mass 40 were simulated. Central collisions were selected to isolate the effects due to geometrical considerations and no uncertainty due to binning in experimental measurements. In Fig. 6.3, the average multiplicity of fragments for mass bins (a) $[3 \leq A \leq 5]$, (b) $[5 \leq A \leq 7]$, and

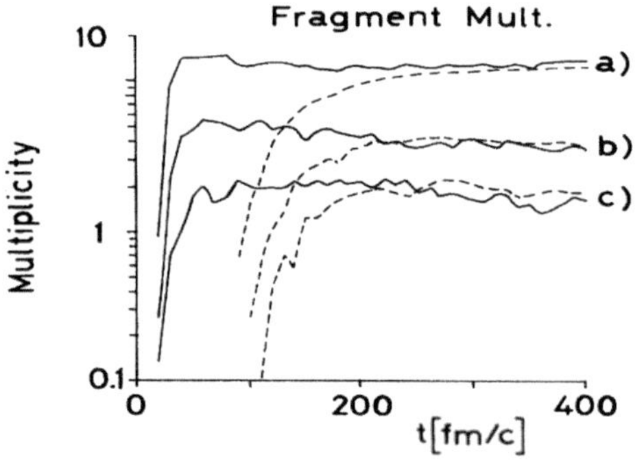

Figure 6.3 The fragment multiplicity as a function of the times calculated for ECRA analysis (solid lines) and the conventional MST procedure (dashed lines) for mass bins (a) $[3 \leq A \leq 5]$, (b) $[5 \leq A \leq 7]$, and (c) $[7 \leq A \leq 9]$. The figure has been taken from Ref. [Balonga (1995)].

(c) $[7 \leq A \leq 9]$ as a function of the times obtained using ECRA (full lines) and MST procedures (dashes lines) is shown.

One can see from the figure that for shorter times ($t < 200$ fm/c), the MST procedure does not recognize a realistic fragmentation pattern. The ECRA type analysis is, however, better suited for earlier times. After the violent phase is over, the fragmentation pattern is already realized long before the fragments are supposed to be formed in thermal models.

6.5 Simulated Annealing Clusterization Algorithm (SACA): A Faster Approach

The ECRA analysis has an inherent limitation that the nucleons inside a fragment are assumed to be bound even if the binding energy of the fragments is extremely small in magnitude though negative. Further, the simulation of heavier nuclear systems becomes a problem, where the number of possible fragment configurations increases tremendously.

Santini *et al.* investigated fragmentation dynamics for the excited nuclear matter formed in central head-on collisions at an intermediate energy of 400 AMeV [Santini (2005)]. Their approach to investigate HI collisions was the transport equation of Boltzmann type,

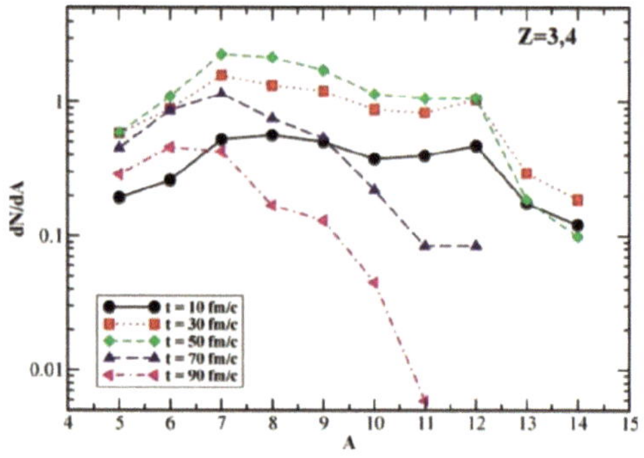

Figure 6.4 The mass distributions of $Z = 3, 4$ clusters calculated at different times as obtained in head-on $^{197}Au +^{197} Au$ collisions at 400 AMeV incident energy. The figure has been taken from Ref. [Santini (2005)].

namely *Relativistic Boltzmann-Uhlenbeck-Uheling* (RBUU) equation, which can be found in detail in Refs. [Botermans (1990), Blattel (1993)]. The phase space coalescence method was applied at several stages of central $^{197}Au +^{197} Au$ collision. The mass distributions of $Z = 3, 4$ fragments were calculated at different times, as shown in Fig. 6.4.

The heavier clusters are observed to originate quite early ($t \approx$ 10–30 fm/c) during the high-density phase. The evaporation of single nucleons and light fragments started at the freeze-out stage only. One possible answer for fast clusterization may be the onset of radial flow very early during the expansion phase. Thus, the transfer of thermal energy to collective motion makes the early emission of heavy clusters possible.

This signature of an early heavy cluster formation also confirms the possibility to explore compressional and high-density isospin effects in fragmentation dynamics at relativistic energies [Bass (1998), Reisdorf (2004)].

On the basis of these earlier investigations on theoretical as well as experimental fronts, one can infer that nuclear fragmentation is a fast process. In peripheral geometry, the origin of fragments should be articulated with excitation energy deposition in the spectator matter [Gossiaux (1997)]. There is no doubt that dynamical approaches are capable of reproducing yields of intermediate mass fragments at peripheral geometries especially at an incident energy of 100 AMeV [Tsang (1993)]. However, the validity of molecular dynamics approaches seemed to be questionable at relativistic energies from 400 to 1000 AMeV. This raises the question, what, if not the phase space coalescence, controls the mechanism behind multifragmentation.

Puri and collaborators proposed that one possibility is to study the time evolution of heavy-ion reactions within dynamical models and look when will be the first time the fragments can be identified. This requires a fragment recognition mechanism at an early stage of the reaction unlike the conventional MST procedure in which one has to simulate HI reactions for about 250 fm/c [Aichelin (1991)] using the QMD approach. They developed a novel approach for the faster recognition of clusters in the phase space based on the simulated annealing technique [Metropolis (1953), Laarhoven (1987)]. This algorithm was dubbed as *Simulated Annealing Clusterization Algorithm* (SACA) [Puri (2000)]. The efficiency of SACA and results obtained with this algorithm are found to be quite promising for low- and high-incident energy regimes.

SACA is also found to reproduce successfully the experimental charge yields obtained in asymmetric O+Ag/Br reactions at incident energies between 25 and 200 AMeV [Puri (2002)]. In another attempt, $^{129}Xe +^{120} Sn$ reactions were analyzed for fragment observables such as charge yields, proton-like fragments, IMF yields, angular distribution, and average kinetic energy of fragments. The SACA method was able to explain all these observables quite nicely, whereas the MST method fails to explain these experimental trends [Nebauer (1999)].

In the SACA approach, the following assumptions are made:

1. The nucleons from target and projectile are grouped into fragments (of any size) and into free nucleons.

2. Though the nucleons inside a fragment can interact with each other, they do not interact with the nucleons from other fragments or free nucleons.

3. That pattern of nucleons and fragments is realized in nature, which maximizes the fragments' total binding energy.

We employ a binding energy check to avoid the unnecessary formation of too many fragments:

$$\zeta_a = \frac{1}{N_f}\sum_{i=1}^{N_f}\left[\sqrt{\left(\mathbf{p}_i - \mathbf{P}_{N_f}^{cm}\right)^2 + m_i^2} - m_i + \frac{1}{2}\sum_{j\neq i}^{N_f} V_{ij}(\mathbf{r}_i, \mathbf{r}_j)\right] < -E_{bind}, \tag{6.6}$$

with E_{bind} = 4.0 MeV if $N_f \geq 3$, else $E_{bind} = 0$. In this equation, E_{bind} is the fragments' binding energy per nucleon, N_f is the number of nucleons in a fragment, and $P^c_{N_f}m$ is the center-of-mass momentum of the fragment. Such binding energy check is very useful to identify the most bound fragment configuration out of a huge number of possible configurations.

The new binding energy check of 4 MeV has the advantage that it excludes the formation of loosely bound fragments that would decay later on. This new definition of binding energy check is different from that used in ECRA analysis [Dorso (1994), Dorso (1995)], where nucleons are assumed to be bound even if the binding energy of the fragment is extremely small. An iterative procedure is followed, where a transition to the fragment configuration with higher binding energy is always accepted. The transitions leading to lower binding energies are also accepted, but with a certain probability. Since this procedure is called simulated annealing [Laarhoven (1987)], this clusterization method is dubbed as *Simulated Annealing Clusterization Algorithm* (SACA). This name has been derived from the "annealing" process used in the solidification of liquid metals. It is a sequence of Metropolis algorithm with gradually decreasing control parameter "ϑ," which can be interpreted as a *temperature*. For each metropolicity at a given temperature, one has to perform a sequence of steps until the binding energy does not change anymore. Each step is executed as follows:

1. Given some initial fragment configuration "*a*" with energy ζ_a, a new configuration "*b*" with energy ζ_b is generated in the neighborhood of "*a*" using the Monte-Carlo procedure.

2. Let the energy difference between *a* and *b* be $\Delta\zeta = \zeta_b - \zeta_a$.

3. If $\Delta\zeta$ is negative, the new configuration is always accepted. If $\Delta\zeta$ is positive, it is accepted with a probability $e^{-\Delta\zeta/\vartheta}$.

At the start, the temperature ϑ is taken to be large enough so that almost all attempted transitions are accepted. This is to overcome any kind of local minima. After the binding energy remains constant, a gradual decrease in the control temperature ϑ is made and the Metropolis algorithm is repeated.

The first attempt to use the Metropolis algorithm in this direction was made by Dorso *et al.* in the ECRA method [Dorso (1993)]. The Metropolis algorithm was used in ECRA so as to identify the optimal fragment configuration that maximizes the binding energy of the system. But the main limitation of ECRA analysis was the prohibitively slow algorithm as discussed earlier in Sec. 6.4.

SACA is executed in two steps. The first step involves an exchange of nucleons only among the fragments. The second step involves an exchange of sub-clusters, i.e., a group of nucleons among the fragments.

I. Nucleon Exchange Procedure

To start with, a random configuration "*a*" (which consists of fragments and free nucleons) is chosen. The total energy associated with the configuration "*a*" is given by:

$$\begin{aligned}
\zeta_a = & \sum_{i=1}^{N_f^l}\left\{\sqrt{(\mathbf{p}_i - \mathbf{P}^{cm}_{N_f^l})^2 + m_i^2} - m_i + \frac{1}{2}\sum_{j\neq i}^{N_f^l} V_{ij}(\mathbf{r}_i, \mathbf{r}_j)\right\}_1 \\
& + \cdots \sum_{i=1}^{N_f^\nu}\left\{\sqrt{(\mathbf{p}_i - \mathbf{P}^{cm}_{N_f^\nu})^2 + m_i^2} - m_i + \frac{1}{2}\sum_{j\neq i}^{N_f^\nu} V_{ij}(\mathbf{r}_i, \mathbf{r}_j)\right\}_\nu \\
& + \sum_{i=1}^{N_f^\mu}\left\{\sqrt{(\mathbf{p}_i - \mathbf{P}^{cm}_{N_f^\mu})^2 + m_i^2} - m_i + \frac{1}{2}\sum_{j\neq i}^{N_f^\mu} V_{ij}(\mathbf{r}_i, \mathbf{r}_j)\right\}_\mu \\
& + \cdots \sum_{i=1}^{N_f^n}\left\{\sqrt{(\mathbf{p}_i - \mathbf{P}^{cm}_{N_f^n})^2 + m_i^2} - m_i + \frac{1}{2}\sum_{j\neq i}^{N_f^n} V_{ij}(\mathbf{r}_i, \mathbf{r}_j)\right\}_n .
\end{aligned}$$

Here, N_f^μ is the number of nucleons in a fragment μ, $\mathbf{P}^{cm}_{N_f^\mu}$ is the center-of-mass momentum of the fragment μ, and $V_{ij}(\mathbf{r}_i, \mathbf{r}_j)$ is the interaction energy between nucleons i and j in the given fragment μ. Note that the total energy is the sum of the energies of individual fragments in their respective center-of-mass system. Therefore, ζ_a differs from the (conserved) total energy of the system because (i) the kinetic energies of fragments calculated in their center of masses and (ii) the interactions between fragments/free nucleons are neglected.

A new configuration is generated using the Monte-Carlo procedure by either (i) transferring a nucleon from some randomly chosen fragment to another fragment or by (ii) setting a nucleon of a fragment free, or (iii) absorbing a free nucleon into a fragment. Let the new configuration "b" be generated by transferring a nucleon from a fragment ν to a fragment μ. Then the energy of the new configuration "b" is given by:

$$\zeta_b = \sum_{i=1}^{N_f^l}\left\{\sqrt{(\mathbf{p}_i - \mathbf{P}_{N_f^l}^{cm})^2 + m_i^2} - m_i + \frac{1}{2}\sum_{j\neq i}^{N_f^l} V_{ij}(\mathbf{r}_i, \mathbf{r}_j)\right\}_1$$
$$+ \cdots \sum_{i=1}^{N_f^\nu - 1}\left\{\sqrt{(\mathbf{p}_i - \mathbf{P}_{N_f^\nu - 1}^{cm})^2 + m_i^2} - m_i + \frac{1}{2}\sum_{j\neq i}^{N_f^\nu - 1} V_{ij}(\mathbf{r}_i, \mathbf{r}_j)\right\}_\nu$$
$$+ \sum_{i=1}^{N_f^\mu + 1}\left\{\sqrt{(\mathbf{p}_i - \mathbf{P}_{N_f^\mu + 1}^{cm})^2 + m_i^2} - m_i + \frac{1}{2}\sum_{j\neq i}^{N_f^\mu + 1} V_{ij}(\mathbf{r}_i, \mathbf{r}_j)\right\}_\mu$$
$$+ \cdots \sum_{i=1}^{N_f^n}\left\{\sqrt{(\mathbf{p}_i - \mathbf{P}_{N_f^n}^{cm})^2 + m_i^2} - m_i + \frac{1}{2}\sum_{j\neq i}^{N_f^n} V_{ij}(\mathbf{r}_i, \mathbf{r}_j)\right\}_n$$

Note that in this procedure, the individual energies of all fragments except for the donor fragment (ν) and the receptor fragment (μ) remain the same. The change in the energy when going from configuration "a" to the new configuration "b" is:

$$\Delta\zeta = \zeta_b - \zeta_a. \tag{6.7}$$

Between the Metropolis algorithms, the system is cooled by decreasing the control parameter ϑ. A decrease in the temperature means that we narrow the energy difference, which is accepted in a

Metropolis step. After many Metropolis steps, one would arrive at a minimum, i.e., the most bound configuration.

II. Fragment Exchange Procedure

The problem with the above nucleon-exchange procedure is that one usually arrives at a local minimum only. Between the two local minima, we find a huge maximum. Let us give an example: Assume we have two fragments, but the most bound configuration would be one single fragment that combines both. Now each exchange of a single nucleon raises the binding energy, and only the exchange of all nucleons at the same time lowers the total binding energy. This effect is well known in chemistry, where it is termed activation energy. In order to avoid this, one adds, therefore, a second simulated annealing algorithm in which the nucleons are not anymore considered entities to be exchanged in each Metropolis step (as in the first step of simulated annealing), but also fragments or nucleons obtained after the first step. This second stage of minimization is called fragment exchange procedure. Note that even in this stage, the free nucleons can be exchanged as before. The total energy associated with any configuration "c" during the second stage of iterations is given by:

$$\begin{aligned}\zeta_c = &\left\{\sum_{i=1}^{N_{S_1}}\left[\sqrt{(\mathbf{p}_i-\mathbf{P}^{cm}_{N_{S_1}})^2+m_i^2}-m_i+\frac{1}{2}\sum_{j\neq i}^{N_{S_1}}V_{ij}(\mathbf{r}_i,\mathbf{r}_j)\right]\right\}_1\\
&+\cdots\left\{\sum_{i=1}^{N_{S_\nu}}\left[\sqrt{(\mathbf{p}_i-\mathbf{P}^{cm}_{N_{S_\nu}})^2+m_i^2}-m_i+\frac{1}{2}\sum_{j\neq i}^{N_{S_\nu}}V_{ij}(\mathbf{r}_i,\mathbf{r}_j)\right]\right\}_\nu\\
&+\left\{\sum_{i=1}^{N_{S_\mu}}\left[\sqrt{(\mathbf{p}_i-\mathbf{P}^{cm}_{N_{S_\mu}})^2+m_i^2}-m_i+\frac{1}{2}\sum_{j\neq i}^{N_{S_\mu}}V_{ij}(\boldsymbol{r}_i,\mathbf{r}_j)\right]\right\}_\mu\\
&+\cdots\left\{\sum_{i=1}^{N_{S_n}}\left[\sqrt{(\mathbf{p}_i-\mathbf{P}^{cm}_{N_{S_n}})^2+m_i^2}-m_i+\frac{1}{2}\sum_{j\neq i}^{N_{S_n}}V_{ij}(\mathbf{r}_i,\mathbf{r}_j)\right]\right\}_n.\end{aligned}$$

Here N_{S_μ} is the number of nucleons in a super-fragment $S_\mu = \sum_{k=1}^{N^f_{S_\mu}} N^k_{S_\mu}$, where $N^k_{S_\mu}$ is the number of nucleons in the kth fragment contained in the super-fragment S_μ, and $N^f_{S_\mu}$ is the number of pre-fragments contained in the super-fragment S_μ. The $\mathbf{P}^{cm}_{N_{S_\mu}}$ is the center-of-mass momentum of the super fragment S_μ, and $V_{ij}(\mathbf{r}_i, \mathbf{r}_j)$ is the interaction energy between nucleons i and j in a given super-fragment. Note that now the particle i interacts with its fellow nucleons in the same pre-fragment and also with the nucleons of other pre-fragments, which are contained in a given super-fragment S_μ. The new configuration is generated using the Monte-Carlo procedure either by (i) transferring a pre-fragment from some randomly chosen super-fragment to another super-fragment or by (ii) setting a pre-fragment free, or (iii) absorbing a single isolated pre-fragment into a super-fragment. Let us suppose that a new configuration "d" is generated by transferring a pre-fragment "k" (with mass $N^k_{S_\nu}$) from a super-fragment ν to a super-fragment μ. The associated energy of new configuration "d" reads as:

$$\begin{aligned}
\zeta_d = & \left\{ \sum_{i=1}^{N_{S_1}} \left[\sqrt{(\mathbf{p}_i - \mathbf{P}^{cm}_{N_{S_1}})^2 + m_i^2} - m_i + \frac{1}{2} \sum_{j \neq i}^{N_{S_1}} V_{ij}(\mathbf{r}_i, \mathbf{r}_j) \right] \right\}_1 \\
& + \cdots \left\{ \sum_{i=1}^{N_{S_\nu} - N^k_{S_\nu}} \left[\sqrt{(\mathbf{p}_i - \mathbf{P}^{cm}_{N_{S_\nu} - N^k_{S_\nu}})^2 + m_i^2} - m_i + \frac{1}{2} \sum_{j \neq i}^{N_{S_\nu} - N^k_{S_\nu}} V_{ij}(\mathbf{r}_i, \mathbf{r}_j) \right] \right\}_\nu \\
& + \left\{ \sum_{i=1}^{N_{S_\mu} + N^k_{S_\nu}} \left[\sqrt{(\mathbf{p}_i - \mathbf{P}^{cm}_{N_{S_\mu} + N^k_{S_\nu}})^2 + m_i^2} - m_i + \frac{1}{2} \sum_{j \neq i}^{N_{S_\mu} + N^k_{S_\nu}} V_{ij}(\mathbf{r}_i, \mathbf{r}_j) \right] \right\}_\mu \\
& + \cdots \left\{ \sum_{i=1}^{N_{S_n}} \left[\sqrt{(\mathbf{p}_i - \mathbf{P}^{cm}_{N_{S_n}})^2 + m_i^2} - m_i + \frac{1}{2} \sum_{j \neq i}^{N_{S_n}} V_{ij}(\mathbf{r}_i, \mathbf{r}_j) \right] \right\}_n .
\end{aligned}$$

The only difference between the nucleon and the fragment exchange procedures occurs for the bound nucleons. Now the bound nucleons cannot change their identity neither by being absorbed by other pre-fragments nor by becoming free. They will remain bound in a pre-fragment. The pre-fragment itself can change its identity by either getting transferred to a new super-fragment, or be set free. As in the first stage, one calculates the energy difference between the new and the old configurations $\Delta\zeta$, and the Metropolis procedure is continued till the most bound configuration is obtained.

For a specific problem of cluster identification, one needs to generate neighborhood solutions and reduce the control temperature "ϑ" gradually. This cooling schedule is designed to execute the SA algorithm properly with the following parameters [Metropolis (1953), Laarhoven (1987)]:

1. The initial temperature ϑ_i.
2. Cooling function to decrease the temperature ϑ gradually.
3. The length of the Markov chain M_{ch}. The SA algorithm will repeat the process M_{ch} times at a given temperature.
4. Final temperature ϑ_f to terminate the SA algorithm.

6.5.1 Time Evolution of Fragments Using SACA and MST Approaches

Let us now discuss the fragmentation process in $^{197}Au +^{197} Au$ collisions as analyzed by Puri *et al.* [Puri (1996)] at 600 AMeV and at impact parameter $b = 8$ fm. This reaction system has been extensively studied by ALADiN collaboration for its results presenting direct evidence for liquid–gas phase transition [Pochodzalla (1995)]. Figure 6.5 displays the evolution of $^{197}Au +^{197} Au$ collision, analyzed within QMD+MST and QMD+SACA approaches. The first row shows the time evolution of the collision rate dN_{coll}/dt and mean nucleonic density $\rho(t)$. The mean nucleonic density is obtained as:

$$\rho(t) = \langle \frac{1}{N} \sum_{i=1, j\neq i}^{N} \frac{1}{(2\pi L)^{3/2}} e^{-(\mathbf{r}_i(t)-\mathbf{r}_j(t))^2/2L} \rangle, \tag{6.8}$$

with $L = 1.08\ fm^2$ and r_i are centroids of gaussian wave packets of nucleons. The mean nucleonic density reaches its maximum at 30 fm/c. With an increase in time, the collision rate attains its peak value at about 60 fm/c and then decreases sharply. Free nucleons keep emitting from the clusters very slowly and this continues for long till the end of HI reaction. This happens due to a small transfer of energy to spectator matter, and as a result, the emission time scale becomes

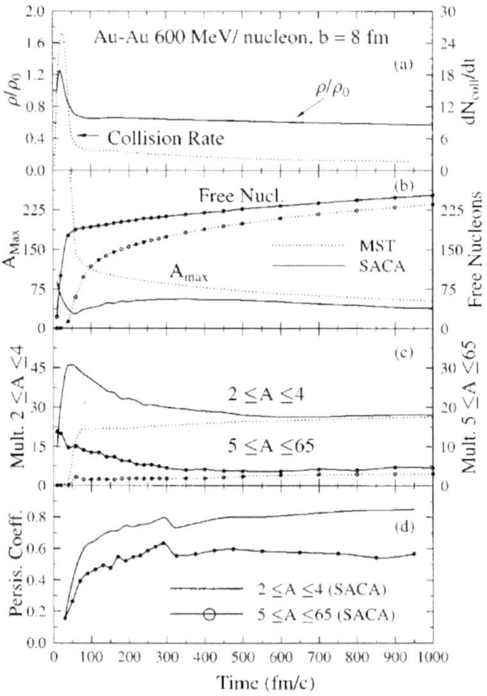

Figure 6.5 Evolution of $^{197}Au +^{197} Au$ collision at 600 AMeV and at an impact parameter $b = 8$ fm. Top to bottom rows display time evolution of density and of collision rate, the size of the heaviest fragment and of the number of emitted nucleons, multiplicity of fragments with mass $2 \leq A \leq 4$ and with $5 \leq A \leq 65$, and the persistence coefficient. The figure has been taken from Ref. [Puri (1996)]. Reprinted with permission from [Rajeev K. Puri, Christoph Hartnack, and Jorg Aichelin, Physical Review C, 54, R28, 1996] Copyright (2020) from the American Physical Society.

quite large. Even the persistence coefficient is shown to attain the value 0.8 after a very long time as displayed in the bottom row. From the second row, we see that SACA is not able to identify the size of the heaviest fragment $\langle A_{max} \rangle$ before 150 fm/c and MST needs much longer time. Before that time, one observes a rise in the number of IMFs due to the fact that the largest fragment size falls into the mass bracket of IMFs $[5 \leq A \leq 65]$ as clear from the figure. How can this conjecture be interpreted?

The additional fragments predicted by SACA at an earlier time $t \sim 60$ fm/c are obtained because this configuration gives the lowest binding energy. This seems contradictory to the thermalization of spectator matter that takes place at a time longer than 60 fm/c. From this observation, one may conjecture that realistic and most bound fragment structure can be obtained using SACA at the moment where the size of the largest fragment is minimum. As soon as the violent phase of reaction is over, the final fragment structure is already determined in the phase space. A faster recognition of fragment structure with the QMD+SACA theoretical approach obviously seems to be a better choice than the QMD+MST approach. At an earlier time $t \sim$ 60fm/c, one obtains a minimum in size of the heaviest fragment $\langle A_{max} \rangle$, particularly in peripheral collisions where the QMD model was earlier unable to transfer the excitation energy to the spectator zones [Gossiaux (1997)].

Based on the extensive multifragmentation data available from the ALADiN collaboration [Schüttauf (1996)], it remained to be seen whether the QMD+SACA approach can also reproduce spectator matter fragmentation of Au-projectiles or not. In this direction, Puri and collaborators planned to systematically analyze the spectator matter fragmentation in $^{197}Au +^{197} Au$ collisions at relativistic bombarding energies of 400, 600, and 1000 AMeV [Vermani (2009)]. Model calculations using the original SACA version and the MST approach are finally confronted with ALADiN multifragmentation data. Figure 6.6 shown below gives the time evolution of binding energy per nucleon for the fragments with mass $A = 4$ and IMFs $[5 \leq A \leq 65]$. In MST calculations, the binding energy turns negative around 100 fm/c. However, with the SACA method, one obtains a stable fragment configuration at much earlier times.

It is clear from the figure that bound fragments can be identified as early as 60 fm/c, just after the violent phase is over. All the fragments at this time posses binding energy greater than 4 MeV per nucleon. Strikingly, an earlier detection of the bound fragments at all incident energies up to 1000 AMeV gives us the possibility to look into the n–n interactions when nuclear matter is still hot and dense. Further, one is also free from the problem of stability of fragments. The failure of the MST method to detect the final fragment pattern also questions its validity at incident energies as high as 1000 AMeV.

Finally, we confronted our model calculations using SACA and MST versions of the clusterization algorithm with $^{197}Au +^{197} Au$ multifragmentation data available at all the three incident energies.

Figure 6.7 displays the mean IMF multiplicity $\langle N_{IMF} \rangle$ as a function of the impact parameter b in $^{197}Au +^{197} Au$ reactions at bombarding energies of 400, 600, and 1000 AMeV. Our model calculations for $^{197}Au +^{197} Au$ reactions with the SACA method are in nice agreement with the ALADiN data [Schüttauf (1996)] at all incident energies. As is clear from the figure, we achieved a reasonable reproduction of the shape of impact parameter dependence of $\langle N_{IMF} \rangle$. One main feature of ALADiN data on the Au-projectile fragmentation is the universality behavior, that is, the yield of IMFs is independent of the target and the bombarding energy. This universality characteristics is nicely reproduced at all the three bombarding energies using the SACA approach. Due to the shallow minima in the size of $\langle A_{max} \rangle$, sometimes we also display the mean IMF multiplicity observed at the second local minimum (earlier than 60 fm/c) for peripheral geometries. The calculated values at these minima using the SACA method are marked as black filled squares. These values are further closer to the experimental data reflecting the conception behind the SACA

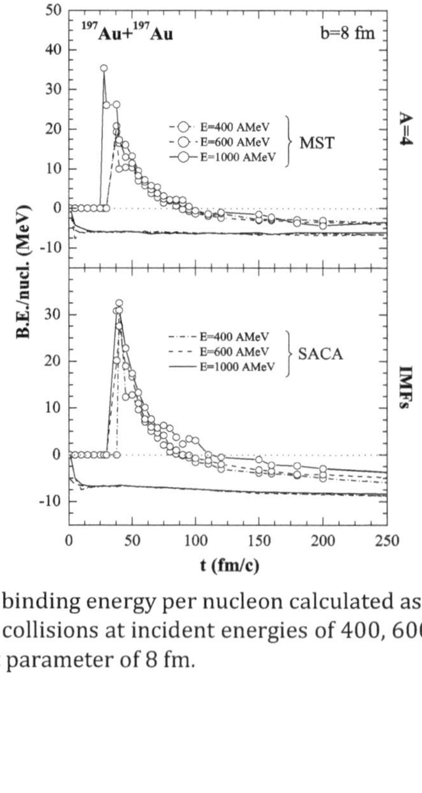

Figure 6.6 Mean binding energy per nucleon calculated as a function of time in $^{197}Au +^{197} Au$ collisions at incident energies of 400, 600, and 1000 AMeV and at an impact parameter of 8 fm.

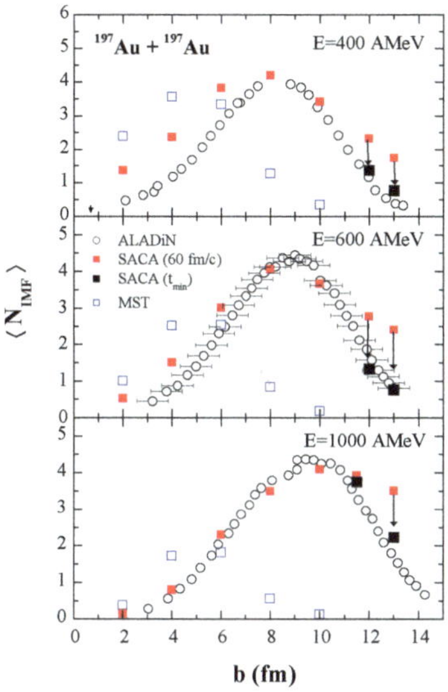

Figure 6.7 Mean IMF multiplicity $\langle N_{IMF} \rangle$ *vs.* impact parameter b calculated using SACA and MST approaches.

technique to recognize the fragment yield faster. Further, the peak value of $\langle N_{IMF} \rangle$ and the corresponding impact parameter b is also well estimated with the QMD+SACA method. The prominent feature of the spectator decay is the invariant nature of the IMF distribution with respect to the bombarding energy. The SACA method successfully reproduced this universal nature of spectator fragmentation at all the three bombarding energies. It is worth interesting to mention that these universal features observed in the multifragmentation of gold nuclei persist up to much higher bombarding energies than explored in this work [Adamovich (1997)].

The QMD transport model clubbed with the SACA method also allowed us to study the time scale of multifragmentation. This confrontation of model calculations with experimental data again confirms the validity of molecular dynamics approaches to study the multifragmentation phenomenon. The secondary clusterization algorithm one chooses remains, however, the key tenet in the understanding of multifragmentation dynamics in heavy-ion collisions.

References

Aguilera, E. F. *et al.* (2013). Near-barrier fusion and total reaction of a proton-rich projectile: ^{3}He+^{58}Ni, *Phys. Rev. C* **87**, 1, pp. 014613 (7).

Aichelin, J. (1991). "Quantum" molecular dynamics: A dynamical microscopic n-body approach to investigate fragment formation and the nuclear equation of state in heavy-ion collisions, *Phys. Rep.* **202**, 5, pp. 233–360.

Aichelin, J. and Stöcker, H. (1986). Quantum molecular dynamics: A novel approach to N-body correlations in heavy ion collisions, *Phys. Lett. B* **176**, 1, pp. 14–19.

Aichelin, J. *et al.* (1987). Importance of momentum-dependent interactions for the extraction of the nuclear equation of state from high-energy heavy-ion collisions, *Phys. Rev. Lett.* **58**, 19, pp. 1926–1929.

Balonga, P. E. and Dorso, C. O. (1995). Intermittency in microscopic simulations of multifragmentation, *Phys. Rev. C* **52**, 2, pp. 915–918.

Bass, S. A. *et al.* (1998). Microscopic models for ultrarelativistic heavy ion collisions, *Prog. Part. Nucl. Phys.* **41**, pp. 255–369.

Bauer, W., Bertsch, G. F., and Das Gupta, S. (1987). Fluctuations and clustering in heavy-ion collisions, *Phys. Rev. Lett.* **58**, 9, pp. 863–866.

Bertsch, G. F. and Das Gupta, S. (1988). A guide to microscopic models for intermediate energy heavy ion collisions, *Phys. Rep.* **160**, 4, pp. 189–233.

Blaich, Begemann M. *et al.* (1993). Quantum molecular dynamics simulation of multifragment production in heavy-ion collisions at E/A=600 MeV, *Phys. Rev. C* **48**, pp. 610–617.

Blattel, B., Koch, V., and Mosel, U. (1993). Transport-theoretical analysis of relativistic heavy-ion collisions, *Rep. Prog. Phys.* **56**, 1, pp. 1–62.

Boal, D. H. and Glosli, J. N. (1988). Quasiparticle model for nuclear dynamics, *Phys. Rev. C* **38**, 4, pp. 1870–1878.

Boal, D. H., Glosli, J. N., and Wicentowich, C. (1989). Temperatures in heavy-ion reactions: Simulation via quasiparticle dynamics, *Phys. Rev. C* **40**, 2, pp. 601–609.

Botermans, W. and Malfliet, R. (1990). Quantum transport theory of nuclear matter, *Phys. Rep.* **198**, 3, pp. 115–194.

Bowman, D. R. *et al.* (1991). Multifragment disintegration of the ^{129}Xe+^{197}Au system at E/A=50 MeV, *Phys. Rev. Lett.* **67**, 12, pp. 1527–1530.

Cugnon, J. and Volant, C. (1989). Multifragmentation in the intranuclear cascade, *Z. Phys. A* **334**, 4, pp. 435–442.

Das, A., Mishra, M., Satpathy, M., and Satpathy, L. (1993). Effects of interfragment nuclear interaction and statistical approach to multifragmentation phenomena, *J. Phys. G: Nucl. Part. Phys.* **19**, 2, pp. 319–338.

Das, C. B., Das, A., Satpathy, L., and Satpathy, M. (1996). Statistical simultaneous multifragmentation model for heavy-ion collisions with entrance channel characteristics, *Phys. Rev. C* **53**, 4, pp. 1833–1839.

Das, C. B., Das, A., Satpathy, M., and Satpathy, L. (1997). Entrance channel dependence of fragmentation dynamics in heavy-ion collisions, *Phys. Rev. C* **56**, 3, pp. 1444–1451.

Dhawan, J. K. and Puri, R. K. (2007). On the momentum correlations in the fragmentation of ^{197}Au + ^{197}Au reactions, *Eur. Phys. J. A* **33**, 1, pp. 57–64.

Dorso, C. O. and Balonga, P. E. (1994). Fluctuation dynamics of fragmenting spherical nuclei, *Phys. Rev. C* **50**, 2, pp. 991–995.

Dorso, C. O. and Aichelin, J. (1995). When and how are fragments formed in heavy ion collisions? *Phys. Lett. B* **345**, 3, pp. 197–202.

Dorso, C. O. and Randrup, J. (1993). Early recognition of clusters in molecular dynamics, *Phys. Lett. B* **301**, 4, pp. 328–333.

Dutt, I. *et al.* (2010). Comparison of different proximity potentials for asymmetric colliding nuclei, *Phys. Rev. C* **81**, 6, pp. 064609 (12).

EMU-01/12 – collaboration, (1997). Multifragmentation of gold nuclei in the interactions with photoemulsion nuclei at 10.7 GeV/nucleon, *Z. Phys. A* **359**, pp. 277–290.

FOPI Collaboration, Reisdorf, W. (2004). Nuclear stopping from 0.09*A* to 1.93*A* GeV and its correlation to flow, *Phys. Rev. Lett.* **92**, 23, pp. 232301 (4).

Friedman, W. A. (1990). Rapid massive cluster formation, *Phys. Rev. C* **42**, 2, pp. 667–673.

Friedman, W. A. and Lynch, W. G. (1983). Statistical formalism for particle emission, *Phys. Rev. C* **28**, 1, pp. 16–23.

Gossiaux, P. B., Puri, R., Hartnack, Ch., and Aichelin, J. (1997). The multifragmentation of spectator matter, *Nucl.Phys. A* **619**, 3–4, pp. 379–390.

Hubele, J. *et al.* (1991). Fragmentation of gold projectiles: From evaporation to total disassembly, *Z. Phys. A* **340**, pp. 263–270.

Ko, C. M. and Li, G. Q. (1996). Medium effects in high energy heavy-ion collisions, *J. Phys. G: Nucl. Part. Phys.* **22**, 12, pp. 1673–1725.

Kreutz, P. *et al.* (1993). Charge correlations as a probe of nuclear disassembly, *Nucl. Phys. A* **556**, 4, pp. 672–696.

Kumar, S. and Puri, R. K. (1998). Role of momentum correlations in fragment formation, *Phys. Rev. C* **58**, 1, pp. 320–325.

Kunde, G. J. *et al.* (1995). Fragment flow and the multifragmentation phase space, *Phys. Rev. Lett.* **74**, 1, pp. 38–41.

Laarhoven, P. J. M. and Aarts, E. H. L. (1987). *Simulated Annealing: Theory and Applications* (Reidel, Dordrecht).

Lynch, W. G. (1995). Fragmentation in exclusive measurements, *Nucl. Phys. A* **583**, 6, pp. 471–479.

Metropolis, N., Rosenbluth, A. W., Rosenbluth, M. N., Teller, A. H., and Teller, E. (1953). Equation of state calculations by fast computing machines, *J. Chem. Phys.* **21**, 6, pp. 1087–1092.

Naderi, D. (2013). A dynamical interpretation of fusion–fission reactions using four-dimensional Langevin equations, *J. Phys. G: Nucl. Part. Phys.* **40**, 12, pp. 125103.

Nebauer, R. *et al.* (1999). *Proceedings of the International Workshop XXVII, Hirschegg, Austria, 1999*, edited by Feldmeier, H., Knoll, J., Noerenberg, W., and Wambach, J. (GSI, Darmstadt), pp. 43–61.

Ogilvie, C. A. *et al.* (1991). Rise and fall of multifragment emission, *Phys. Rev. Lett.* **67**, pp. 1214–1217.

Pan, J. and Das Gupta, S. (1995). Unified description for the nuclear equation of state and fragmentation in heavy-ion collisions, *Phys. Rev. C* **51**, 3, pp. 1384–1392.

Peilert, G. *et al.* (1989). Multifragmentation, fragment flow, and the nuclear equation of state, *Phys. Rev. C* **39**, 4, pp. 1402–1419.

Peilert, G. *et al.* (1992). Dynamical treatment of Fermi motion in a microscopic description of heavy ion collisions, *Phys. Rev. C* **46**, 4, pp. 1457–1473.

Pochodzalla, J. *et al.* (1995). Probing the nuclear liquid–gas phase transition, *Phys. Rev. Lett.* **75**, 6, pp. 1040–1043.

Pratt, S., Montoya, C., and Ronning, F. (1995). Balancing nuclear matter between liquid and gas, *Phys. Lett. B* **349**, 3, pp. 261–266.

Puri, R. K. and Aichelin, J. (2000). Simulated annealing clusterization algorithm for studying the multifragmentation, *J. Comput. Phys.* **162**, 1, pp. 245–266.

Puri, R. K., Hartnack, C., and Aichelin, J. (1996). Early fragment formation in heavy-ion collisions, *Phys. Rev. C* **54**, 1, pp. R28–R31.

Puri, R. K., Singh, J., and Kumar, S. (2002). Fragment production in $^{16}O +^{80} Br$ reaction within dynamical microscopic theory, *Pramana J. Phys.* **59**, 1, pp. 19–31.

Santini, E., Gaitanos, T., Colonna, M., and Di Toro, M. (2005). Fragment formation in central heavy-ion collisions at relativistic energies, *Nucl. Phys. A* **756**, 3–4, pp. 468–484.

Satpathy, L., Mishra, M., Das, A., and Satpathy, M. (1990). Fragment interactions in nuclear multifragmentation phenomena, *Phys. Lett. B* **237**, 2, pp. 181–186.

Schüttauf, A. *et al.* (1996). Universality of spectator fragmentation at relativistic bombarding energies, *Nucl. Phys. A* **607**, 4, pp. 457–486.

Singh, J. and Puri, R. K. (2001). Study of the formation of fragments with different clusterization methods, *J. Phys. G: Nucl. Part. Phys.* **27**, 10, pp. 2091–2108.

Sisan, D. *et al.* (2001). Intermediate mass fragment emission in heavy-ion collisions: Energy and system mass dependence, *Phys. Rev. C* **63**, 2, pp. 027602 (4).

Strachan, A. and Dorso, C. O. (1997). Fragment recognition in molecular dynamics, *Phys. Rev. C* **56**, 2, pp. 995–1001.

Tsang, M. B. *et al.* (1993). Onset of nuclear vaporization in $^{197}Au +^{197} Au$ collisions, *Phys. Rev. Lett.* **71**, 10, pp. 1502–1505.

Vermani, Y. K. and Puri, R. K. (2009). Mass dependence of the onset of multifragmentation in low energy heavy-ion collisions, *J. Phys. G: Nucl. Part. Phys.* **36**, pp. 105103 (8).

Vermani, Y. K. and Puri, R. K. (2009). Microscopic approach to the spectator matter fragmentation from 400 to 1000 AMeV, *Europhys. Lett.* **85**, pp. 62001 (6).

Wang, N. and Guo, Lu (2016). New neutron-rich isotope production in $^{154}Sm +^{160} Gd$, *Phys. Lett. B* **760**, pp. 236–241.

Chapter 7

Symmetry Energy of Finite Nuclei Using Relativistic Mean Field Densities within Coherent Density Fluctuation Model

Manpreet Kaur,[a,b] Ankit Kumar,[a,b] Abdul Quddus,[c]
M. Bhuyan,[d,e] and S. K. Patra[a,b]

[a]*Institute of Physics, Bhubaneswar 751005, India*

[b]*Homi Bhabha National Institute, Anushakti Nagar, Mumbai 400085, India*

[c]*Aligarh Muslim University, Aligarh 202002, India*

[d]*Department of Physics, Faculty of Science, University of Malaya, Kuala Lumpur 50603, Malaysia*

[e]*Institute of Research Development, Duy Tan University, Da Nang 550000, Vietnam*

patra@iopb.res.in

7.1 Introduction

The nuclear symmetry energy is one of the key issues of contemporary nuclear physics, which is associated with the isospin nature of the nuclear systems. An imbalance between neutron and proton composition of the system leads to the emergence of the symmetry energy term in the equation of state (EOS) of nuclear matter. The EOS imparts the information about the properties of nuclei and the isovector part of the nuclear interaction, analogous to

Multifragmentation in Heavy-Ion Reactions: Theory and Experiments
Edited by Rajeev K. Puri, Yu-Gang Ma, and Arun Sharma

ISBN 978-981-4968-69-0 (Hardcover), 978-1-003-38513-4 (eBook)
www.jennystanford.com

ordinary matter where van der Waals forces yield phase diagrams shedding light on different phase transitions with pressure and temperature variation. The symmetry energy plays a crucial role in neutron-skin formation in finite nuclei, estimation of neutron star radius and pressure of neutron star matter, which lead to the determination of other properties such as moments of inertia, tidal polarizability, and quadrupole of a (binary) neutron star [Lattimer (2001), Lattimer (2007)]. In other words, it exhibits a pervasive role extending from the study of the structure of nuclei [Niksic (2008), Van Giai (2010)], dynamics of isospin asymmetric heavy-ion collisions [Li (2008), Colonna (2009)] in terrestrial laboratory to neutron star, and stellar nucleosynthesis [Steiner (2009), Lattimer (2007)] in the cosmos. Several astrophysical observations and the availability of radioactive ion beams (RIBs) at FAIR (Germany), RIKEN (Japan), CSR (CHINA), GANIL (France), and upcoming FRIB (USA) have provided the opportunities to explore the nuclei under extreme conditions of isospin asymmetry.

The different physical properties of nuclei, like their masses, radii, density distributions, occupation probabilities, and fission properties are influenced by the isospin dependence of strong interactions among nucleons. In order to interpret the properties of neutron-rich nuclei and the neutron star matter, the determination of the symmetry energy is an important step. But experimentally it is not a directly measurable quantity and is extracted from the observables related to it. A lot of efforts have been undertaken to probe the symmetry energy and its coefficient [Tsang (2012)]. Different theoretical approaches such as the liquid drop model [Myers (1966), Moller (1995), Pomorski (2003)], energy density functional of Skyrme force [Chen (2005), Yoshida (2006), Chen (2010)], random phase approximation based on Hartree–Fock (HF) approach [Carbone (2010)], relativistic nucleon–nucleon interaction [Lee (1998), Agrawal (2010)], and the effective relativistic Lagrangian with density-dependent meson–nucleon vertex function [Vretenar (2003)] have been advanced to investigate the symmetry energy and related observables. Some empirical correlations reveal the significance of isospin interactions in finite nuclei and neutron stars [Steiner (2009)]. Numerous investigations have found a nearly linear correlation among the values of symmetry energy, its slope and curvature [Holt (2018)]. The neutron-skin thickness of nuclei is interrelated to the radius of a neutron star [Horowitz (2001)] and pressure of neutron star matter at sub-nuclear density [Brown (2000)]. There are some signatures of correlation of symmetry energy slope parameter (L) with the neutron-skin thickness of ^{208}Pb [Brown (2000), Furnstahl (2002), Centelles (2009), Roca-Maza (2011)] and the radii of neutron stars having low masses ($M_{NS} \sim 0.6$–$1.2\ M_{\odot}$). The neutron pressure is related to L, which is sensitive to the EOS of nuclear matter. Recently, we have shown a correlation between neutron pressure and the neutron-skin thickness of neutron-rich thermally fissile nuclei [Quddus (2019), Quddus (2019a)]. The symmetry energy and neutron pressure, collectively termed the effective surface properties, are defined extensively in Refs. [Gaidarov (2011), Bhuyan (2018)].

In addition to the above-mentioned studies, earlier investigations by Feenberg, Cameron, and Green [Feenberg (1947), Cameron (1957), Green (1958)] have shown that surface symmetry energy contribution needs to be taken into account along with volume symmetry energy for an accurate description of symmetry energy. Danielwicz has predicted that the ratio of volume to surface symmetry energy is related with the neutron-skin thickness [Danielewicz (2003), Tsang (2009), Danielewicz (2014)]. Many theoretical studies are devoted to determine the volume and surface symmetry energy components in finite nuclei with and without inclusion of temperature [Myers (1966), Danielewicz (2003), Tsang (2009), Danielewicz (2014), Warda (2010), Bhuyan (2018), Quddus (2019), Antonov (2016)]. It is worth mentioning that the symmetry energy can be correlated with the magic numbers for light and heavy nuclei corresponding to the neutron number N = Z = 8, 20, 28, 50 [Antonov (2016), Quddus (2019a)]. In the view of the above facts, here we analyze the symmetry energy in the isotopic chains of light, medium as well as heavy nuclei. Also, we discuss the mass dependence of volume and surface contributions in the symmetry energy and their possible correlation with shell closure/magicity in the rare earth mass region. It is important to note that both the symmetry energy as well as shell structure affect the r-process nucleosynthesis mechanism. For this, we intend to explore an enigmatic peak at A $\sim$ 160 in the abundance curve, although being less pronounced, which is supposed to be due to mid-shell nuclear deformation. Here, we also emphasize on a new deformed magic number at N $\sim$ 100, which is in line with our earlier work of stability of ^{162}Sm nuclei [Ghorui (2012)] and later on got the experimental conformation [Patel (2014)].

In the present work, the bulk properties and densities of the nuclei are calculated within the effective field theory motivated relativistic mean field (E-RMF) approach. The calculated densities of the nuclei are used as an input to the coherent density fluctuation model (CDFM) to investigate the surface properties. The advantages of CDFM [Antonov (1982), Gaidarov (2011), Bhuyan (2018), Gaidarov (2012)] over other methods are that this method takes care of (i) the fluctuation arising in the nuclear density distribution via weight function $|F(x)|^2$, and (ii) the momentum distributions through the mixed density matrix (i.e., the Wigner distribution function) [Bhuyan (2018), Gaidarov (2011), Gaidarov (2012)]. In other words, in the CDFM approach, the relative uncertainty arising in the estimation of the surface properties for finite nuclei due to density and momentum is taken care of.

The chapter is organized as follows: in Section 7.2, we present the formalism. In subsection 7.2.1, we outline the E-RMF model, which has been used to calculate the ground-state bulk properties and densities of the nuclei. Subsection 7.2.2 contains the general idea of calculating symmetry energy and relevant quantities like neutron pressure and symmetry energy curvature. The effective surface properties of nuclei are calculated within the CDFM, discussed in subsection 7.2.3. The calculated results are discussed in Section 7.3 and the work is summarized in Section 7.4.

7.2 Formalism

7.2.1 Effective Field Theory Motivated Relativistic Mean Field Model (E-RMF)

Relativistic mean field (RMF) theory is one of the microscopic approaches to solve the many-body problem of nuclear systems. In the RMF model, the nucleons are assumed to interact through the exchange of mesons. The model predicts ground as well as intrinsic excited state properties of nuclei such as the binding energy, root mean square radius, density distributions, deformation parameter, and single particle energies remarkably well throughout the nuclear landscape. The details of RMF models and their parametrizations can be found in Refs. [Furnstahl (1996), Del Estal (2001), Kumar (2018), Quddus (2018)]. For the sake of completeness, here we present the E-RMF formalism briefly. The effective mean field approximated Lagrangian density has, in principle, several terms with all possible types of self- and cross-couplings of mesons. In this work, we have used the E-RMF Lagrangian having contributions of δ meson and photon up to 2nd order exponent and the rest up to 4th order of exponents, which has been shown to be a good approximation to predict the finite nuclei and the nuclear matter observables up to considerable satisfaction [Furnstahl (1996)]. The energy density, obtained within the E-RMF Lagrangian by applying mean field approximation, is given as:

$$
\begin{aligned}
\mathcal{E}(r) = \sum_i \varphi_i^\dagger(r) \Bigg\{ & -i\boldsymbol{\alpha}\cdot\boldsymbol{\nabla} + \beta\left[M - \Phi(r) - \tau_3 D(r)\right] + W(r) \\
& + \frac{1}{2}\tau_3 R(r) + \frac{1+\tau_3}{2} A(r) - \frac{i\beta\boldsymbol{\alpha}}{2M}\cdot \\
& \left(f_\omega \boldsymbol{\nabla} W(r) + \frac{1}{2} f_\rho \tau_3 \boldsymbol{\nabla} R(r) \right) \Bigg\} \varphi_i(r) \\
& + \left(\frac{1}{2} + \frac{\kappa_3}{3!}\frac{\Phi(r)}{M} + \frac{\kappa_4}{4!}\frac{\Phi^2(r)}{M^2} \right) \frac{m_s^2}{g_s^2}\Phi^2(r) \\
& - \frac{\zeta_0}{4!}\frac{1}{g_\omega^2} W^4(r) + \frac{1}{2g_s^2}\left(1 + \alpha_1 \frac{\Phi(r)}{M} \right) \left(\boldsymbol{\nabla}\Phi(r)\right)^2 \\
& - \frac{1}{2g_\omega^2}\left(1 + \alpha_2 \frac{\Phi(r)}{M} \right) \left(\boldsymbol{\nabla} W(r)\right)^2 \\
& - \frac{1}{2}\left(1 + \eta_1 \frac{\Phi(r)}{M} + \frac{\eta_2}{2}\frac{\Phi^2(r)}{M^2} \right) \frac{m_\omega^2}{g_\omega^2} W^2(r) - \frac{1}{2e^2}\left(\boldsymbol{\nabla} A(r)\right)^2 \\
& - \frac{1}{2g_\rho^2}\left(\boldsymbol{\nabla} R(r)\right)^2 - \frac{1}{2}\left(1 + \eta_\rho \frac{\Phi(r)}{M} \right) \frac{m_\rho^2}{g_\rho^2} R^2(r) - \Lambda_\omega \\
& \left(R^2(r) \times W^2(r) \right) + \frac{1}{2g_\delta^2}\left(\boldsymbol{\nabla} D(r)\right)^2 + \frac{1}{2}\frac{{m_\delta}^2}{g_\delta^2}\left(D^2(r) \right), \qquad (7.1)
\end{aligned}
$$

where Φ, W, R, D, and A are the fields that have been redefined as $\phi = g_\sigma \sigma$, $W = g_\omega \omega^0$, $R = g_\rho \rho^0$, and $A = eA^0$. m_σ, m_ω, m_ρ, and m_δ are the masses and g_σ, g_ω, g_ρ, g_δ, $\frac{e^2}{4\pi}$ are the coupling constants for σ, ω, ρ, δ mesons and photon, respectively.

7.2.2 Nuclear Matter Parameters

The energy per nucleon of nuclear matter $\mathcal{E}/A$=$e(\rho, \alpha)$ (where ρ is the baryon density) can be expanded by the Taylor series expansion method in terms of isospin asymmetry parameter $\alpha \left(= \frac{\rho_n - \rho_p}{\rho_n + \rho_p}\right)$:

$$e(\rho, \alpha) = \frac{\mathcal{E}}{\rho_B} - M = e(\rho) + S(\rho)\alpha^2 + \mathcal{O}(\alpha^4), \tag{7.2}$$

where $e(\rho)$, $S(\rho)$, and M are the energy density of symmetric nuclear matter (SNM) (α = 0), the symmetry energy, and the mass of a nucleon, respectively. The odd powers of α are forbidden by the isospin Symmetry, and the terms proportional to α^4 and higher orders have negligible contribution. The symmetry energy $S(\rho)$ is defined by:

$$S(\rho) = \frac{1}{2}\left[\frac{\partial^2 e(\rho, \alpha)}{\partial \alpha^2}\right]_{\alpha=0}. \tag{7.3}$$

Near the saturation density ρ_0, the symmetry energy can be expanded through the Taylor series expansion method as:

$$S(\rho) = J + L\mathcal{Y} + \frac{1}{2}K_{sym}\mathcal{Y}^2 + \frac{1}{6}Q_{sym}\mathcal{Y}^3 + \mathcal{O}[\mathcal{Y}^4], \tag{7.4}$$

where $J = S(\rho_0)$ is the symmetry energy at saturation and $\mathcal{Y} = \frac{\rho - \rho_0}{3\rho_0}$. The slope parameter ($L$-coefficient), the symmetry energy curvature (K_{sym}), and the skewness parameter (Q_{sym}) are defined as

$$L = 3\rho \left.\frac{\partial S(\rho)}{\partial \rho}\right|_{\rho=\rho_0}, \tag{7.5}$$

$$K_{sym} = 9\rho^2 \frac{\partial^2 S(\rho)}{\partial \rho^2}\bigg|_{\rho=\rho_0}, \tag{7.6}$$

and

$$Q_{sym} = 27\rho^3 \frac{\partial^3 S(\rho)}{\partial \rho^3}\bigg|_{\rho=\rho_0}, \tag{7.7}$$

respectively. The neutron pressure of asymmetric nuclear matter can also be evaluated from the L-coefficient by using the relation:

$$L^{NM} = \frac{3P_0^{NM}}{\rho_0}. \tag{7.8}$$

It is to be noted that these nuclear matter properties at saturation (i.e., ρ_0, $S(\rho_0)$, $L(\rho_0)$), $K_{sym}(\rho_0)$, and $Q_{sym}(\rho_0)$ are model dependent and vary with certain uncertainties. More details of these quantities and their values along with the uncertainties for the non-relativistic and relativistic mean field models with various force parameters can be found in Refs. [Dutra (2012), Dutra (2014)].

7.2.3 Coherent Density Fluctuation Model

The coherent density fluctuation model (CDFM) was suggested and developed in Refs. [Antonov (1982)]. In the CDFM, the one-body density matrix ρ $(\mathbf{r}, \mathbf{r}')$ of a nucleus can be written as a coherent superposition of an infinite number of one-body density matrices ρ_x $(\mathbf{r}, \mathbf{r}')$ for spherical pieces of the nuclear matter called *Fluctons* [Antonov (1982), Gaidarov (2011), Bhuyan (2018)],

$$\rho_x(\mathbf{r}) = \rho_0(x)\Theta(x - |\mathbf{r}|), \tag{7.9}$$

with $\rho_o(x) = \frac{3A}{4\pi x^3}$. The generator coordinate x is the spherical radius of the nucleus contained in a uniformly distributed spherical Fermi gas. In a finite nuclear system, the one-body density matrix can be given as [Antonov (1982), Gaidarov (2011), Gaidarov (2012),

Bhuyan (2018)]

$$\rho(\mathbf{r},\mathbf{r'}) = \int_0^\infty dx|F(x)|^2\rho_x(\mathbf{r},\mathbf{r'}), \tag{7.10}$$

where $|F(x)|^2$ is the weight function (Eq. (7.14)). The term $\rho_x(\mathbf{r},\mathbf{r'})$ is the coherent superposition of the one-body density matrix and defined as

$$\rho_x(\mathbf{r},\mathbf{r'}) = 3\rho_0(x)\frac{J_1\left(k_f(x)|\mathbf{r}-\mathbf{r'}|\right)}{\left(k_f(x)|\mathbf{r}-\mathbf{r'}|\right)} \times\Theta\left(x-\frac{|\mathbf{r}+\mathbf{r'}|}{2}\right) \tag{7.11}$$

Here, J_1 is the first-order spherical Bessel function. The Fermi momentum of nucleons in the Fluctons with radius x is expressed as $k_f(x) = (3\pi^2/2\rho_0(x))^{1/3} = \gamma_a/x$, where $\gamma_a = (9\pi A/8)^{1/3} \approx 1.52A^{1/3}$. The Wigner distribution function of the one-body density matrices in Eq. (7.11) is

$$W(\mathbf{r},\mathbf{k}) = \int_0^\infty dx|F(x)|^2 W_x(\mathbf{r},\mathbf{k}) \tag{7.12}$$

Here, $W_x(\mathbf{r},\mathbf{k}) = \frac{4}{8\pi^3}\Theta(x-|\mathbf{r}|)\Theta(k_F(x)-|\mathbf{k}|)$. Similarly, the density ρ (r) within CDFM can be expressed in terms of the same weight function as

$$\rho(r) = \int d\mathbf{k}W(\mathbf{r},\mathbf{k}) = \int_0^\infty dx|F(x)|^2\frac{3A}{4\pi x^3}\Theta(x-|\mathbf{r}|) \tag{7.13}$$

and it is normalized to the nucleon numbers of the nucleus, $\int \rho(\mathbf{r})d\mathbf{r} = A$. By taking the δ-function approximation to the Hill–Wheeler integral equation, we can obtain the differential equation for the weight function in the generator coordinate [Bhuyan (2018)]. The weight function for a given density distribution ρ (r) can be expressed as

$$|F(x)|^2 = -\left(\frac{1}{\rho_0(x)}\frac{d\rho(r)}{dr}\right)_{r=x}, \tag{7.14}$$

with $\int_0^\infty dx|F(x)|^2 = 1$. For a detailed analytical derivation, one can follow Refs. [Antonov (1994), Fuchs (1995), Antonov (2018), Bhuyan (2018)]. The symmetry energy, neutron pressure, and symmetry energy curvature for a finite nucleus are defined below by weighting the corresponding quantities for infinite nuclear matter within the CDFM. The CDFM allows us to make a transition from the properties of nuclear matter to those of finite nuclei. Following the CDFM approach, the expression for the effective symmetry energy S, pressure P, and curvature ΔK for a nucleus can be written as [Antonov (1994), Fuchs (1995), Gaidarov (2011), Gaidarov (2012), Bhuyan (2018), Antonov (2017)],

$$S = \int_0^\infty dx|F(x)|^2 S_0^{NM}(\rho(x)), \tag{7.15}$$

$$P = \int_0^\infty dx|F(x)|^2 P_0^{NM}(\rho(x)), \tag{7.16}$$

and

$$\Delta K = \int_0^\infty dx|F(x)|^2 \Delta K_0^{NM}(\rho(x)). \tag{7.17}$$

Here, the quantities on the left-hand side of Eqs. (7.15–7.17) are surface-weighted average of the corresponding nuclear matter quantities with local density approximation, which have been determined within the method of Brückner *et al.* [Brückner (1968), Brückner (1969)].

In the present work, considering the pieces of nuclear matter with density ρ_0(x), we have used the matrix element V(x) of the nuclear Hamiltonian, corresponding to the energy of nuclear matter from the method of Brückner *et al.* [Brückner (1968), Brückner (1969)]. In the Brückner energy density functional method, V(x) is given by

$$V(x) = AV_0(x) + V_C + V_{CO}, \tag{7.18}$$

where

$$V_0(x) = 37.53[(1+\delta)^{5/3} + (1-\delta)^{5/3}]\rho_0^{2/3}(x)$$
$$+b_1\rho_0(x) + b_2\rho_0^{4/3}(x) + b_3\rho_0^{5/3}(x)$$
$$+\delta^2[b_4\rho_0(x) + b_5\rho_0^{4/3}(x) + b_6\rho_0^{5/3}], \quad (7.19)$$

with $b_1 = -741.28, b_2 = 1179.89, b_3 = -467.54, b_4 = 148.26, b_5 = 372.84$, and $b_6 = -769.57$. The V_0(x) in Eq. 7.18 is the energy per particle of nuclear matter (in MeV), which accounts for the neutron–proton asymmetry. V_C is the Coulomb energy of charge particle (proton) in a Flucton:

$$V_C = \frac{3}{5}\frac{Z^2e^2}{x}, \quad (7.20)$$

and V_{CO} is the Coulomb exchange energy given by

$$V_{CO} = 0.7386Ze^2(3Z/4\pi x^3)^{1/3}. \quad (7.21)$$

On substituting V_0(x) in Eq. 7.3 and taking its second-order derivative, the symmetry energy S_0^{NM}(x) of nuclear matter with density ρ_0(x) is obtained:

$$S_0^{NM}(x) = 41.7\rho_0^{2/3}(x) + b_4\rho_0(x) + b_5\rho_0^{4/3}(x) + b_6\rho_0^{5/3}(x). \quad (7.22)$$

The corresponding parametrized expressions for the pressure P_0^{NM}(x) and the symmetry energy curvature ΔK_0^{NM}(x) for such a system within the Brückner energy density functional method have the forms

$$P_0^{NM}(x) = 27.8\rho_0^{5/3}(x) + b_4\rho_0^2(x) + \frac{4}{3}b_5\rho_0^{7/3}(x)$$
$$+\frac{5}{3}b_6\rho_0^{8/3}(x), \quad (7.23)$$

and

$$\Delta K_0^{NM}(x) = -83.4\rho_0^{2/3}(x) + 4b_5\rho_0^{4/3}(x) + 10b_6\rho_0^{5/3}(x), \quad (7.24)$$

respectively. These quantities are folded in Eqs. (7.15–7.17) with the weight function to find the corresponding quantities of finite nuclei within the CDFM.

7.2.4 Volume and Surface Components of the Nuclear Symmetry Energy

The expression for nuclear energy given in the liquid droplet model, which is an extension of the Bethe–Weizsãcker liquid drop model incorporating the volume and surface asymmetry, can be written as [Steiner (2009), Myers (1966)]:

$$E(A,Z) = -B.A + E_S A^{2/3} + S_V A \frac{(1-2Z/A)^2}{1 + S_S A^{-1/3}/S_V}$$
$$+E_C \frac{Z^2}{A^{1/3}} + E_{dif} \frac{Z^2}{A} + E_{ex} \frac{Z^{4/3}}{A^{1/3}} + a\Delta A^{-1/2}. \quad (7.25)$$

In Eq. 7.25, B $\sim$ 16 MeV is the binding energy per particle of bulk symmetric matter at saturation. E_S, E_C, E_{dif}, and E_{ex} are coefficients that correspond to the surface energy of symmetric matter, the Coulomb energy of a uniformly charged sphere, the diffuseness correction, and the exchange correction to the Coulomb energy, respectively. The last term gives the pairing corrections, which is essential for the open shell nuclei. S_V is the volume symmetry energy parameter, and S_S is the modified surface symmetry energy parameter in the liquid drop model (see Ref. [Steiner (2009)]). The symmetry energy at temperature T is rewritten [as the third term on the right-hand side of Eq. 7.25] in the form $S(T) = (N-Z)^2/A$, where

$$S(T) = \frac{S_V(T)}{1 + \frac{S_S(T)}{S_V(T)} A^{-1/3}} = \frac{S_V(T)}{1 + A^{-1/3}/\kappa(T)} \quad (7.26)$$

From Eq. 7.26, the individual components can be written as

$$S_V(T) = S(T)\left(1 + \frac{1}{\kappa(T) A^{1/3}}\right) \quad (7.27)$$

and

$$S_S(T) = \frac{S(T)}{\kappa(T)}\left(1 + \frac{1}{\kappa(T)A^{1/3}}\right) \tag{7.28}$$

The temperature-dependent symmetry energy $S(T)$ is calculated by

$$S(T) = \int_0^\infty dx|F(x,T)|^2 S[\rho(x,T)], \tag{7.29}$$

where the weight function $|F(x,T)|^2$ depends on the temperature-dependent density distribution $\rho(r,T)$ as

$$|F(x,T)|^2 = -\left(\frac{1}{\rho_0(x)}\frac{d\rho(r,T)}{dr}\right)_{r=x} \tag{7.30}$$

Following Refs. [Danielewicz (2003), Danielewicz (2004)], an approximate expression for the ratio $\kappa(T) \equiv \frac{S^V}{S^S}$ can be written within the CDFM:

$$\kappa(T) = \frac{3}{R\rho_0}\int_0^\infty dx|F(x,T)|^2 x\rho_0(x)\left[\left(\frac{S(\rho_0)}{S(\rho(x,T)}\right) - 1\right], \tag{7.31}$$

where $|F(x,T)|^2$ is determined by Eq. (7.30), and $S(\rho_0)$ is the nuclear symmetry energy at equilibrium nuclear matter density ρ_0 and T = 0 MeV. Employing the density dependence of symmetry energy [Danielewicz (2003)], we have

$$S[\rho(x,T)] = S^V(T)\left(\frac{\rho(x,T)}{\rho_0}\right)^\gamma. \tag{7.32}$$

There exist various estimations for the value of the parameter γ. In the present work, we use γ = 0.3 in reference to [Antonov (2018)]. Using the above equation and $S(\rho_0) = S_V$, Eqs. (7.29) and (7.31) can be rewritten as follows:

$$S(T) = S(\rho_0)\int_0^\infty dx|F(x,T)|^2\left(\frac{\rho(x,T)}{\rho_0}\right)^\gamma, \tag{7.33}$$

and

$$\kappa(T) = \frac{3}{R\rho_0} \int_0^\infty dx |F(x, T)|^2 x \rho_0(x) \left(\left(\frac{\rho_0}{\rho(x, T)} \right)^\gamma - 1 \right). \quad (7.34)$$

It is to be noted that the present work has been carried out using the above equations at T = 0 MeV. Moreover, in the calculations of volume and surface components of the symmetry energy in search of magic properties of neutron-rich rare earth nuclei, we have used the deformed RMF densities, which are converted into spherical equivalent ones using the procedure mentioned in Refs. [Shukla (2007), Patra (2009), Panda (2014)].

7.3 Results and Discussions

Here, we present our results of the calculations obtained from the E-RMF formalism and subsequently report the effective surface properties of finite nuclei within the framework of CDFM. Then we will use the knowledge to locate the deformed magic number in the neutron-rich region of the periodic table.

7.3.1 Bulk Properties of Finite Nuclei within E-RMF Formalism

Before investigating the effective surface properties of finite nuclei, we would like to see the applicability of the E-RMF formalism for finite nuclei. In this context, we present the bulk properties such as binding energy, charge radius, and the neutron-skin ($\triangle R = R_n - R_p$=difference of neutron and proton distribution radius) thickness for some selected finite nuclei over the periodic table. The calculated results are shown in Table 7.1, and the comparison with experimental data is made wherever available. From the table, one can notice that the calculated results are as per the experimental data. After this the calculated densities of the nuclei under study are used to explore the effective surface properties within the CDFM formalism.

7.3.2 The Effective Surface Properties of the Nuclei

The density profile and weight function are calculated within the spherically symmetric E-RMF formalism corresponding to the NL3, IOPB-I, G3, and FSUGarnet parameter sets (shown in Fig. 7.1).

Table 7.1 The calculated binding energy per particle (B/A) and charge radius (R_{ch}) of nuclei using NL3, IOPB-I, G3, and FSUGranet parameters and comparison with the available experimental data [Wang (2012), Angeli (2013)]. The predicted neutron-skin thickness $\Delta r = R_n - R_p$ is also shown.

Nucleus	Obs.	Expt.	NL3	FSUGarnet	G3	IOPB-I
^{16}O	B/A	7.976	7.917	7.876	8.037	7.977
	R_{ch}	2.699	2.714	2.690	2.707	2.705
	R_n-R_p	-	-0.026	-0.029	-0.028	-0.027
^{40}Ca	B/A	8.551	8.540	8.528	8.561	8.577
	R_{ch}	3.478	3.466	3.438	3.459	3.458
	R_n-R_p	-	-0.046	-0.051	-0.049	-0.049
^{48}Ca	B/A	8.666	8.636	8.609	8.671	8.638
	R_{ch}	3.477	3.443	3.426	3.466	3.446
	R_n-R_p	-	0.229	0.169	0.174	0.202
^{68}Ni	B/A	8.682	8.698	8.692	8.690	8.707
	R_{ch}	-	3.870	3.861	3.892	3.873
	R_n-R_p	-	0.262	0.184	0.190	0.223
^{90}Zr	B/A	8.709	8.695	8.693	8.699	8.691
	R_{ch}	4.269	4.253	4.231	4.276	4.253
	R_n-R_p	-	0.115	0.065	0.068	0.091
^{100}Sn	B/A	8.253	8.301	8.298	8.266	8.284
	R_{ch}	-	4.469	4.426	4.497	4.464
	R_n-R_p	-	-0.073	-0.078	-0.079	-0.077
^{132}Sn	B/A	8.355	8.371	8.372	8.359	8.352
	R_{ch}	4.709	4.697	4.687	4.732	4.706
	R_n-R_p	-	0.349	0.224	0.243	0.287
^{208}Pb	B/A	7.867	7.885	7.902	7.863	7.870
	R_{ch}	5.501	5.509	5.496	5.541	5.521
	R_n-R_p	-	0.283	0.162	0.180	0.221

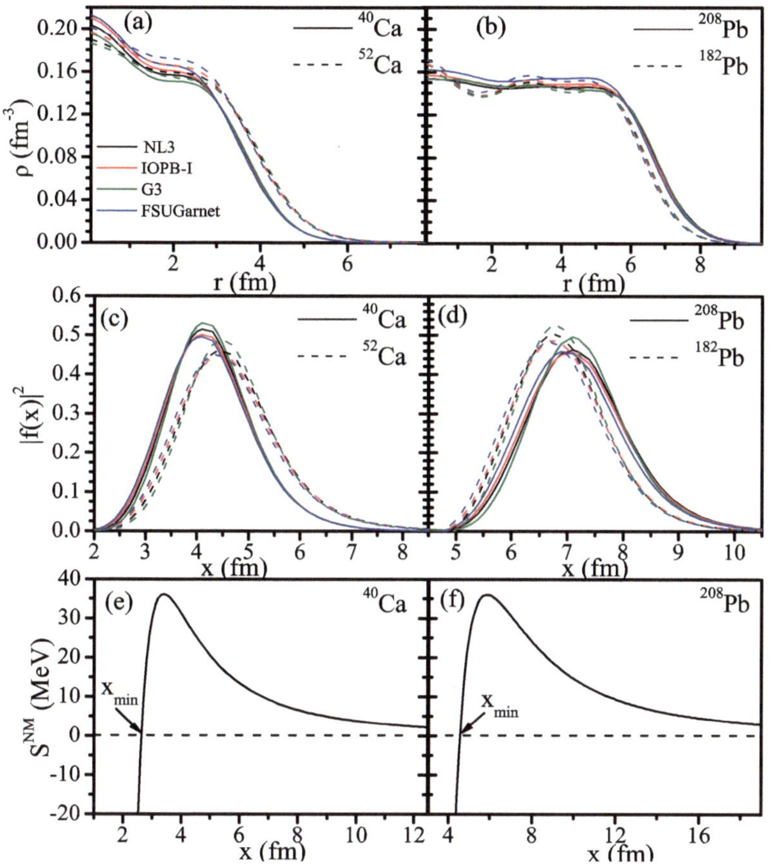

Figure 7.1 The density profiles for (a)^{40,52}Ca, (b)^{182,208}Pb and weight functions for (c)^{40,52}Ca, (d)^{182,208}Pb corresponding to NL3, IOPB-I, G3, and FSUGarnet parameters sets. Symmetry energy at the fluctuation density of (e)^{40}Ca, (f)^{208}Pb within Brückner energy density functional (BEDF) as a function of the local coordinate x. The x_{min} represents the lower limit of the integrations.

The nuclear matter symmetry energy obtained from the Brückner [Brückner (1968), Brückner (1969)] energy density functional is also shown in the figure (Fig. 7.1e,f). The calculated densities from E-RMF are further used to obtain the weight functions for the corresponding nucleus. The weight functions for the ^{40}Ca and ^{208}Pb nuclei, as the representative cases are shown. From Fig. 7.1c,d, one can notice that the trends of the weight functions are exactly opposite as compared to the density (Fig. 7.1a,b). In other words, the lower value of the central density gives the larger height of the weight function of the nucleus. Further, it is noted that the maxima of the weight functions shift toward the right (larger r) as the size of the nucleus increases. The G3 parameter set predicts the larger weight function, while the lower one corresponds to the FSUGarnet parameter set.

In principle, the limits of integration in Eqs. (7.15–7.17) and (7.33, 7.34) are set from 0 to ∞. But, the symmetry energy of the infinite symmetric nuclear matter within the Brückner energy density functional method has some negative values (unphysical points) in certain regions. In order to avoid the unphysical points of the symmetry energy of nuclear matter, the limits of integration x_{min} and x_{max} are put rather than what mentioned above. In general, x_{min} and x_{max} are the points where the symmetry energy of nuclear matter changes from negative to positive and from positive to negative, respectively [Gaidarov (2012)]. For a better understanding of the concept of finding x_{min} and x_{max}, we present the symmetry energy of nuclear matter within the Brückner energy density functional for ^{40}Ca and ^{208}Pb in Fig. 7.1e,f. As it mentioned earlier, the symmetry energy, neutron pressure, and symmetry energy curvature of nuclear matter at local coordinate are folded with the calculated weight function, which result in the corresponding effective surface properties of the finite nucleus. The significant values of weight functions (its peak value) lie in the range that corresponds to the surface part of the density. That is why these quantities are called surface properties.

The effective surface properties for the isotopic series of O, Ca, Ni, Zr, Sn, and Pb nuclei are shown in Fig. 7.2. The first, second, and third rows of each panel of the figure represent the symmetry energy S, neutron pressure P, and symmetry energy curvature ΔK, respectively. The values of the symmetry energy for finite nuclei lie in the range of 24–31 MeV. It is observed from the figure that the symmetry energy is larger for the FSUGarnet parameter set, whereas it is minimum for the NL3 set. The nature of parameter sets gets reversed in the cases of neutron pressure P and symmetry energy curvature ΔK. For example, the NL3 parameter set predicts a larger value of P and ΔK for all the isotopic series. Furthermore,

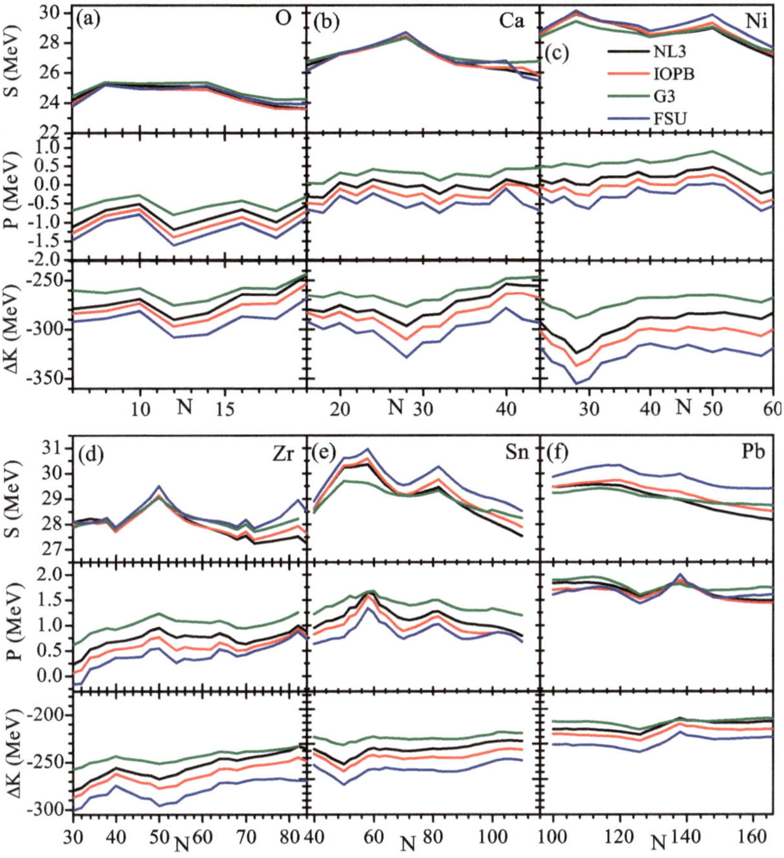

Figure 7.2 The symmetry energy (S), pressure (P), and symmetry energy curvature (ΔK) for the isotopic series of O, Ca, Ni, Zr, Sn, and Pb nuclei corresponding to NL3, IOPB-I, G3, and FSUGarnet parameter sets.

we find several peaks at neutron numbers, which correspond to the magic number and/or shell–sub-shell closure for each isotopic chain. These peaks in the symmetry energy curve imply that the stability of the nuclei at the magic neutron number is more as compared to the neighboring isotopes and correspond to the doubly magic isotopes of nuclei. Moreover, these peaks imply that more energy is required to convert one neutron to proton or vice versa. In addition to some peaks at the magic neutron number, a few small peaks are also seen, which may arise due to the shell structure or the density distribution of the nuclei. The present investigation predicts some neutron magic numbers beyond the known magic number based on the well-known feature of the symmetry energy over an isotopic chain. We find negative neutron pressure for the isotopes of oxygen nuclei for all parameter sets. Also, a few negative values of *P* are estimated for the isotopic series of Ca and Ni corresponding to IOPB-I and FSUGarnet. It is to be noted that the negative value of *P* arises due to the significant value of the weight function in the range of x (fm), where the pressure of nuclear matter is negative.

7.3.3 Correlation of Skin Thickness with the Symmetry Energy

It has been shown that the neutron-skin thickness is correlated with the surface properties in Refs. [Antonov (1982), Gaidarov (2011), Bhuyan (2018), Gaidarov (2012)]. It is found to be linearly correlated with the surface properties except some kinks, which correspond to the magic/semi-magic nuclei in an isotopic chain [Antonov (1982), Gaidarov (2011), Bhuyan (2018), Gaidarov (2012)]. Here, we present the correlation between the symmetry energy and the neutron-skin thickness for the isotopic series of Ca and Sn nuclei

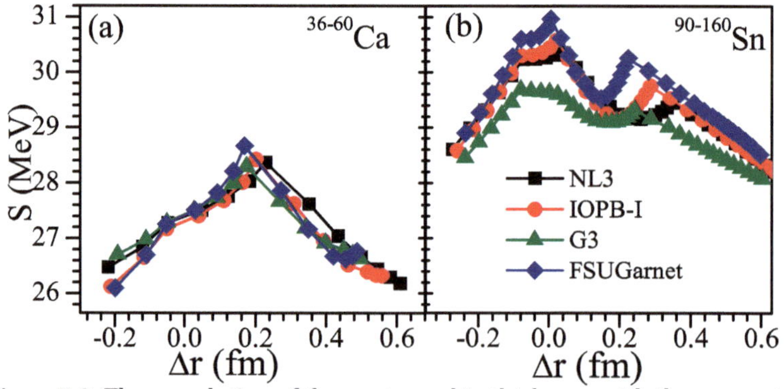

Figure 7.3 The correlation of the neutron-skin thickness with the symmetry energies for the isotopes of (a) Ca, (b) Sn nuclei corresponding to NL3, IOPB-I, G3, and FSUGarnet parameter sets.

for the NL3, IOPB-I, G3, and FSUGarnet parameter sets in Fig. 7.3. In Refs. [Quddus (2018), Kumar (2018)], it is reported that the stiffer EOS of nuclear matter predicts the larger neutron-skin thickness of nuclei. Among the chosen parameter sets, NL3 is the stiffest, which predicts a larger skin thickness, while FSUGarnet being softer estimates a smaller skin thickness. This behavior is clear in Fig. 7.3. It can be noticed from Fig. 7.3 that the symmetry energy predicted by FSUGarnet is higher compared to the other parameter sets. The peaks in the symmetry energy curves (in Fig. 7.3) correspond to the magic or semi-magic neutron numbers. The symmetry energy decreases with a varying neutron number in either direction of the magic/semi-magic number. It implies that for exotic nuclei (nuclei lying at the drip line), a less amount of energy is required to convert one proton to neutron or vice versa, depending on the neutron–proton asymmetry. The behavior of the symmetry energy with skin thickness is monotonous in nature. For Ca and Sn nuclei, the symmetry energy curve is almost linear before and after the peaks.

7.3.4 Volume and Surface Contributions in the Symmetry Energy of Rare Earth Nuclei

Here, we investigate the symmetry energy S, volume symmetry energy (S_V), and surface symmetry energy (S_S) and their ratio κ ($\equiv = S_V/S_V$) in the isotopic chains of rare earth Nd, Sm, Gd, and Dy nuclei with N = 82–126, which is shown in Fig. 7.4. For these nuclei, the deformed densities calculated using NL3 and IOPB-I parameter sets are converted into spherical equivalent by fitting two Gaussian functions for their further use in the calculation of the weight function. A rise and fall trend with mass is observed in symmetry energy and its bulk and surface components and a dip in κ is also seen. The observed peak at neutron number N $\sim$ 100 shows more stability of these nuclei compared to neighboring nuclei in isotopic chain. It is to be noted that the origin of this peak lies in the density distribution ρ(r) of the considered isotopes. The weight function $|F(x,T)|^2$, containing the derivative of ρ(r), is an important term in the evaluation of the integrand in Eq. (7.33). The peculiarities of ρ(r) for the closed shell lead to a peak in the symmetry energy and its components. This peak signifies the persistence of a new magic number/shell closure at N $\sim$ 100 in exotic nuclei near the neutron drip line [Kaur (2020a), Kaur (2020b)].

In Ref. [Quddus (2019a)], it is reported that the symmetry energy has a relation with the magicity, i.e., a peak is observed at the conventional magic numbers N = Z = 8, 20, 28, and 50. It is also noted that there is no peak at N = Z = 82 depicting that it is not a good magic number unlike the other classical magic numbers. It is interesting to point out here that in Ref. [Ghorui (2012)], the large shell gap in ^{162}Sm compared to the neighboring isotopes

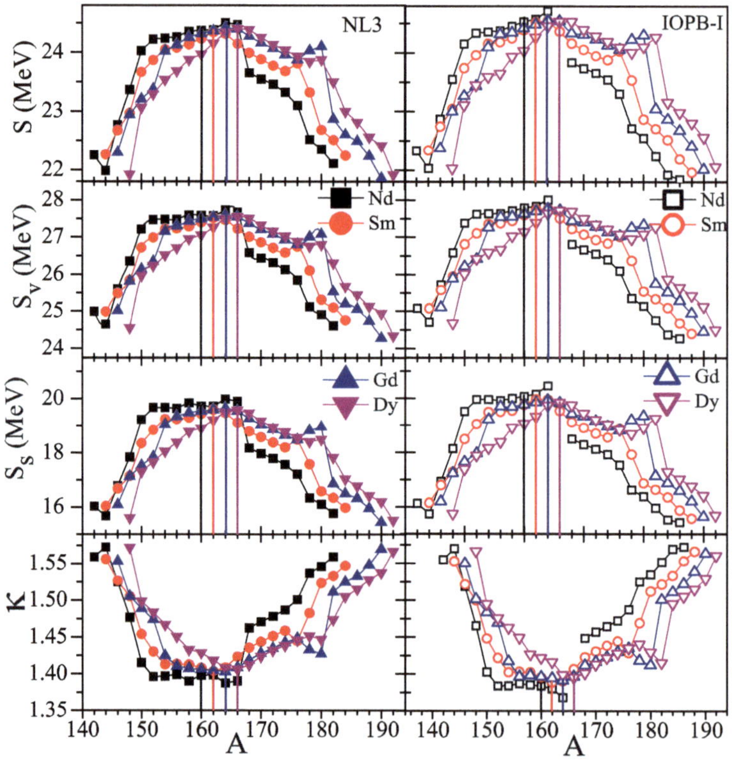

Figure 7.4 Mass dependence of effective nuclear symmetry energy S, its volume S_V, and surface S_S parts and their ratio κ for isotopic chain of Nd, Sm, Gd, Dy with $\gamma = 0.3$ at T = 0 MeV with NL3 parameter set (on the left side) and with IOPB-I parameter set (on the right side).

of the Sm nuclei provided an evidence of magicity N = 100. In addition, the RMF calculations of two neutron separation energy (S_{2n}) and dS_{2n}-differential and calculations within infinite nuclear matter (INM) mass formalism have conjectured the existence of N = 100 magic number in ^{162}Sm [Satpathy (2003)]. The present results also corroborate the proposed "island of stability" near the neutron drip-line at N = 100. Besides, the exploration of high spin excited states in rare earth nucleus ^{169}Tm has shown the back-bending as observed in ^{165}Tm, which is in contrast to its immediate neighbor ^{167}Tm. It was interpreted in terms of the effect of N = 98 deformed shell gap, which has considerable effect in defining the nature of alignment rather than its shape effect [Asgar (2017)]. It is worth mentioning here that in 2014, Patel *et al.* [Patel (2014)] carried out the isomer decay spectroscopy of ^{166}Gd and ^{164}Sm isotonic systems with N = 102. This analysis reported that a decrease in ground band energies of ^{166}Gd and ^{164}Sm in comparison to the energies of ^{164}Gd and ^{162}Sm isotones (with N = 100), which conforms the prediction of N = 100 deformed shell closure in consonance with earlier studies by one of the authors [Satpathy (2003), Ghorui (2012)]. This result has an important implication in the synthesis of heavy nuclei in astrophysical entities, since due to magic properties the nuclei with N = 100 will act as a waiting point in the nucleosynthesis in *r*-process.

It is also noted that there is no peak at N = 82, which portrays that it does not exhibit magicity in neutron-deficient nuclei. It is pointed out here that a new phenomenon of shape change/coexistence is observed in the neutron-deficient Pb nuclei, indicating the disappearance of Z = 82 shell gap in ^{186,188,190,192}Pb isotopes [Toth (1984)]. The theoretical results within RMF also predicted the change in shape for neutron-deficient Pb isotopes, indicating that Z = 82 is no longer a good magic number [Yoshida (1994)]. These results were subsequently verified experimentally, showing the existence of triplet state of spherical, prolate, and oblate shapes as the lowest three states in the energy spectrum of ^{186}Pb [Andreyev (2000)], which demonstrated the weakening of Z = 82 shell closure. Nayak *et al.* [Satpathy (1998)] have predicted the shell quenching at N = 82 within the INM model. Chen *et al.* [Chen (1995)] have also shown that shell quenching for the N = 82 shell is indispensable in order to explain the peak around A $\sim$ 130 while analyzing the elemental abundance in *r*-process nucleosynthesis. In addition, in Fig. 7.4, the peaks near N = 114 and 116 for super-deformed ^{180}Gd (β_2 = 0.379) and ^{180}Dy (β_2 = 0.318) nuclei, respectively, predict the new "island of inversion" in the neutron drip-line region.

7.4 Summary and Conclusion

We have investigated the effective surface properties such as symmetry energy, neutron pressure, and symmetry energy curvature over the nuclear chart using RMF densities (with NL3, IOPB-I, G3, and FSUGarnet parameter sets) within CDFM. For this purpose, the isotopic chains of O, Ca, Ni, Zr, Sn, and Pb nuclei have been chosen. First, their ground-state bulk properties such as binding energies, charge radius, and neutron-skin thickness are calculated and compared with the available experimental data. At the next step, the study of effective surface properties shows that the symmetry energy is maximum for the FSUGarnet parameter set and minimum for the NL3 parameter set. This trend is reversed for neutron pressure and symmetry energy curvature. For each isotopic chain, some kinks are seen at a neutron number that corresponds to the magic number or shell/sub-shell closure and infers the more stability of these nuclei. The analysis also presents a correlation between symmetry energy and neutron-skin thickness of nuclei in an isotopic series.

Furthermore, the exploration of symmetry energy, its volume and surface components, and their ratio (κ) in isotopic series of rare earth Nd, Sm, Gd, and Dy nuclei shows the persistence of a peak at N $\sim$ 100, which is a signature of deformed magic number N $\sim$ 100. These results are in consonance with an earlier study by one of the authors [Ghorui (2012)] and further strengthened the experimental study of the stability of ^{162}Sm and ^{164}Gd nuclei with N = 100 [Patel (2014)]. The considerable significance of this result lies in the fact that these N = 100 nuclei can serve as the waiting point in the nucleosynthesis r-process. Moreover, peaks at N = 114 and 116 for ^{180}Gd and ^{180}Dy nuclei, respectively, show the presence of a new *"Island of Inversion"* and an experimental verification of the same is called for.

References

Agrawal, B. K. (2010). Asymmetric nuclear matter and neutron skin in an extended relativistic mean-field model, *Phys. Rev. C* **81**, pp. 034323.

Andreyev, A. N. *et al.* (2000). A triplet of differently shaped spin-zero states in the atomic nucleus ^{186}Pb, *Nature* **405**, pp. 430.

Angeli, I. and Marinova, K. P. (2013). Table of experimental nuclear ground state charge radii: An update, *At. Data Nucl. Data Tables* **99**, pp. 69.

Antonov, A. N. *et al.* (2018). Temperature dependence of the volume and surface contributions to the nuclear symmetry energy within the coherent density fluctuation model, *Phys. Rev. C* **98**, pp. 054315.

Antonov, A. N., Gaidarov, M. K., Sarriguren, P., and Moya de Guerra, E. (2016). Volume and surface contributions to the nuclear symmetry energy within the coherent density fluctuation model, *Phys. Rev. C*, **94**, pp. 014319.

Antonov, A. N., Kadrev, D. N., and Hodgson, P. E. (1994). Effect of nucleon correlations on natural orbitals, *Phys. Rev. C* **50**, pp. 164.

Antonov, A. N., Kadrev, D. N., Gaidarov, M. K., Sarriguren, P., and Moya de Guerra, E. (2017). Temperature dependence of the symmetry energy and neutron skins in Ni, Sn, and Pb isotopic chains, *Phys. Rev. C* **95**, pp. 024314.

Antonov, A. N., Nikolaev, V. A., and Petkov, I. Zh. (1982). Spectral functions and hole nuclear states, *Z. Phys. A*, **304**, pp. 239.

Asgar, Md. A., Roy, T., Mukherjee, G., and Dhal, A. (2017). Return of backbending in ^{169}Tm and the effect of the N = 98 deformed shell gap, *Phys. Rev. C* **95**, pp. 031304.

Bhuyan, M., Carlson, B. V., Patra, S. K., and Zhou, S.-G. (2018). Surface properties of neutron-rich exotic nuclei within relativistic mean field formalisms, *Phys. Rev. C* **97**, pp. 024322.

Brown, B. A. (2000). Neutron radii in nuclei and the neutron equation of state, *Phys. Rev. Lett.* **85**, pp. 5296; Typel, S., Brown, B. A. (2001). Neutron radii and the neutron equation of state in relativistic models, *Phys. Rev. C* **64**, pp. 027302.

Brueckner, K. A., Buchler, J. R., Jorna, S., and Lombard, R. J. (1968). Statistical theory of nuclei, *Phys. Rev.* **171**, pp. 1188.

Brueckner, K. A., Buchler, J. R., Clark, R. C., and Lombard, R. J. (1969). Statistical theory of nuclei. II. Medium and heavy nuclei, *Phys. Rev.* **181**, pp. 1543.

Cameron, A. G. W. (1957). A revised semiempirical atomic mass formula, *Can. J. Phys.* **35**, pp. 1021.

Carbone, A., Colò, G., Bracco, A., Cao, L.-G., Bortignon, P. F., Camera, F., and Wieland, O. (2010). Constraints on the symmetry energy and neutron skins from pygmy resonances in ^{68}Ni and ^{132}Sn *Phys. Rev. C* **81**, pp. 041301.

Centelles, M., Roca-Maza, X., Viñas, X., and Warda, M. (2009). Nuclear symmetry energy probed by neutron skin thickness of nuclei, *Phys. Rev. Lett.* **102**, pp. 122502.

Chen, B. *et al.* (1995). Influence of shell-quenching far from stability on the astrophysical r-process, *Phys. Lett. B* **355**, pp. 37.

Chen, L.-W., Ko, C. M., and Li, B.-A. (2005). Nuclear matter symmetry energy and the neutron skin thickness of heavy nuclei, *Phys. Rev. C* **72**, pp. 064309.

Chen, L.-W., Ko, C. M., Li, B.-A., and Xu, J. (2010). Density slope of the nuclear symmetry energy from the neutron skin thickness of heavy nuclei, *Phys. Rev. C* **82**, pp. 024321.

Colonna, M. (2009). Nuclear matter and nuclear dynamics, *J. Phys. Conf. Ser.* **168**, pp. 012006.

Danielewicz, P. (2003). Surface symmetry energy, *Nucl. Phys. A*, **727**, pp. 233.

Danielewicz, P. (2004). Nuclear symmetry energy: From nuclear extremes to neutron-star matter, arXiv:nucl-th/0411115; Danielewicz, P. (2006), Symmetry Energy, arXiv:nucl-th/0607030.

Danielewicz, P. and Lee, J. (2014). Symmetry energy II: Isobaric analog states, *Nucl. Phys. A* **922**, pp. 1.

Del Estal, M., Centelles, M., Viñas, X., and Patra, S. K. (2001). Pairing properties in relativistic mean field models obtained from effective field theory, *Phys. Rev. C* **63**, pp. 044321; Del Estal, M., Centelles, M., Viñas, X., and Patra, S. K. (2001). Effects of new nonlinear couplings in relativistic effective field theory, *Phys. Rev C* **63**, pp. 024314.

Dutra, M., Lourenco, O., Avancini, S. S., Carlson, B. V., Delfino, A., Menezes, D. P., Providencia, C., Typel, S., and Stone, J. R. (2014). Relativistic mean-field hadronic models under nuclear matter constraints, *Phys. Rev. C* **90**, pp. 055203.

Dutra, M., Lourenco, O., Sa Martins, J. S., Delfino, A., Stone, J. R., and Stevenson, P. D., (2012). Skyrme interaction and nuclear matter constraints, *Phys. Rev. C* **85**, pp. 035201.

Feenberg, E. (1947). Semi-empirical theory of the nuclear energy surface, *Rev. Mod. Phys.*, **19**, pp. 239.

Fuchs, C., Lenske, H., and Wolter, H. H. (1995). Density dependent hadron field theory, *Phys. Rev. C* **52**, pp. 3043.

Furnstahl, R. J. (2002). Neutron radii in mean-field models, *Nucl. Phys. A* **706**, pp. 85.

Furnstahl, R. J., Serot, B. D., and Tang, H. B. (1996). Analysis of chiral mean-field models for nuclei, *Nucl. Phys. A* **598**, pp. 539; Furnstahl, R. J., Serot, B. D., and Tang, H. B. (1997). A chiral effective Lagrangian for nuclei, *Nucl. Phys. A* **615**, pp. 441.

Gaidarov, M. K., Antonov, A. N., Sarriguren, P., and Moya de Guerra, E. (2011). Surface properties of neutron-rich exotic nuclei: A source for studying the nuclear symmetry energy, *Phys. Rev. C*, **84**, pp. 034316.

Gaidarov, M. K., Antonov, A. N., Sarriguren, P., and Moya de Guerra, E. (2012). Symmetry energy of deformed neutron-rich nuclei, *Phys. Rev. C* **85**, pp. 064319.

Ghorui, S. K., Sahu, B. B., Praharaj, C. R., and Patra, S. K. (2012). Examining the stability of Sm nuclei around N = 100, *Phys. Rev. C* **85**, pp. 064327.

Green, A. E. S. (1958). Nuclear sizes and the Weiszäcker mass formula, *Rev. Mod. Phys.*, **30**, pp. 569.

Holt, J. W. and Lim, Y.(2018). Universal correlations in the nuclear symmetry energy, slope parameter, and curvature, *Phys. Lett. B* **784**, pp. 77.

Horowitz, C. J. and Piekarewicz, J. (2001). Neutron star structure and the neutron radius of ^{208}Pb, *Phys. Rev. Lett.* **86**, pp. 5647.

Kaur, M., Quddus, A., Kumar, A., Bhuyan, M., and Patra, S. K. (2020). Effect of temperature on the volume and surface contributions in the symmetry energy of rare earth nuclei, *Nucl. Phys. A* **1000**, pp. 121871.

Kaur, M., Quddus, A., Kumar, A., Bhuyan, M., and Patra, S. K. (2020). On the symmetry energy and deformed magic number at N = 100 in rare earth nuclei, *J. Phys. G: Nucl. Part. Phys.* **47**, pp. 105102.

Kumar, B., Patra, S. K., and Agrawal, B. K. (2018). New relativistic effective interaction for finite nuclei, infinite nuclear matter, and neutron stars, *Phys. Rev. C* **97** , pp. 045806.

Lattimer, J. M. and Prakash, M. (2001). Neutron star structure and the equation of state, *Astrophys. J.* **550**, pp. 426.

Lattimer, J. M. and Prakash, M. (2007). Neutron star observations: Prognosis for equation of state constraints, *Phys. Rep.* **442**, pp. 109.

Lee, C.-H., Kuo, T. T. S., Li, G. Q., and Brown, G. E. (1998). Nuclear symmetry energy, *Phys. Rev. C* **57**, pp. 3488.

Li, B.-A., Chen, L.-W., and Ko, C. M. (2008). Recent progress and new challenges in isospin physics with heavy-ion reactions, *Phys. Rep.* **464**, pp. 113.

Möller, P. *et al.* (1995). Nuclear ground-state masses and deformations, *At. Data Nucl. Data Tables* **59**, pp. 185.

Myers, W. D. and Swiatecki, W. J. (1966). Nuclear masses and deformations, *Nucl. Phys. A* **81**, pp. 1.

Naik, K. C., Kaur, M., Kumar, A., and Patra, S. K. (2020). Density-dependence of symmetry energy of ^{162}Sm nucleus, *Int. J. Mod. Phys. E* **28**, pp. 1950100.

Nikšić, T. , Vretenar, D., and Ring, P. (2008). Relativistic nuclear energy density functionals: Adjusting parameters to binding energies, *Phys. Rev. C* **78**, pp. 034318.

Panda, R. N., Sharma, M. K., and Patra, S. K. (2014). Nuclear structure and reaction properties of Ne, Mg and Si isotopes with RMF densities, *Mod. Phys. Lett. A* **29**, pp. 1450013.

Patel, Z. *et al.* (2014). Isomer decay spectroscopy of ^{164}Sm and ^{166}Gd: Midshell collectivity around N = 100, *Phys. Rev. Lett.* **113**, pp. 262502.

Patra, S. K., Panda, R. N., Arumugam, P., and Gupta, R. K. (2009). Nuclear reaction cross sections of exotic nuclei in the Glauber model for relativistic mean field densities, *Phys. Rev. C* **80**, pp. 064602.

Pomorski, K. and Dudek, J. (2003). Nuclear liquid-drop model and surface-curvature effects, *Phys. Rev. C* **67**, pp. 044316.

Quddus, A., Bhuyan, M., Ahmad, S., Carlson, B. V., and Patra, S. K. (2019). Temperature-dependent symmetry energy of neutron-rich thermally fissile nuclei, *Phys. Rev. C*, **99**, pp. 044314.

Quddus, A., Bhuyan, M., and Patra, S. K. (2020). Effective surface properties of light, heavy, and superheavy nuclei, *J. Phys. G: Nucl. Part. Phys.*, **47**, pp. 045105.

Quddus, A., Naik, K. C., and Patra, S. K. (2018). Study of hot thermally fissile nuclei using relativistic mean field theory, *J. Phys. G: Nucl. Part. Phys.* **45**, pp. 075102.

Roca-Maza, X., Centelles, M., Viñas, X., and Warda, M. (2011). Neutron skin of ^{208}Pb, nuclear symmetry energy, and the parity radius experiment, *Phys. Rev. Lett.* **106**, pp. 252501.

Sahoo, T., Kaur, M., Panda, R. N., Das, Piyush, R., and Patra, S. K. (2019). Investigation of proton emission using deformed relativistic mean field densities, *Int. J. Mod. Phys. E* **28**, pp. 1950095.

Satpathy, L. and Nayak, R. C. (1998). Study of nuclei in the drip-line regions, *J. Phys. G: Nucl. Part. Phys.* **24**, pp. 1527.

Satpathy, L. and Patra, S. K. (2003). New magic numbers and new islands of stability in drip-line regions in mass model, *Nucl. Phys. A* **771**, pp. 24.

Shukla, A., Sharma, B. K., Chandra, R., Arumugam, P., and Patra, S. K. (2007). Nuclear reaction studies of unstable nuclei using relativistic mean field formalisms in conjunction with the Glauber model, *Phys. Rev. C* **76**, pp. 034601.

Steiner, A. W., Prakash, M., Lattimer, J. M., and Ellis, P. J. (2005). Isospin asymmetry in nuclei and neutron stars, *Phys. Rep.*, **411**, pp. 325.

Toth, K. *et al.* (1984). Evidence from α decay that Z = 82 is not magic for light Pb isotopes, *Phys. Rev. Lett.* **53**, pp. 1623.

Tsang, M. B. *et al.* (2009). Constraints on the density dependence of the symmetry energy, *Phys. Rev. Lett.*, **102**, pp. 122701; Tsang, M. B. *et al.* (2010). Constraints on the density dependence of the symmetry energy, *Int. J. Mod. Phys. E*, **19**, pp. 1631.

Tsang, M. B. *et al.* (2012). Constraints on the symmetry energy and neutron skins from experiments and theory, *Phys. Rev. C* **86**, pp. 015803.

Van Giai, N., Carlson, B. V., Ma, Z., and Wolter, H. (2010). The Dirac–Brueckner–Hartree–Fock approach: From infinite matter to effective Lagrangians for finite systems, *J. Phys. G* **37**, pp. 064043.

Vretenar, D., Nikšić, T. and Ring, P. (2003). A microscopic estimate of the nuclear matter compressibility and symmetry energy in relativistic mean-field models, *Phys. Rev. C* **68**, pp. 024310.

Wang, M., Audi, G., Wapstra, A. H., Kondev, F. G., Mac-Cormick, M., Xu, X., and Pfeiffer, B. (2012). The Ame2012 atomic mass evaluation, *Chin. Phys. C* **36**, pp. 1603.

Warda, M., Viñas, X., Roca-Maza, X., and Centelles, M. (2010). Analysis of bulk and surface contributions in the neutron skin of nuclei, *Phys. Rev. C* **81**, pp. 054309.

Yoshida, S. and Sagawa, H. (2006). Isovector nuclear matter properties and neutron skin thickness, *Phys. Rev. C* **73**, pp. 044320.

Yoshida, S., Patra, S. K., Takigawa, N., and Praharaj, C. R. (1994). Structure of neutron-deficient Pt, Hg, and Pb isotopes, *Phys. Rev. C*, **50**, pp. 1398.

Chapter 8

Nuclear Symmetry Energy in Heavy-Ion Collisions

Gao-Chan Yong

Institute of Modern Physics, Chinese Academy of Sciences, Lanzhou 730000, China

School of Nuclear Science and Technology, University of Chinese Academy of Sciences, Beijing 100049, China

yonggaochan@impcas.ac.cn

8.1 Introduction

Nowadays in the topics of properties of nuclei, far from stability and objects in heaven, an asymmetric nuclear matter is involved and its properties play a crucial role. To describe the property of asymmetric nuclear matter, the equation of state of nuclear matter is frequently mentioned. The equation of state (EoS) of nuclear matter at density ρ and isospin asymmetry δ ($\delta = (\rho_n - \rho_p)/(\rho_n + \rho_p)$) can be expressed as [Bombaci (1991), Li (2008), Baran (2005)]

$$E(\rho, \delta) = E(\rho, 0) + E_{\text{sym}}(\rho)\delta^2 + \mathcal{O}(\delta^4), \tag{8.1}$$

where $E_{\text{sym}}(\rho)$ is the nuclear symmetry energy. The nuclear symmetry energy thus describes the single nucleonic energy of nuclei

Multifragmentation in Heavy-Ion Reactions: Theory and Experiments
Edited by Rajeev K. Puri, Yu-Gang Ma, and Arun Sharma

ISBN 978-981-4968-69-0 (Hardcover), 978-1-003-38513-4 (eBook)
www.jennystanford.com

or nuclear matter changes as one replaces protons in a system with neutrons. Nowadays the EoS of isospin symmetric nuclear matter $E(\rho, 0)$ is relatively well determined [Danielewicz (2002)], but the EoS of isospin asymmetric nuclear matter, especially the high-density behavior of the nuclear symmetry energy, is still very uncertain [Guo (2014)]. Besides its impacts in nuclear physics [Li (2008), Baran (2005)], in a density range of $0.1 \sim 10$ times nuclear saturation density, the symmetry energy determines the birth of neutron stars and supernova neutrinos [Sumiyoshi (1995)], a range of neutron star properties such as cooling rates, the thickness of the crust, the mass–radius relationship, and the moment of inertia [Sumiyoshi (1994), Lattimer (2004), Steiner (2005), Lattimer (2014)]. The nuclear symmetry energy also plays a crucial role in the evolution of core-collapse supernova [Fischer (2014)] and astrophysical r-process nucleosynthesis [Nikolov (2011), Goriely (2011), Bauswein (2013), Wanajo (2014)], the gravitational-wave frequency [Maselli (2013), Bauswein (2014)], and the gamma-ray bursts [Lasky (2014)] in neutron star mergers [Rosswog (2015), Abbott (2017)].

To constrain the symmetry energy in broad density regions, besides the studies in astrophysics, many terrestrial experiments are being carried out or planned using a wide variety of advanced new facilities, such as the Facility for Rare Isotope Beams (FRIB) in the United States [Bollen (2010)], or the Radioactive Isotope Beam Facility (RIBF) in Japan [Yano (2007)]. To unscramble symmetry energy related experimental data, various isospin-dependent transport models are frequently used to probe the symmetry energy below and above the saturation density [Li (2008), Baran (2005)]. With great efforts, the nuclear symmetry energy and its slope around the saturation density of nuclear matter from the analysis of terrestrial nuclear laboratory experiments and astrophysical observations have been roughly pinned down [Li (2013)], while recent interpretations of the FOPI and FOPI-LAND experimental measurements by different groups made the symmetry energy at supra-saturation densities fall into chaos [Guo (2014), Xiao (2009), Feng (2010), Russotto (2011), Cozma (2013), Xie (2013), Wang (2014)]. It does not seem to be clarified why the nuclear symmetry energy at supra-saturation densities is so uncertain; maybe the effects of pion in-medium effects [Guo (2015), Hong (2014), Xu (2013)], the isospin dependence of in-medium nuclear strong interactions [Yong (2011)], and the short-range tensor force [Pandharipande (1972), Xu (2011)] are some factors.

Nowadays, constraints on the high-density symmetry energy by the measurements of pion and nucleon, triton and ^{3}He yields ratio in the isotope reaction systems ^{132}Sn $+^{124}$ Sn and ^{108}Sn $+^{112}$ Sn at about 300 MeV/nucleon, are being carried out at RIBF-RIKEN in Japan [SE Project (2014), Shane (2015)]. Probing the high-density symmetry energy with other heavy systems at higher incident beam energies is also being carried out/planned at FOPI/GSI and CSR/Lanzhou [Russotto (2016), L.M. (2016)] and some progress has been made by measuring nucleon and light-charged cluster flows [Russotto (2016)].

8.2 Sensitive Probes of Nuclear Symmetry Energy in Heavy-Ion Collisions

To study density-dependent symmetry energy using heavy-ion collisions, symmetry-energy-sensitive probes are frequently used to constrain nuclear symmetry energy through comparisons between theoretical simulations with different symmetry energy inputs and experimental data. Because the symmetry potentials have opposite signs for neutrons and protons and the fact that the symmetry potentials are generally smaller compared to the isoscalar potential at the same density, most of the observables proposed so far use differences or ratios of isospin multiplets of baryons, mirror nuclei and mesons, such as, the neutron/proton ratio of nucleon emissions, neutron–proton differential flow, neutron–proton correlation function, $t/^3$He, π^-/π^+, Σ^-/Σ^+ and K^0/K^+ ratios, etc. [Li (2008), Baran (2005)].

We examined in the lower window of Fig. 8.1 the transverse momentum dependence of the neutron/proton (n/p) ratio of midrapidity nucleons emitted in the direction perpendicular to the reaction plane [Yong (2007)]. The n/p ratio is determined mostly by the density dependence of the symmetry energy and almost not affected by the EOS of symmetric nuclear matter. It is interesting to see in the lower window that the symmetry energy effect on the n/p ratio increases with the increasing transverse momentum

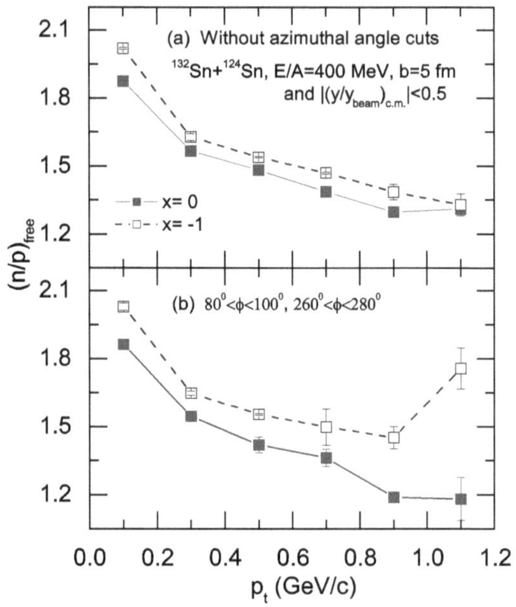

Figure 8.1 Transverse momentum distribution of the ratio of midrapidity neutrons to protons emitted in the reaction of ^{132}Sn+^{124}Sn at the incident beam energy of 400 MeV/nucleon and impact parameter of $b = 5$ fm. For the lower panel, we make an azimuthal angle cut of $80° < \phi < 100°$ and $260° < \phi < 280°$ to make sure that the free nucleons are from the direction perpendicular to the reaction plane, i.e., to analyze the squeezed-out nucleons only. Different x parameters refer different density-dependent symmetry energy. Taken from Ref. [Yong (2007)].

p_t. At a transverse momentum of 1 GeV/c, the effect can be as high as 40%. The high p_t particles most likely come from the high density region in the early stage during heavy-ion collisions, and they are just more sensitive to the high-density behavior of the symmetry energy. Without the cut on the azimuthal angle, the n/p ratio of free nucleons in the midrapidity region is shown in the upper window. This ratio is much less sensitive to the symmetry energy in the whole range of transverse momentum. The n/p ratio of squeezed-out nucleons carries directly information of the symmetry

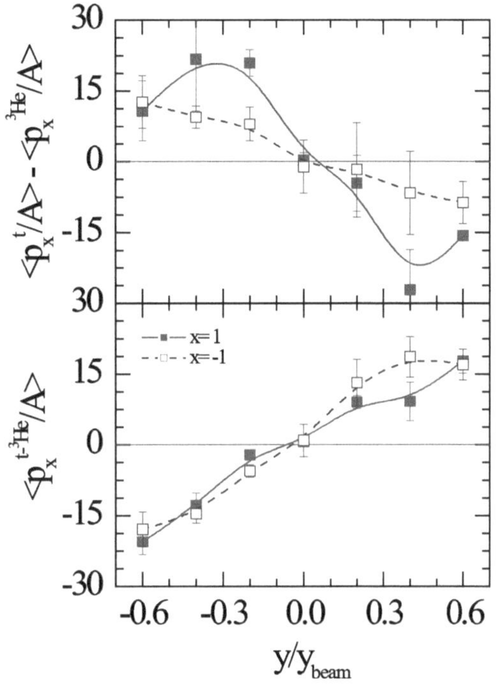

Figure 8.2 Triton-^{3}He relative and differential flows (in MeV) as a function of the reduced C.M. rapidity in the reaction of $^{132}Sn + ^{124}Sn$ at a beam energy of 400 MeV/nucleon and an impact parameter of 5 fm with the symmetry energy parameter $x = 1$ (soft) and $x = -1$ (stiff), respectively. Taken from [Yong (2009)].

potential/energy since it acts directly on nucleons. The sensitivity to the high-density behavior of the nuclear symmetry energy observed in the n/p ratio of squeezed-out nucleons is probably the highest found so far among all observables studied within the same transport model [Yong (2007)].

It was predicted that the neutron–proton differential flow is sensitive to the high-density behavior of the nuclear symmetry energy. However, it is difficult to measure observables involving neutrons. One question often asked by some experimentalists is whether the triton–^{3}He (t–^{3}He) pair may carry the same information as the neutron–proton one. The transverse flow is a result of the actions of several factors, including the isoscalar, symmetry and Coulomb potentials and nucleon–nucleon scatterings. It is well known that the transverse flow is sensitive to the isoscalar potential. Given the remaining uncertainties associated with the isoscalar potential and the small size of the symmetry energy effects, it would be very difficult to extract any reliable information about the symmetry energy from the individual flows of triton and ^{3}He clusters. Thus, techniques of reducing effects of the isoscalar potential while enhancing the effects of the isovector potential are very helpful. We studied in Fig. 8.2 the triton–^{3}He relative and differential flows [Yong (2009)]. From the upper panel of Fig. 8.2, it is seen that the triton–^{3}He relative flow is very sensitive to the symmetry energy. Because of the larger slope of the ^{3}He flow, the triton–^{3}He relative flow shows a negative slope at midrapidity. Effects of the symmetry energy on the differential flow shown in the lower panel, however, is relatively small. The slope $F(x) \equiv d < p_x/A > /d(y/y_{beam})$ of the transverse flow at midrapidity can be used to characterize more quantitatively the symmetry energy effects. We found that for the t–^{3}He relative flow, $F(x=1) \approx -74$ MeV/c and $F(x=-1) \approx -22$ MeV/c, respectively. For the t–^{3}He differential flow, $F(x=1) \approx 21$ MeV/c and $F(x=-1) \approx 42$ MeV/c, respectively.

Another sensitive probe of nuclear symmetry energy is the charged π^-/π^+ ratio. Shown in Fig. 8.3 is the quantity $(\pi^-/\pi^+)_{like}$ as a function of time [Li (2005)]. First, it is seen that the $(\pi^-/\pi^+)_{like}$ ratio reaches a very high value in the early stage of the reaction. This is due to the abundant neutron–neutron scatterings when the two neutron skins start overlapping at the beginning of the reaction. Second, the $(\pi^-/\pi^+)_{like}$ ratio saturates after about 25 fm/c, indicating that a chemical freeze-out stage has been reached. Finally, the sensitivity to the symmetry energy is clearly shown in the final π^-/π^+ ratio.

Hadronic probes, especially if used for studying the high-density behavior of the symmetry energy, inevitably suffer from distortions due to the strong interactions in the final state. Ideally one would like to have more clean ways to study the symmetry energy especially at supranormal densities. Shown in Fig. 8.4 is the spectra ratio of hard photons calculated for four cases using both the p_γ^a and p_γ^b production forms [Yong (2008)]. First of all, it is seen clearly that the full calculations with the p_γ^a and p_γ^b and the in-medium NN cross

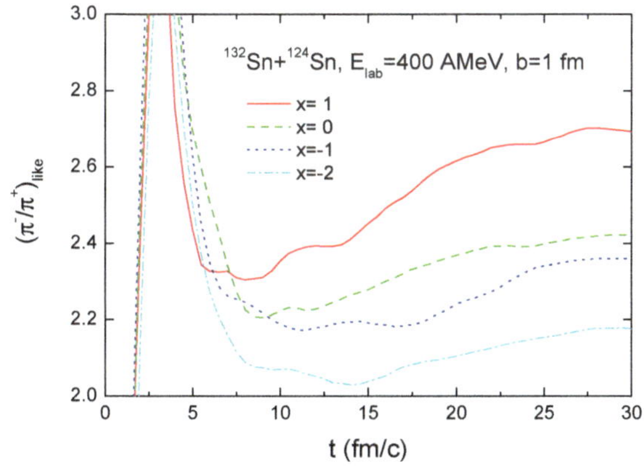

Figure 8.3 Evolution of the π^-/π^+ ratio in the reaction of $^{132}Sn + ^{124}Sn$ at a beam energy of 400 MeV/nucleon and an impact parameter of 1 fm with different symmetry energy parameter *x*. Taken from Ref. [Li (2005)].

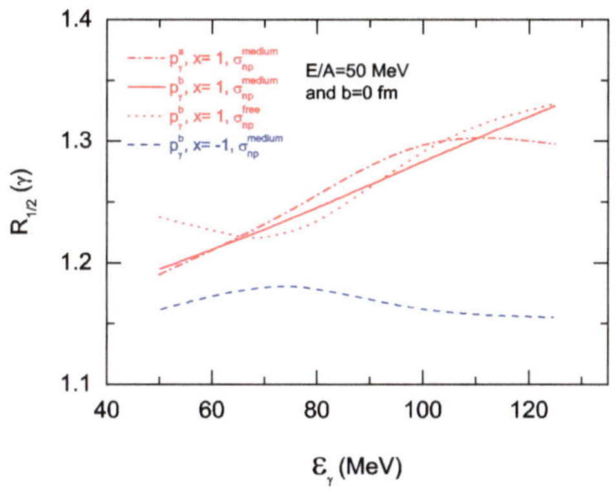

Figure 8.4 The spectra ratio of hard photons in the reactions of $^{132}Sn+$ ^{124}Sn and $^{112}Sn + ^{112}$ Sn reactions at a beam energy of 50 MeV/A with the symmetry energies of x=1, x=-1. Taken from Ref. [Yong (2008)].

sections indeed lead to about the same $R_{1/2}(\gamma)$ within statistical errors as expected. It is also clearly seen that the effects of the in-medium NN cross sections get almost completely cancelled out. These observations thus verify numerically the advantage of using

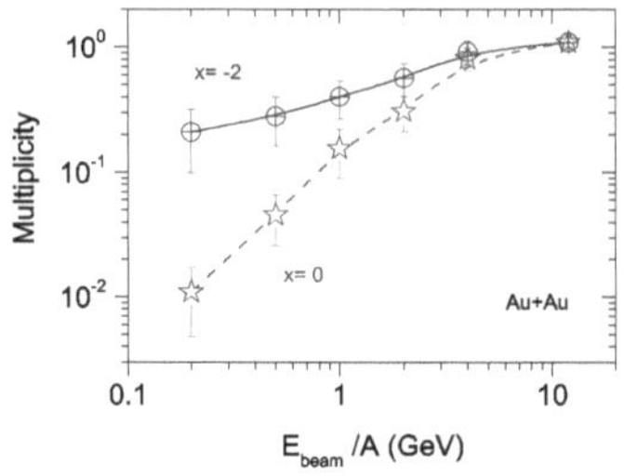

Figure 8.5 Multiplicity of inclusive η production as a function of incident beam energy in $^{197}Au +^{197} Au$ reactions with the two different values for the symmetry energy parameter x. Taken from [Yong (2013)].

the $R_{1/2}(\gamma)$ as a robust probe of the symmetry energy essentially free of the uncertainties associated with both the elementary photon production and the NN cross sections. Most interestingly, comparing the calculations with $x = 1$ and $x = -1$ both using the p_{γ}^{b}, it is clearly seen that the $R_{1/2}(\gamma)$ remains sensitive to the symmetry energy especially for very energetic photons. Again, the symmetry energy is varied by at most 20% in the reaction considered by varying the parameter x from 1 to -1. Thus, the approximately 15% maximum change in the spectra ratio represents a relatively significant sensitivity.

It is well known that the π^-/π^+ ratio is more sensitive to the symmetry energy at lower beam energies especially in the subthreshold region where the mean-field dominates the reaction dynamics and has longer time to modify the momentum of nucleons. Creating an η at subthreshold energies requires even longer reaction time for both multiple collisions and the mean-field to act coherently in order to accumulate enough energy on the two colliding baryons or their resonances. Thus, it is interesting to examine the multiplicity of η production as a function of the incident beam energy. And because of the hidden strangeness, η mesons experience weaker final state interactions compared to pions. It is thus interesting to know if the η meson might be useful for exploring the symmetry energy. Shown in Fig. 8.5 is the η multiplicity as a function of beam energy in inclusive $^{197}Au +^{197} Au$ reactions with both $x = 0$ (softer) and $x = -2$ (stiffer) [Yong (2013)]. It is seen that the η multiplicity decreases rapidly with decreasing beam energy, especially for the soft symmetry energy. The η multiplicity saturates at an incident energy of about 10 GeV/nucleon. It is interesting to see that indeed the effect of nuclear symmetry energy is stronger in the deeper subthreshold region as one expects. In fact, it is more sensitive to the symmetry energy than the π^-/π^+ ratio in the same energy region although the pion yields are relatively more abundant.

8.3 Blind Spots of Probing the High-Density Symmetry Energy in Heavy-Ion Collisions

To probe the symmetry energy by heavy-ion collisions, one usually varies the density dependence of the symmetry energy in transport model simulations and makes comparisons with experimental data. This method is not a problem for those symmetry energies that evidently depend on the density. But for the symmetry energy that has less dependence on the density, the above method is questionable. Because according to the chemical equilibrium condition of nuclear matter formed in heavy-ion collisions [Li (2002)], only if the symmetry energy changes with density, the liquid–gas phase transition can occur. Therefore, if the value of the symmetry energy is less density dependent, a freeze-out observable in heavy-ion collisions cannot prove the symmetry energy effectively.

To demonstrate the blind spots of probing the high-density symmetry energy in heavy-ion collisions, we studied the symmetry-energy-sensitive observable neutron to proton ratio (n/p ratio) in the central $^{197}Au +^{197} Au$ reaction at 300 MeV/nucleon [Yong (2018a)]. The used isospin-dependent Boltzmann–Uehling–Uhlenbeck (IBUU) transport model originates from the IBUU04 model [Li (2005)]. The effects of neutron–proton short-range correlations are appropriately taken into account in the mean-field potential and the initialization of colliding nuclei, respectively [Yong (2017a), Yong (2017b)]. And the transition momentum of the minority is set to be the same as that of the majority in asymmetric matter [Yong (2018b)]. The in-medium inelastic baryon–baryon collisions and the in-medium pion transport are also considered [Yong (2016), Yong (2015)]. The

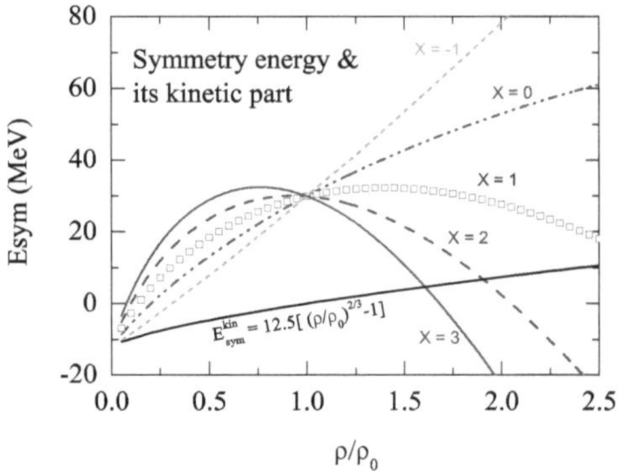

Figure 8.6 Density-dependent symmetry energy and its kinetic part with different x parameters. Taken from [Yong (2016)].

neutron and proton density distributions in nucleus are given by the Skyrme–Hartree–Fock with Skyrme M* force parameters [Friedrich (1986)]. In projectile and target nuclei, the proton and neutron momentum distributions with high-momentum tails are employed [Yong (2017a), Yong (2017b), Subedi (2008), Hen (2014), Yong (2014)].

The isospin- and momentum-dependent single-nucleon mean-field potential reads as [Yong (2017a)]

$$
\begin{aligned}
U(\rho, \delta, \vec{p}, \tau) = & A_u(x)\frac{\rho_{\tau'}}{\rho_0} + A_l(x)\frac{\rho_\tau}{\rho_0} \\
& + B\left(\frac{\rho}{\rho_0}\right)^\sigma (1 - x\delta^2) - 8x\tau\frac{B}{\sigma+1}\frac{\rho^{\sigma-1}}{\rho_0^\sigma}\delta\rho_{\tau'} \\
& + \frac{2C_{\tau,\tau}}{\rho_0}\int d^3p'\frac{f_\tau(\vec{r}, \vec{p'})}{1 + (\vec{p} - \vec{p'})^2/\Lambda^2} \\
& + \frac{2C_{\tau,\tau'}}{\rho_0}\int d^3p'\frac{f_{\tau'}(\vec{r}, \vec{p'})}{1 + (\vec{p} - \vec{p'})^2/\Lambda^2},
\end{aligned} \tag{8.2}
$$

where ρ_0 denotes the saturation density, τ, τ'=1/2 (-1/2) is for neutron (proton). $\delta = (\rho_n - \rho_p)/(\rho_n + \rho_p)$ is the isospin asymmetry, and ρ_n, ρ_p denote neutron and proton densities, respectively. The parameter values are $A_u(x)$ = 33.037 - 125.34x MeV, $A_l(x)$ = -166.963 + 125.34x MeV, B = 141.96 MeV, $C_{\tau,\tau}$ = 18.177 MeV, $C_{\tau,\tau'}$ = -178.365 MeV, $\sigma = 1.265$, and $\Lambda = 630.24$ MeV/c. With these settings, the empirical values of nuclear matter at normal density are reproduced, i.e., the saturation density ρ_0 = 0.16 fm^{-3}, the binding energy E_0 = -16 MeV, the incompressibility K_0 = 230 MeV, the isoscalar effective mass $m_s^* = 0.7m$, the single-particle potential U_∞^0 = 75 MeV at infinitely large nucleon momentum at saturation density in symmetric nuclear matter, and the symmetry energy $E_{\text{sym}}(\rho_0) = 30$ MeV [Yong (2017a)]. Note here that at nuclear densities studied here, the average kinetic symmetry energy just accounts for less than 20% of the total symmetry energy; thus the symmetry potential plays a major role in the constitution of the high-density symmetry energy. In Eq. (8.2), different symmetry energies' stiffness parameters x can be used in different density regions to mimic different density-dependent symmetry energies.

The isospin-dependent baryon–baryon (BB) scattering cross section in medium σ_{BB}^{medium} is reduced compared with their free-space value σ_{BB}^{free} by a factor of

$$R_{medium}^{BB}(\rho,\delta,\vec{p}) \equiv \sigma_{BB_{elastic,inelastic}}^{medium} / \sigma_{BB_{elastic,inelastic}}^{free}$$
$$= (\mu_{BB}^*/\mu_{BB})^2, \tag{8.3}$$

where μ_{BB} and μ_{BB}^* are the reduced masses of the colliding baryon pairs in free space and medium, respectively. The effective mass of baryon in isospin asymmetric nuclear matter is expressed as

$$\frac{m_B^*}{m_B} = 1/\left(1 + \frac{m_B}{p}\frac{dU}{dp}\right). \tag{8.4}$$

More details on the presently used model can be found in Ref. [Yong (2017a)].

Figure 8.6 shows the used symmetry energy derived from the single-particle potential Eq. (8.2) with different x parameters at low and high densities. Because below the saturation density, the symmetry energy is roughly constrained [Horowitz (2014), Chen (2017)], we fix the form of the low-density symmetry energy with parameter x = 1. From Fig. 8.7, it is seen that the low-density

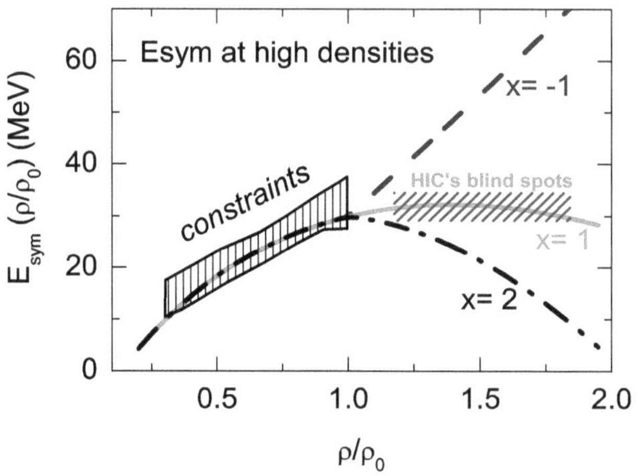

Figure 8.7 The used density-dependent symmetry energy at high densities. The experimental constraints at low densities is also shown. Taken from Ref. [Yong (2018a)].

symmetry energy with parameter x = 1 is well consistent with the current constraints. Since the symmetry energy at high densities is still not well constrained [Guo (2014)], in the present study, we vary the high-density symmetry energy parameter x in the range of x = -1, 1, 2. These choices cover the current uncertainties of the high-density behavior of the symmetry energy [Guo (2014), Chen (2017)]. In the study, we also make transport simulations without the high-density symmetry energy. This is achieved by setting $\rho_n = \rho_p = (\rho_n + \rho_p)/2$ and $U_n = U_p = (U_n + U_p)/2$.

To probe the high-density symmetry energy in dense matter formed in heavy-ion collisions, it is useful to first see the asymmetry of the formed dense matter. Figure 8.8 shows the evolution of the neutron to proton ratio (n/p) of dense matter formed in the central $^{197}Au +^{197} Au$ reaction at 300 MeV/nucleon with different high-density symmetry energies. For each case, we use the same low-density symmetry energy as shown in Fig. 8.7. In each panel of Fig. 8.8, the solid line denotes the case without high-density

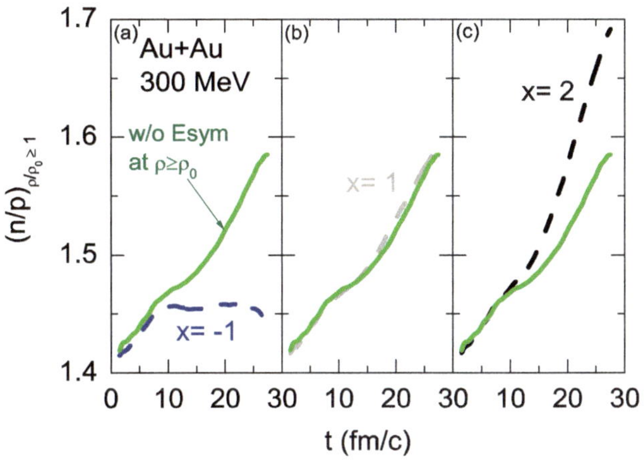

Figure 8.8 Neutron to proton ratio n/p of dense matter ($\rho/\rho_0 \geq 1$) as a function of time in the central $^{197}Au +^{197} Au$ reaction at 300 MeV/nucleon with different high-density symmetry energies. The solid line denotes the case without high-density symmetry energy. Taken from Ref. [Yong (2018a)].

symmetry energy. From Fig. 8.8 (a) and (c), compared with the case without high-density symmetry energy, one sees that the stiffer high-density symmetry energy (x= -1) causes a smaller asymmetry of dense matter, while a softer high-density symmetry energy (x= 2) causes a larger asymmetry. Both cases are understandable since the high-density symmetry energy is repulsive/attractive for neutrons with the stiff/soft high-density symmetry energy, thus leaving a small/large proportion of neutrons in the dense matter for the stiff/soft high-density symmetry energy.

While from Fig. 8.8 (b), compared with the case without high-density symmetry energy, it is interesting to see that the less density-dependent high-density symmetry energy (with parameter x= 1, shown in Fig. 8.7) almost has no effect on the asymmetry of dense matter formed in heavy-ion collisions. That is to say, the less density-dependent high-density symmetry energy almost does not affect the isospin fractionation of dense matter formed in heavy-ion collisions [Li (2002)]. In fact, for the isospin fractionation of nuclear matter, there is a chemical equilibrium condition [Li (2002), Müller (1995),

Li (1997), Baran (1998), Shi (2000)]

$$E_{sym}(\rho_1)\delta_1 = E_{sym}(\rho_2)\delta_2, \tag{8.5}$$

where $E_{sym}(\rho_1), E_{sym}(\rho_2)$ are the symmetry energies at different density regions and δ_1, δ_2 are, respectively, the asymmetries of the two different parts of nuclear matter. $\delta \equiv (\rho_n - \rho_p)/(\rho_n + \rho_p)$ with ρ_n and ρ_p being the neutron and proton densities. It is energetically favorable in a dynamical process to have a migration of nucleons with its direction determined by Eq. (8.5) according to the density dependence of the symmetry energy. Since the high-density symmetry energy with parameter x= 1 almost does not change its value with the increase in density, i.e., $E_{sym}(\rho_1) \approx E_{sym}(\rho_2)$, the asymmetry of dense matter would not be affected by such density-dependent symmetry energy, thus in dense matter the isospin-fractionation $\delta_1 \approx \delta_2$. Here the asymmetry n/p = $(1+\delta)/(1-\delta)$. Therefore, compared with the case without high-density symmetry energy, one sees in Fig. 8.8 (b) that the less density-dependent high-density symmetry energy with parameter x= 1 almost does not affect the asymmetry of dense matter formed in heavy-ion collisions. Alternatively, in heavy-ion collisions, the less density-dependent high-density symmetry energy may not be observable, since it cannot cause isospin fractionation.

Because the symmetry potential has opposite signs for neutron and proton and the fact that the symmetry potential is generally smaller compared to the isoscalar potential, and also because the symmetry potential acts directly on nucleons and normally nucleon emissions are rather abundant in typical heavy-ion reactions, the neutron/proton ratio n/p of nucleon emissions may be one of the best observables to probe the symmetry energy [Li (2006), Li (1997)].

Shown in Fig. 8.9 is the free neutron to proton ratio n/p as a function of kinetic energy in the central $^{197}Au + ^{197}Au$ reaction at 300 MeV/nucleon with different high-density symmetry energies. The free neutron to proton ratio n/p at low kinetic energies is not shown since the low-energy nucleons may be from cluster decays of hot fragments, which complicate the question studied here. From Figs. 8.9 (a) and (c), one sees that compared with the case without high-density symmetry energy, the stiffer high-density symmetry

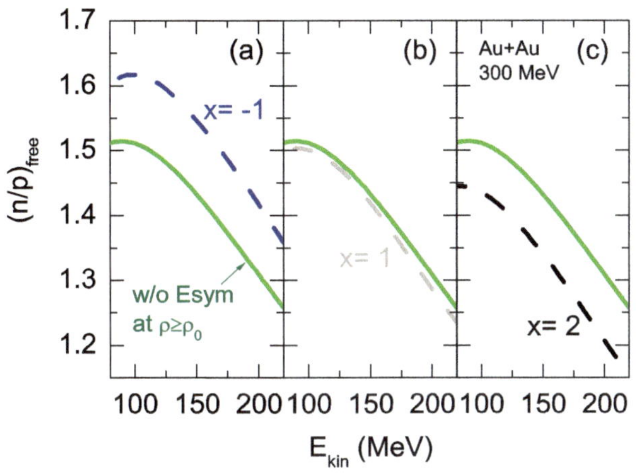

Figure 8.9 Free neutron to proton ratio n/p as a function of kinetic energy in the central $^{197}Au + ^{197}Au$ reaction at 300 MeV/nucleon with different high-density symmetry energies. The solid line denotes the case without high-density symmetry energy. Taken from Ref. [Yong (2018a)].

energy (x= -1) causes a larger free n/p ratio, while the softer high-density symmetry energy (x= 2) gives a smaller free n/p ratio. This is understandable since the stiffer/softer high-density symmetry energy push more/less neutrons to be free in heavy-ion collisions. As expected, From Fig. 8.9 (b) one again sees that the less density-dependent high-density symmetry energy (x= 1) does not affect the free n/p ratio evidently. This is because the less density-dependent high-density symmetry energy cannot cause isospin fractionation in dense matter formed in heavy-ion collisions as discussed previously. Therefore, one can conclude that the less density-dependent high-density symmetry energy may not be probed directly in heavy-ion collisions.

One may argue that if the experimental data lie between the results with the stiff (x= -1) and the soft (x= 2) symmetry energies, the true high-density symmetry energy should also lie between the stiff and the soft symmetry energies. This deduction may not be reliable because the high-density symmetry energy may, for instance, have an abrupt change at a certain density point due to the possible chiral phase transition [Xia (2016)]. If the symmetry energy is less density dependent in a certain density region, its effectiveness in heavy-ion collisions should be roughly the same as that without the symmetry energy.

In the nuclear physics community, especially in the recent 10 years, one usually tries to constrain the high-density behavior of the symmetry energy by rare isotope reactions worldwide, such as at the Facility for Rare Isotope Beams (FRIB) in the United States [Bollen (2010)], the Radioactive Isotope Beam Facility (RIBF) at RIKEN in Japan [Shane (2015)], or the GSI Facility for Antiproton and Ion Research (FAIR) in Germany [Russotto (2016)], the Cooling Storage Ring on the Heavy Ion Research Facility at IMP (HIRFL-CSR) in China [L.M. (2016)], and the Rare Isotope Science Project (RISP) in Korea [Tshoo (2013)]. However, as we discussed above, one may not directly probe the high-density symmetry energy in heavy-ion collisions in case the high-density symmetry energy is less density dependent. From the above studies, it is shown that the effectiveness of the less density-dependent high-density symmetry energy is roughly equal to the case without high-density symmetry energy. One, therefore, has to find some other ways to probe the less density-dependent high-density symmetry energy.

The blind spots of probing the high-density symmetry energy in heavy-ion collisions only occur in case the high-density symmetry energy in a certain density region is less density dependent. The blind spots do not appear in the density regions if the high-density symmetry energy is evidently density dependent.

8.4 Model Dependence of Symmetry-Energy-Sensitive Probes and Qualitative Probe

There are many factors affecting nuclear reaction transport simulation, e.g., the initialization, the nucleon–nucleon interaction potential, the nucleon–nucleon scattering cross sections, and the framework of transport models. It is thus necessary to make a study among different models to see how large the differences are on the values of isospin-sensitive observables.

Within the frameworks of isospin-dependent transport models Boltzmann–Uehling–Uhlenbeck (IBUU) and Ultrarelativistic Quantum Molecular Dynamics (UrQMD), the model dependences of frequently used isospin-sensitive observables π^-/π^+ ratio and n/p ratio of free nucleons have been demonstrated [Guo (2013)]. Figure 8.10 shows the n/p ratio of free nucleons and π^-/π^+ ratio as

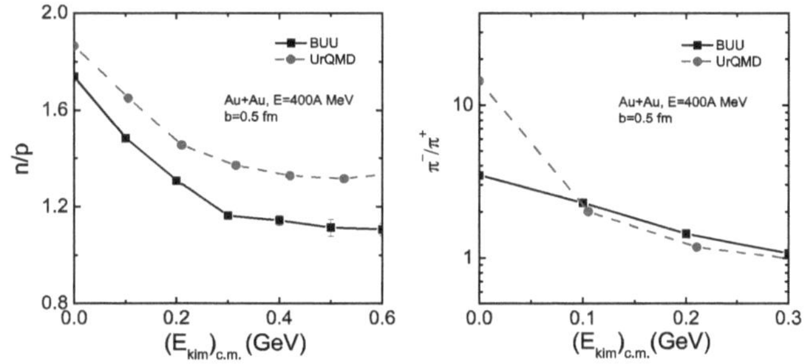

Figure 8.10 The n/p ratio of free nucleons and π^-/π^+ ratio as a function of kinetic energy in $^{197}Au +^{197} Au$ at a beam energy of 400 MeV/nucleon with the same density-dependent symmetry energy simulated by IBUU and UrQMD models. Taken from Ref. [Guo (2013)].

a function of kinetic energy in the $^{197}Au +^{197} Au$ reaction at a beam energy of 400 MeV/nucleon simulated by the IBUU and the UrQMD models. From the left panel, one can see that both models give the same trend of n/p ratio as a function of nucleonic kinetic energy. The result of the UrQMD model is overall larger than that of the IBUU model due to their different strengths of symmetry potential [Guo (2013)]. From the right panel of Fig. 8.10, one can see that there is a cross between the π^-/π^+ ratios from the UrQMD model and that from the IBUU model. At lower kinetic energies, the value of the π^-/π^+ ratio from the UrQMD model is much larger than that from the IBUU model. And from the left panel of Fig. 8.11, one can see that the nucleonic transverse flow given by the IBUU model shows a large isospin effect than that with the UrQMD model. From the right panel of Fig. 8.11, it is seen that the slope of the neutron–proton differential

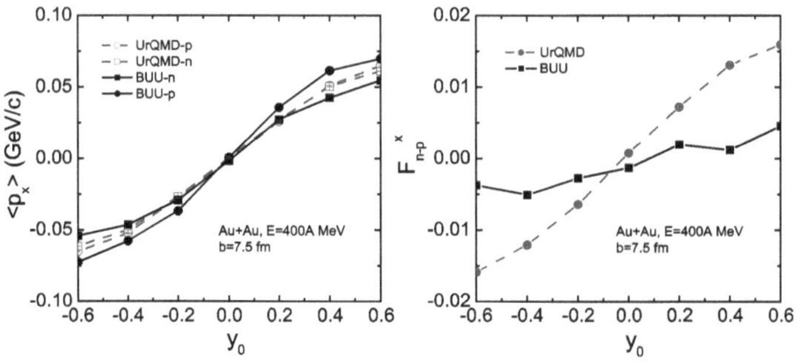

Figure 8.11 Same as Fig. 8.10, but for the rapidity distributions of the transverse flow $<p_x(y)>$ for neutrons and protons and the neutron–proton differential transverse flow F^x_{n-p}. Taken from Ref. [Guo (2013)].

flow is evidently larger for the UrQMD model than that for the IBUU model.

To study the model dependence of quantitatively sensitive probes of nuclear symmetry energy, what is crucially needed for the first step is a qualitative observable to probe whether the symmetry energy at high densities is stiff or soft, i.e., whether values of the symmetry energy increase or decrease with densities above saturation density.

Figure 8.12 shows nucleon kinetic energy dependence of the n/p ratios of free (gas) and bound (liquid) nucleons in the central $^{197}Au + ^{197}Au$ reaction at a beam energy of 400 MeV/A with positive and negative symmetry potentials at supra-saturation densities simulated by the IBUU transport model [Guo (2014)]. From the left panel, one can see that the value of n/p of nuclear gas phase is larger than that of the liquid phase in the whole kinetic energy distribution region with the positive symmetry potential. However, it is interesting to see that the value of n/p of gas phase is smaller than that of the liquid phase at higher kinetic energies with the negative symmetry potential at supradensities. The same thing is seen in Fig. 8.13 with the UrQMD model. One discrepancy is that the kinetic energy of transition point with the UrQMD model is lower

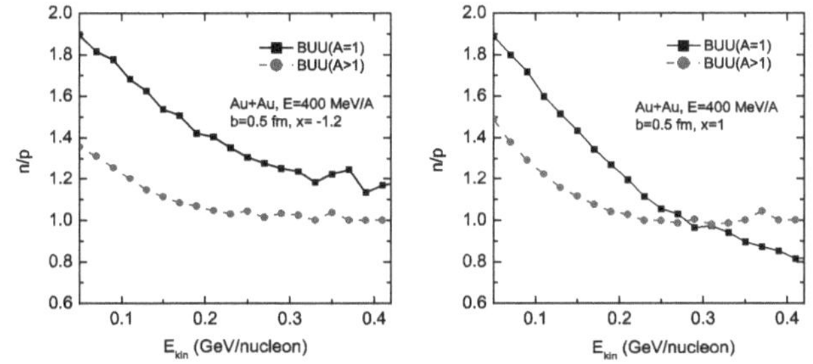

Figure 8.12 The neutron to proton ratio n/p of free and bound nucleons as a function of nucleon kinetic energy in the central $^{197}Au +^{197} Au$ reaction at a beam energy of 400 MeV/A with positive (left) and negative (right) symmetry potentials at supra-saturation densities. Simulated with the IBUU transport model. Taken from Ref. [Guo (2014)].

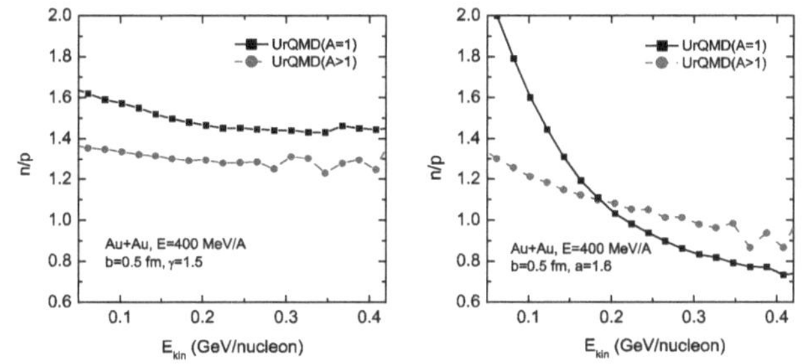

Figure 8.13 Same as Fig. 8.12 but with the UrQMD transport model. Taken from Ref. [Guo (2014)].

than that with the IBUU model. With positive/negative symmetry potential at supradensities, for energetic nucleons, the value of neutron to proton ratio of free nucleons is larger/smaller than that of bound nucleon fragments. Compared with the extensively studied quantitative observables of nuclear symmetry energy, the normal or abnormal isospin fractionation of energetic nucleons can be a qualitative probe of nuclear symmetry energy at supradensities.

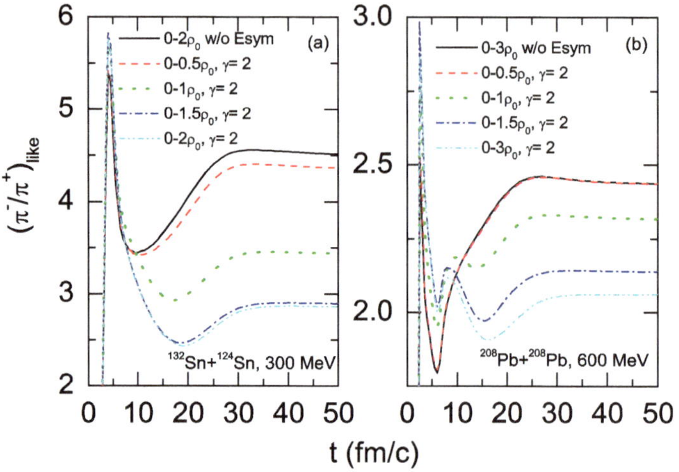

Figure 8.14 Comparison of the relative sensitivity of the symmetry-energy-sensitive observable $(\pi^-/\pi^+)_{like}$ ratio as a function of time in the ^{132}Sn+^{124}Sn at 300 MeV/nucleon with an impact parameter b= 3 fm and the ^{208}Pb+^{208}Pb at 600 MeV/nucleon with an impact parameter b= 0 fm. Taken from Ref. [Yong (2019)].

8.5 Determination of the Density Region of the Symmetry Energy Probed by the π^-/π^+ Ratio and Nucleon Observables

To probe the symmetry energy at high densities, the π^-/π^+ ratio has been proposed in the literature [Li (2002a), Li (2005), Yong (2006)]. Aiming at probing the symmetry energy at high densities by the π^-/π^+ ratio, Sn+Sn reactions at 300 or 200 MeV/nucleon experiments are being carried out at RIKEN in Japan [Shane (2015), Otsu (2016), Jhang (2016)]. Does the π^-/π^+ ratio in heavy-ion collisions at intermediate energies always probe the high-density symmetry energy? And in what conditions does the π^-/π^+ ratio probe the symmetry energy at high densities?

To answer the above questions, the decomposition of the sensitivity of the symmetry energy observables was studied. From

Fig. 8.14 (a), it is seen that in the $^{132}Sn+^{124}Sn$ reaction at 300 MeV/nucleon, the effects of the symmetry energy in the density region 0–0.5ρ_0 are larger than that in the density region 1.5ρ_0–2ρ_0. The effects of the symmetry energy below the saturation density are comparable to that above saturation density. However, the situation is changed when one moves to Fig. 8.14 (b), evolution of the π^-/π^+ ratio in the $^{208}Pb+^{208}Pb$ reaction at 600 MeV/nucleon. From Fig. 8.14 (b), it is seen that the effects of the symmetry energy in the density region 0–0.5ρ_0 are obviously smaller than that in the density region 1.5ρ_0–3ρ_0. The effects of the symmetry energy below saturation density are also obviously smaller than that above the saturation density. Figure 8.14 clearly shows that to use the π^-/π^+ ratio as the probe of the high-density symmetry energy, using a heavy system and at relatively higher incident beam energies is a preferable way.

Similar studies were also carried out for nucleon observables [Fan (2018)]. It is found that the symmetry-energy-sensitive observable n/p ratio in the $^{132}Sn+^{124}Sn$ reaction at 0.3 GeV/nucleon just probes the density-dependent symmetry energy below the density of 1.5ρ_0 and effectively probes the density-dependent symmetry energy around or somewhat below the saturation density. The nucleon elliptic flow can probe the symmetry energy from the low-density region to the high-density region when changing the incident beam energies from 0.3 to 0.6 GeV/nucleon in the semi-central $^{132}Sn+^{124}Sn$ reaction. Nucleon transverse and elliptic flows in the semi-central $^{197}Au+^{197}Au$ reaction at 0.6 GeV/nucleon are in fact more sensitive to the high-density behavior of the nuclear symmetry energy.

8.6 Effects of Short-Range Correlations in Transport Model

The picture of nucleons having maximal momentum—so called Fermi momentum p_F—in a nuclear system and roughly moving independently in the mean field created by their mutually attractive interactions has been established since the 1950s. However, recent proton-removal experiments using electron beams with energies of several hundred MeVs showed that only about 80% nucleons participate in this type of independent particle motion [Lapikas (1993), Kelly (1996), Subedi (2008)]. And high-momentum

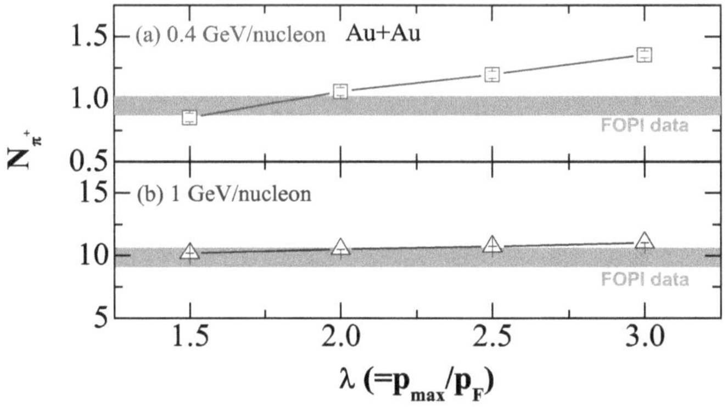

Figure 8.15 The number of produced π^+ meson as a function of high-momentum cut-off parameter λ in the $^{197}Au + ^{197}Au$ collisions at, respectively, 0.4 and 1 GeV/nucleon beam energies. Taken from [Yong (2014)].

transfer measurements have shown that nucleons in nuclei can form pairs with larger relative momenta and smaller center-of-mass momenta [Piasetzky (2006), Shneor (2007)]. This is interpreted by the nucleon–nucleon tensor interaction in short range [Sargsian (2005), Schiavilla (2007)]. The nucleon–nucleon short-range correlations (SRC) in nuclei leads to a high-momentum tail (HMT) in single-nucleon momentum distribution above 300 MeV/c [Bethe (1971), Antonov (1971), Rios (2009), Rios (2013), Ciofi (2015)]. And interestingly, the high-momentum tail's shape caused by two-nucleon SRC is almost identical for all nuclei from deuteron to the heavier nuclei [Ciofi (1996), Fantoni (1984), Pieper (1992), Egiyan (2003)], i.e., roughly exhibits a C/k^4 tail [Hen (2015), Hen (2014), Hen (2015a), Cai (2015)].

The high-momentum tail of nucleon distribution in nuclei or nuclear matter surely affects the yields of π, K, η, and nucleon emission in heavy-ion collisions at intermediate energies. Figure 8.15 shows π^+ production as a function of high-momentum cut-off parameter λ of the colliding nuclei in the $^{197}Au + ^{197}Au$ collisions at 0.4 and 1 GeV/nucleon incident beam energies, respectively [Yong (2014)]. One can clearly see that as the high-momentum cutoff parameter increases, more π^+'s are produced. A larger high-momentum cut-off parameter λ causes a larger nucleon average kinetic energy, especially proton average kinetic energy; thus, the average center-of-mass energy of proton–proton collision also becomes larger. As a consequence, more π^+'s are produced in nucleus–nucleus collision. As incident beam energy increases, the initial movement of nucleons in nuclei becomes less important in nucleus–nucleus collisions.

Figure 8.16 shows the kinetic energy distributions of π^-, π^+ as well as π^-/π^+ ratio with and without the HMT in the same semi-central reaction $^{197}\text{Au} + ^{197}\text{Au}$ at a beam energy of 400 MeV/nucleon [Zhang (2016)]. It is seen that the kinetic energy distributions of both π^- and π^+ are very sensitive to the HMT. The ratio of π^-/π^+ is also very sensitive to the HMT except in the high kinetic energy region. The neutron–proton short-range corrections increase the kinetic energies of a certain proportion of neutrons and protons; thus, more pion mesons are produced. Because in isospin asymmetric reaction system, protons have a larger probability than neutrons to have larger momenta and proton–proton collision mainly produces π^+, one sees lower value of the π^-/π^+ ratio with the HMT than that without the HMT. And from Fig. 8.17, it is seen that the effect of the HMT on the difference of neutron and proton elliptic flows ($v_2^n - v_2^p$) is larger than that of the total nucleon elliptic flow [Zhang (2016)].

It is generally considered that below the Fermi momentum, protons or neutrons have independent movements, while above their respective Fermi momenta, they respectively have $1/k^4$ distributions starting from their respective Fermi momenta. This naive opinion, however, is not consistent with the correlation picture of a neutron–proton pair. In asymmetric nuclear matter or neutron matter, such as the neutron stars, neutron and proton may have very different Fermi momenta. If each correlated neutron and proton has $1/k^4$ distributions starting from their respective Fermi momenta, then the correlated neutron and proton would have very different momenta. This point evidently contradicts the thought of the n-p dominance model [Hen (2014)]. It is thus reasonable that in neutron-rich matter, the minority proton has the same transition momentum as that of the majority neutron.

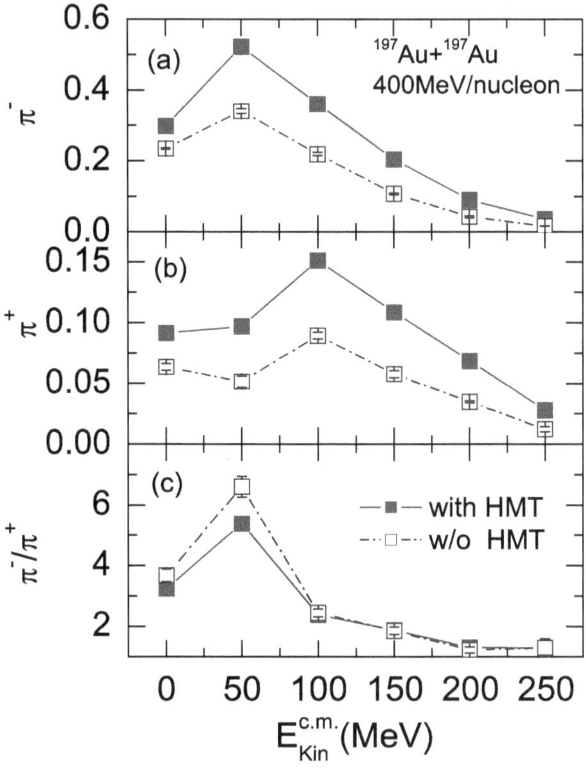

Figure 8.16 Kinetic energy distributions of π^- (upper panel), π^+ (middle panel) as well as π^-/π^+ (bottom panel) ratio with and without the HMT in the reaction of ^{197}Au $+^{197}$ Au at incident beam energy of 400 MeV/nucleon. Taken from Ref. [Zhang (2016)].

Figure 8.18 shows the nucleon momentum distribution of ^{48}Ca. It is clearly seen that there is an HMT above the nuclear Fermi momentum. Proton has a greater probability than neutron to have momenta greater than the nuclear Fermi momentum. Comparing case A with case B, it is seen that with the starting point of majority Fermi momentum, a proton has even greater probability to have high momenta. This consequence may affect the dynamics of heavy-

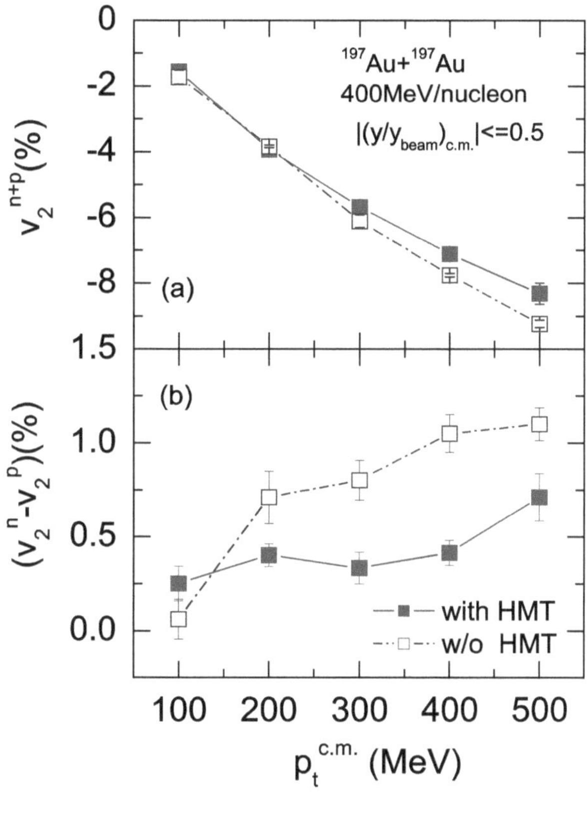

Figure 8.17 Total nucleon elliptic flow (upper panel) and the difference of neutron and proton elliptic flow (lower panel) in the semi-central reaction of ^{197}Au+ ^{197}Au at a beam energy of 400 MeV/nucleon with and without the HMT, respectively. Taken from Ref. [Zhang (2016)].

ion collisions at intermediate energies. Figure 8.19 shows the ratio of π^-/π^+ in ^{132}Sn+^{124}Sn reactions at 300 MeV/nucleon incident beam energy with different proton starting momenta in the HMT. As expected, there is a clear decrease in the value of π^-/π^+ ratio when changing the starting momentum of proton in the HMT from the proton Fermi momentum to the majority neutron Fermi momentum.

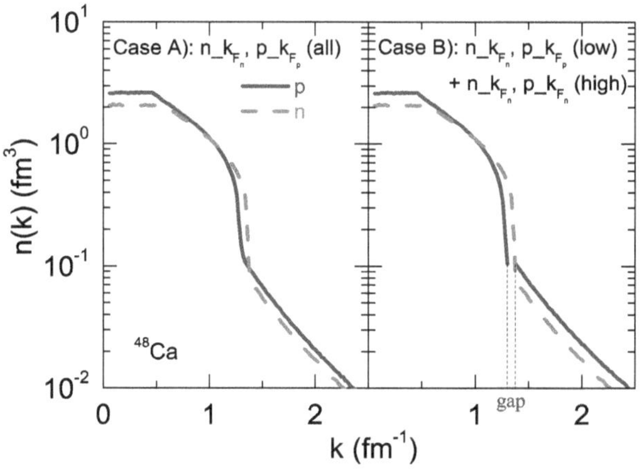

Figure 8.18 Momentum distribution n (k) of nucleon with different starting points of proton $1/k^4$ distribution in nucleus ^{48}Ca with normalization condition $\int_0^{k_{max}} n(k)k^2 dk = 1$. Case A stands for the HMT starting point of minority proton is its own Fermi-momentum while Case B starts the HMT of proton from majority neutron Fermi-momentum. Taken from Ref. [Yong (2018)].

The effect reaches nearly 20%. From the inserted figure, it is shown that such effect is mainly caused by the π^+ production. Since related pion measurements in ^{132}Sn+^{124}Sn at 300 MeV/nucleon incident beam energy are ongoing at Radioactive Isotope Beam Facility (RIBF) at RIKEN in Japan [Shane (2015), Otsu (2016)], to interpret related experimental data more scientifically, it is very necessary to study how the π^-/π^+ ratio is affected by the starting point of the HMT in colliding nuclei.

8.7 Cross-Checking the Symmetry Energy at High Densities

Since constraints of the symmetry energy are usually model dependent, it is necessary to cross-check the symmetry energy using different probes. Figure 8.20 shows the π^-/π^+ ratio predicted by

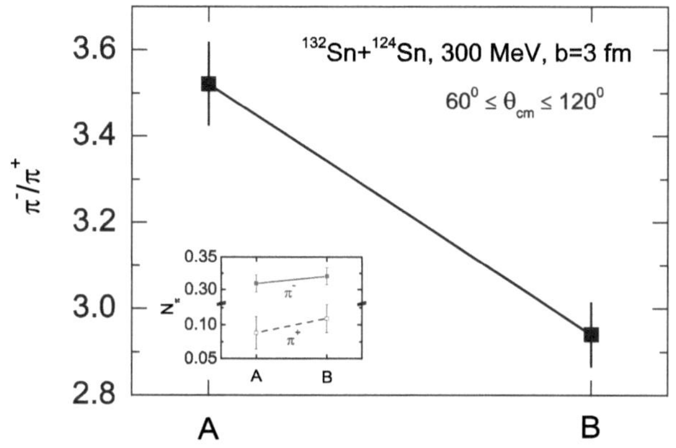

Figure 8.19 The ratio of π^-/π^+ in ^{132}Sn+^{124}Sn reactions at 300 MeV/nucleon incident beam energy with different proton starting momenta in the HMT. θ_{cm} is the polar angle relative to the incident beam direction. The inserted figure shows corresponding numbers of π^- and π^+. Taken from Ref. [Yong (2018)].

our IBUU model with different symmetry energies. Because softer symmetry energy causes more neutron-rich dense matter and π^-'s are mainly from neutron–neutron collision whereas π^+'s are mainly from proton–proton collision, it is not surprising that one sees larger π^-/π^+ ratio with softer symmetry energy. To see the effects of the SRC of nucleon–nucleon and the reduction in the in-medium inelastic baryon–baryon scattering cross section, with same x parameters, we made calculations by turning off the SRC and the reduction in the in-medium inelastic baryon–baryon scattering cross section, respectively. From Fig. 8.20, we can see that both of them affect the value of the π^-/π^+ ratio evidently. Both the SRC of nucleon–nucleon and the reduction in the in-medium inelastic baryon–baryon scattering cross section decrease the value of the π^-/π^+ ratio evidently. Proton–proton collision is also affected by the Coulomb action, so the π^+ production, which is mainly from proton–proton collision, is relatively less affected by the reduction in the in-medium inelastic baryon–baryon scattering cross section. From Fig. 8.20,

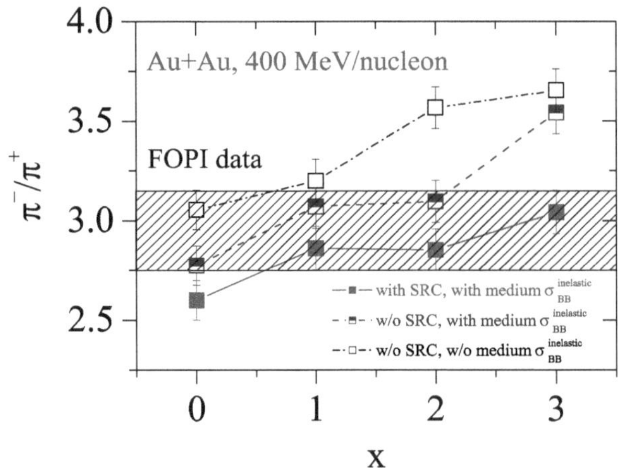

Figure 8.20 π^-/π^+ ratio in Au+Au reaction at 400 MeV/nucleon with different symmetry energies as well as effects of short-range correlations and in-medium cross section on the π^-/π^+ ratio. Taken from Ref. [Yong (2016)].

one can see that the FOPI pion experimental data support a softer symmetry energy ($x = 1, 2$, even $x = 3$).

Shown in Fig. 8.21 is the predicted elliptic flow ratios of neutron and proton V_2^n/V_2^p with different symmetry energies as well as experimental data. Since stiffer symmetry energy/symmetry potential causes relatively more neutrons to emit in the direction perpendicular to the reaction plane, one sees larger values of elliptic flow ratios of neutron and proton V_2^n/V_2^p with stiffer symmetry energies. With the SRC of nucleon–nucleon in the transport model, values of the V_2^n/V_2^p ratio are larger than that without the SRC of nucleon–nucleon. This is because the SRC of nucleon–nucleon causes neutron and proton to be correlated together, the value of the V_2^n/V_2^p ratio trends to unity. Owing to the competing effects of the SRC and the symmetry energy, for $x = 0$ case, the effects of symmetry energy on the trend of the ratio of V_2^n/V_2^p with the SRC change compared with that without the SRC. Figure 8.21 indicates that the FOPI-LAND elliptic flow experimental data do not favor very soft symmetry

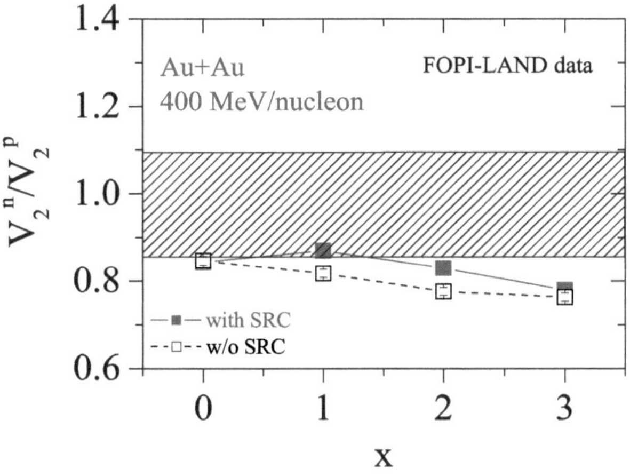

Figure 8.21 The ratio of V_2^n/V_2^p in $^{197}Au + ^{197}Au$ reaction at 400 MeV/nucleon with different symmetry energies. Taken from [Yong (2016)].

energy (x = 2, 3). Combining the studies of nucleon elliptic flow and previous π^-/π^+ ratio, one may roughly obtain the symmetry energy stiffness parameter x = 1. It, in fact, corresponds a mildly soft density-dependent symmetry energy at supra-saturation densities, which is shown in Fig. 8.6.

8.8 Probing the Curvature of Nuclear Symmetry Energy K_{sym} around Saturation Density

The density-dependent symmetry energy around saturation can be Taylor expanded in terms of a few bulk parameters [Vidaña (2009)]:

$$E_{sym}(\rho) = E_{sym}(\rho_0) + L\left(\frac{\rho-\rho_0}{3\rho_0}\right) + \frac{K_{sym}}{2}\left(\frac{\rho-\rho_0}{3\rho_0}\right)^2, \quad (8.6)$$

where $E_{sym}(\rho_0)$ is the value of the symmetry energy at the saturation point and the quantities L, K_{sym} are related to its slope and curvature, respectively, at the same density point,

$$L = 3\rho_0 \frac{\partial E_{sym}(\rho)}{\partial \rho}\bigg|_{\rho=\rho_0}, \quad K_{sym} = 9\rho_0^2 \frac{\partial^2 E_{sym}(\rho)}{\partial \rho^2}\bigg|_{\rho=\rho_0}. \tag{8.7}$$

Over the last two decades, the most probable magnitude $E_{\text{sym}}(\rho_0) = 31.7 \pm 3.2$ MeV and slope $L = 58.7 \pm 28.1$ MeV of the nuclear symmetry energy at saturation density have been found through surveys of 53 analyses [Oertel (2017), Li (2013)]. The curvature K_{sym} is known probably to be in the range of $-400 \leq K_{\text{sym}} \leq 100$ MeV [Tews (2017), Zhang (2017)]. In Eq. (8.6), if we assume $E_{\text{sym}}(\rho_0) = 30$ MeV and slope $L = 60$ MeV, the question of the determination of the high-density symmetry energy is directly converted into the constraints of the coefficient K_{sym}. The curvature of nuclear symmetry energy is, in fact, also closely related to the incompressibility of neutron-rich matter, which is difficult to determine by the properties of neutron-rich nuclei, thus remaining an important open problem [Vidaña (2009), Piekarewicz (2009)].

It is known that the squeezed-out nucleons (emitted in the direction perpendicular to the reaction plane) in semi-central collisions carry more direct information about the high-density phase of the reaction [Stöcker (1986), Bertsch (1988), Cassing (1990), Aichelin (1991), Reisdorf (1997), Danielewicz (2002), Yong (2007)]. Figure 8.22 shows the squeezed-out neutron to proton ratio of nucleons emitted in the direction perpendicular to the reaction plane in $^{197}Au +^{197}Au$ at 400 (left panel) and 600 (right panel) MeV/nucleon [Guo (2019)]. It is clearly seen that the effects of the curvature of the symmetry energy on the squeezed-out neutron to proton ratio n/p are quite evident, especially at high transverse momenta. Since the slope of the high-density symmetry energy with $K_{sym} = 0$ is larger than that with $K_{sym} = -400$ MeV, one sees that the values of the squeezed-out n/p with $K_{sym} = 0$ are higher than that with $K_{sym} = -400$ MeV. From both panels, it is seen that around p_t = 0.8 or 1.3 GeV/c, the effects of the curvature of the symmetry energy on the squeezed-out n/p reach about 40–60%, which are both much larger than the ratio of integrating neutron and proton elliptic flows [Cozma (2018)]. The squeezed-out n/p thus can be one of the most potential probes of the curvature of the symmetry energy and can be carried out on the rare isotope reactions worldwide.

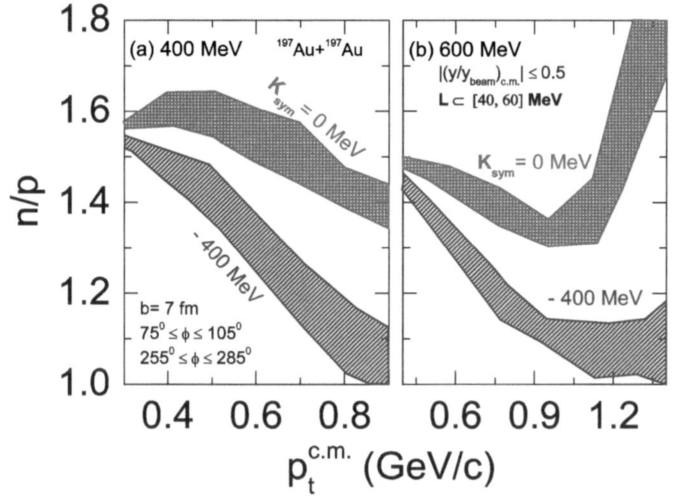

Figure 8.22 Effects of the curvature of nuclear symmetry energy at a saturation density on the transverse momentum distribution of the ratio of midrapidity ($|(y/y_{beam})_{c.m.}| \leq 0.5$) neutrons to protons emitted in the reaction of $^{197}Au + ^{197}Au$ at incident beam energy of 400 (panel (a)) and 600 (panel (b)) MeV/nucleon and impact parameter of b= 7 fm. The azimuthal angle cut is $75° \leq \phi \leq 105°$ and $255° \leq \phi \leq 285°$, to make sure that the free nucleons are from the direction perpendicular to the reaction plane. The bandwidths denote uncertainties of the slope L from 60 to 40 MeV. Taken from Ref. [Guo (2019)].

8.9 Perspective and Acknowledgments

Besides constraining from heavy-ion collisions [Yong (2016), Russotto (2016)], nowadays the high-density symmetry energy can also be determined through the studies of Neutron Stars within a minimum model for neutron stars consisting of nucleons, electrons, and muons [Zhang (2019), Tsang (2019)]. It is fascinating to see the match of constraints of the high-density symmetry energy from heaven [Abbott (2017)] and earth [US LRP (2015), US LRP (2017)] in the near future.

I thank Lie-Wen Chen, Xia-Hua Fan, Yuan Gao, Ya-Fei Guo, Wei-Mei Guo, Bao-An Li, Qing-Feng Li, Gao-Feng Wei, Yong-Jia Wang, Zu-Xing Yang, Fang Zhang, Hong-Fei Zhang, and Wei Zuo for collaborations. This work is supported in part by the National Natural Science Foundation of China under Grants No. 11775275 and No. 11435014 and the 973 Program of China (Program No. 2007CB815004).

References

Abbott, B. P. *et al.* (2017). [LIGO+Virgo Collaborations], GW170817: Observation of gravitational waves from a binary neutron star inspiral, *Phys. Rev. Lett.* **119**, pp. 161101.

Abbott, B. P. *et al.* (2017). Observation of gravitational waves from a binary neutron star Inspiral (LIGO Scientific Collaboration and Virgo Collaboration), *Phys. Rev. Lett.* **119**, pp. 161101.

Aichelin, J. (1991). "Quantum" molecular dynamics: A dynamical microscopic n-body approach to investigate fragment formation and the nuclear equation of state in heavy ion collisions, *Phys. Rep.* **202**, pp. 233.

Antonov, A. N., Hodgson, P. E., and Petkov, I. Z. (1988). *Nucleon Momentum and Density Distributions in Nuclei* (Clarendon Press, Oxford).

Baran, V. *et al.* (2005). Reaction dynamics with exotic beams, *Phys. Rep.* **410**, pp. 335–466.

Baran, V., Larionov, A., Colonna, M., and Di Toro, M. (1998). Spinodal decomposition of low-density asymmetric nuclear matter, *Nucl. Phys. A* **632**, pp. 287.

Bauswein, A., Goriely, S., and Janka, H. T. (2013). *Astrophysics* **773**, pp. 78.

Bauswein, A., Stergioulas, N., and Janka, H. T. (2014). Revealing the high-density equation of state through binary neutron star mergers, *Phys. Rev. D* **90**, pp. 023002.

Bertsch, G. F. and Das Gupta, S. (1988). A guide to microscopic models for intermediate energy heavy ion collisions, *Phys. Rep.* **160**, pp. 189.

Bethe, H. A. (1971). Theory of nuclear matter, *Ann. Rev. Nucl. Part. Sci.* **21**, pp. 93.

Bollen, G. (2010). FRIB–Facility for Rare Isotope Beams, *AIP Conf. Proc.* **1224**, pp.432.

Bombaci, I. and Lombardo, U. (1991). Asymmetric nuclear matter equation of state, *Phys. Rev. C* **44**, pp. 1892.

Cai, B.J. and Li, B.A. (2015). Isospin quartic term in the kinetic energy of neutron-rich nucleonic matter, *Phys. Rev. C* **92**, pp. 011601.

Cassing, W., Metag, V., Mosel, U., and Niita, K. (1990). Production of energetic particles in heavy-ion collisions, *Phys. Rep.* **188**, pp. 363.

Chen, L. W. (2017). Probing high-density symmetry energy using heavy-ion collisions at intermediate energies. *Nucl. Phys. Rev.* **34**, pp. 20.

Ciofi degli Atti, C. (2015). In-medium short-range dynamics of nucleons: Recent theoretical and experimental advances, *Phys. Rep.* **590**, pp. 1 .

Ciofi degli Atti, C. and Simula, S. (1996). Realistic model of the nucleon spectral function in few- and many-nucleon systems, *Phys. Rev. C* **53**, pp. 1689.

Cozma, M. D. (2018). Feasibility of constraining the curvature parameter of the symmetry energy using elliptic flow data, *Eur. Phys. J. A* **54**, pp. 40.

Cozma, M. D., Leifels, Y., Trautmann, W., Li, Q., and Russotto, P. (2013). Toward a model-independent constraint of the high-density dependence of the symmetry energy, *Phys. Rev. C* **88**, pp. 044912.

Danielewicz, P., Lacey, R., and Lynch, W. G. (2002). Determination of the equation of state of dense matter, *Science* **298**, pp. 1592–1596.

Danielewicz, P., Lacey, R., and Lynch, W. G. (2002). Determination of the equation of state of dense matter, *Science* **298**, pp. 1592.

Fan, X. H., Yong, G. C., and Zuo, W. (2018). Probing the density dependence of the symmetry energy by nucleon flow, *Phys. Rev. C* **97**, pp. 034604.

Fantoni, S. and Pandharipande, V. R. (1984). Momentum distribution of nucleons in nuclear matter, *Nucl. Phys. A* **427**, pp. 473.

Feng, Z. Q. and Jin, G. M. (2010). Probing high-density behavior of symmetry energy from pion emission in heavy-ion collisions, *Phys. Lett. B* **683**, pp. 140.

Fischer, T. *et al.* (2014). Symmetry energy impact in simulations of core-collapse supernovae, *Eur. Phys. J. A* **50**, pp. 1.

Friedrich, J. and Reinhard, P. G. (1986). Skyrme-force parametrization: Least-squares fit to nuclear ground-state properties, *Phys. Rev. C* **33**, pp. 335.

Goriely, S. *et al.* (2011). The r-process nucleosynthesis during the decompression of neutron star crust material, *Astrophysics* **738**, pp. 32.

Guo, W. M. *et al.* (2013). Model dependence of isospin sensitive observables at high densities, *Phys. Lett. B* **726**, pp. 211.

Guo, W. M., *et al.* (2014). Normal or abnormal isospin-fractionation as a qualitative probe of nuclear symmetry energy at supradensities, *Phys. Lett. B* **738**, pp. 397–400.

Guo, W. M., *et al.* (2014). Normal or abnormal isospin-fractionation as a qualitative probe of nuclear symmetry energy at supradensities, *Phys. Lett. B* **738**, pp. 397.

Guo, W. M., Yong, G. C., and Zuo, W. (2015). Effects of pion potential and nuclear symmetry energy on the pion ratio in heavy-ion collisions at beam energies around the pion production threshold, *Phys. Rev. C* **91**, pp. 054616.

Guo, W. M., Yong, G. C., Liu, H., and Zuo, W. (2015). Effects of pion potential and nuclear symmetry energy on the ratio in heavy-ion collisions at beam energies around the pion production threshold, *Phys. Rev. C* **91**, pp. 054616.

Guo, Y. F. and Yong, G. C. (2019). High squeezed-out n/p ratio as a probe of Ksym of the symmetry energy, *Phys. Rev. C* **100**, pp. 014617.

Hen O., *et al.* (2015). Symmetry energy of nucleonic matter with tensor correlations, *Phys. Rev. C* **91**, pp. 025803.

Hen, O. *et al.* (2014). Momentum sharing in imbalanced Fermi systems (the CLAS Collaboration), *Science* **346**, pp. 614.

Hen, O. *et al.* (2015). Correlated fermions in nuclei and ultracold atomic gases, *Phys. Rev. C* **92**, pp. 045205.

Hong, J. and Danielewicz, P. (2014). Subthreshold pion production within a transport description of central Au + Au collisions, *Phys. Rev. C* **90**, pp. 024605.

Horowitz, C. J. *et al.* (2014). A way forward in the study of the symmetry energy: Experiment, theory, and observation, *J. Phys. G: Nucl. Part. Phys.* **41**, pp. 093001.

Jhang, G. *et al.* (2016). Beam commissioning of the SpRIT time projection chamber, *J. Kor. Phys. Soc.* **69**, pp. 144.

Kelly, J. (1996). Nucleon knockout by intermediate energy electrons, *Adv. Nucl. Phys.* **23**, pp. 75.

Lapikas, L. (1993). Quasi-elastic electron scattering off nuclei, *Nucl. Phys. A.* **553**, pp. 297.

Lasky, P. D. *et al.* (2014). Nuclear equation of state from observations of short gamma-ray burst remnants, *Phys. Rev. D* **89**, pp. 047302.

Lattimer, J. M. and Prakash, M. (2004). The physics of neutron stars, *Science* **304**, pp. 536–542.

Lattimer, J. M. and Steiner, W. A. (2014). Constraints on the symmetry energy using the mass-radius relation of neutron stars, *Eur. Phys. J. A* **50**, pp. 240.

Li, B. A. (2002). High density behaviour of nuclear symmetry energy and high energy heavy-ion collisions, *Nucl. Phys. A* **708**, pp. 365.

Li, B. A. (2002). Probing the high density behavior of the nuclear symmetry energy with high energy heavy-ion collisions, *Phys. Rev. Lett.* **88**, pp. 192701.

Li, B. A. and Han X. (2013). Constraining the neutron–proton effective mass splitting using empirical constraints on the density dependence of nuclear symmetry energy around normal density, *Phys. Lett. B* **727**, pp. 276

Li, B. A. and Ko, C. M. (1997). Chemical and mechanical instability in hot isospin-asymmetric nuclear matter, *Nucl. Phys. A* **618**, pp. 498.

Li, B. A., Chen, L. W., and Ko, C. M. (2008). Recent progress and new challenges in isospin physics with heavy-ion reactions, *Phys. Rep.* **464**, pp. 113–282.

Li, B. A., Chen, L. W., Yong, G. C., and Zuo, W. (2006). Double neutron/proton ratio of nucleon emissions in isotopic reaction systems as a robust probe of nuclear symmetry energy, *Phys. Lett. B* **634**, pp. 378.

Li, B. A., Ko, C. M., and Ren, Z. Z. (1997). Equation of state of asymmetric nuclear matter and collisions of neutron-rich nuclei, *Phys. Rev. Lett.* **78**, pp. 1644.

Li, B. A., Yong, G. C., and Zuo, W. (2005). Near-threshold pion production with radioactive beams, *Phys. Rev. C* **71**, pp. 014608.

Müller, H. and Serot, B. D. (1995). Phase transitions in warm, asymmetric nuclear matter, *Phys. Rev. C* **52**, pp. 2072.

L. M. Lü *et al.* (2017). Conceptual design of the HIRFL-CSR external-target experiment, *Sci. China-Phys. Mech. Astron.* **60**, pp. 012021.

Maselli, A., Gualtieri, L., and Ferrari, V. (2013). Constraining the equation of state of nuclear matter with gravitational-wave observations: Tidal deformability and tidal disruption, *Phys. Rev. D* **88**, pp. 104040.

Nikolov, N. *et al.* (2011). Surface symmetry energy of nuclear energy density functionals, *Phys. Rev. C* **83**, pp. 034305.

Oertel, M., Hempel, M., Klähn, T., and Typel, S. (2017). Equations of state for supernovae and compact stars, *Rev. Mod. Phys.* **89**, pp. 015007.

Otsu, H. *et al.* (2016). SAMURAI in its operation phase for RIBF users, *Nucl. Instr. Methods Phys. Res. B* **376**, pp. 175.

Pandharipande, V. R. and Garde, V. K. (1972). Effect of tensor forces on composition of dense matter, *Phys. Lett. B* **39**, pp. 608.

Piasetzky, E., Sargsian, M., Frankfurt, L., Strikman, M., and Watson, J. W. (2006). Evidence for strong dominance of proton-neutron correlations in nuclei, *Phys. Rev. Lett.* **97**, pp. 162504.

Piekarewicz, J. and Centelles, M. (2009). Incompressibility of neutron-rich matter, *Phys. Rev. C* **79**, pp. 054311.

Pieper, S. C., Wiringa, R. B., and Pandharipande, V. R. (1992). Variational calculation of the ground state of ^{16}O, *Phys. Rev. C* **46**, pp. 1741.

Reisdorf, W. and Ritter, H. G. (1997). Collective flow in heavy-ion collisions, *Annu. Rev. Nucl. Part. Sci.* **47** pp. 663.

Rios, A., Polls, A., and Dickhoff, W. H. (2009). Depletion of the nuclear Fermi sea, *Phys. Rev. C* **79**, pp. 064308.

Rosswog, S. (2015). The multi-messenger picture of compact binary mergers, *Int. J. Mod. Phys. D* **24**, pp. 1530012.

Russotto, P. *et al.* (2010). Symmetry energy from elliptic flow in ^{197}Au + ^{197}Au, *Phys. Lett. B* **697**, pp. 471.

Russotto, P. *et al.* (2016). Results of the ASY-EOS experiment at GSI: The symmetry energy at supra-saturation density, *Phys. Rev. C* **94**, pp. 034608.

Sargsian, M. M., Abrahamyan, T. V., Strikman, M. I., and Frankfurt, L. L. (2005). Exclusive electrodisintegration of ^{3}He at high Q^2. Decay function formalism, *Phys. Rev. C* **71**, pp. 044615.

Schiavilla, R., Wiringa, R. B., Pieper, S. C., and Carlson, J. (2007). Tensor forces and the ground-state structure of nuclei, *Phys. Rev. Lett.* **98**, pp. 132501.

Sh. Egiyan, K. *et al.* (2003). Observation of nuclear scaling in the A (e, e') reaction at x_B $>$, 1, *Phys. Rev. C* **68**, pp. 014313.

Shane, R. *et al.* (2015). SπRIT: A time-projection chamber for symmetry-energy studies, *Nucl. Instrum. Methods Phys. Res. A* **784**, pp. 513.

Shi, L. and Danielewicz, P. (2000). Effect of cluster formation on isospin asymmetry in the liquid-gas phase transition region, *Euro. Phys. Lett.* **49**, pp. 34.

Shneor, R. *et al.* (2007). Investigation of proton–proton short-range correlations via the 12C (e,e'p p) reaction, *Phys. Rev. Lett.* **99**, pp. 072501.

Stöcker, H. and Greiner, W. (1986). High energy heavy ion collisions–probing the equation of state of highly excited hardronic matter, *Phys. Rep.* **137**, pp. 277.

Steiner, A. W. *et al.* (2005). Isospin asymmetry in nuclei and neutron stars, *Phys. Rep.* **411**, pp. 325.

Subedi, R. *et al.* (2008). Probing cold dense nuclear matter (Hall A. Collaboration), *Science* **320**, pp. 1476.

Sumiyoshi *et al.* (1995). *Astrophys.* **303**, pp. 475.

Sumiyoshi, K. and Toki, H. (1994). *Astrophys.* **422**, pp. 700.

Symmetry Energy Project, `https://groups.nscl.msu.edu/hira/sepweb/pages/home.html`.

Tews, I., Lattimer, J. M., Ohnishi, A., and Kolomeitsev, E. E. (2017). Symmetry parameter constraints from a lower bound on neutron-matter energy, *Astrophys. J.* **848**, pp. 105.

The 2015 U.S. Long Range Plan for Nuclear Science, Reaching for the Horizon, `https://science.energy.gov/~/media/np/nsac/pdf/2015LRP/2015_LRPNS_091815.pdf`

The Nuclear Physics European Collaboration Committee (NuPECC) Long Range Plan 2017, Perspectives in Nuclear Physics, `http://www.esf.org/fileadmin/user_upload/esf/Nupecc-LRP2017.pdf`.

Tsang, C. Y., Tsang, M. B., Danielewicz, P., Lynch, W. G., and Fattoyev, F. J. (2019). Towards a better understanding of the symmetry energy within neutron stars, arXiv:1901.07673.

Tshoo, K. *et al.* (2013). Experimental systems overview of the Rare Isotope Science Project in Korea, *Nucl. Instrum. Methods Phys. Res., Sect. B* **317**, pp. 242.

Vidaña, I., Constança P., Polls, A., and Rios, A. (2009). Density dependence of the nuclear symmetry energy: A microscopic perspective, *Phys. Rev. C* **80**, pp. 045806.

Wanajo, S. *et al.* (2014). Production of all the r-process nuclides in the dynamical ejecta of neutron star mergers, *Astrophysics* **789**, pp. 39.

Wang, Y. J. *et al.* (2014). Constraining the high-density nuclear symmetry energy with the transverse-momentum-dependent elliptic flow, *Phys. Rev. C* **89**, pp. 044603.

Xia, Y. H., Xu, C., and Zong, H. S. (2016). A new approach for calculating nuclear symmetry energy, arXiv:1608.01724.

Xiao, Z. G., Li, B. A., Chen, L. W., Yong, G. C., and Zhang, M. (2009). Circumstantial evidence for a soft nuclear symmetry energy at supra-saturation densities, *Phys. Rev. Lett.* **102**, pp. 062502.

Xie, W. J., Su, J., Zhu, L., and Zhang, F. S. (2013). Symmetry energy and pion production in the Boltzmann–Langevin approach, *Phys. Lett. B* **718**, pp. 1510.

Xu, C. and Li, B. A. (2011). `http://arxiv.org/abs/1104.2075`, arXiv:1104.2075.

Xu, J., Chen, L. W., Ko, C. M., Li, B. A., and Ma, Y. G. (2013). Energy dependence of pion in-medium effects on the pion ratio in heavy-ion collisions, *Phys. Rev. C* **87**, pp. 067601.

Yasushige, Y. (2007). The RIKEN RI beam factory project: A status report, *Nucl. Instrum. Methods Phys. Res. B* **261**, pp. 1009.

Yin, P., Li, J. Y., Wang, P., and Zuo, W. (2013). Three-body force effect on nucleon momentum distributions in asymmetric nuclear matter within the framework of the extended Brueckner–Hartree–Fock approach, *Phys. Rev. C* **87**, pp. 014314.

Yong, G. C. (2016). Cross-checking the symmetry energy at high densities, *Phys. Rev. C* **93**, pp. 044610.

Yong, G. C. (2017). Constraining nucleon high momentum in nuclei, *Phys. Lett. B* **765**, pp. 104.

Yong, G. C. (2018). Blind spots of probing the high-density symmetry energy in heavy-ion collisions, *Phys. Lett. B* **786**, pp. 422.

Yong, G. C. (2018). Probing proton transition momentum in neutron-rich matter, *Phys. Lett. B* **776**, pp. 447.

Yong, G. C. and Li, B. A. (2013). Effects of nuclear symmetry energy on meson production and its rare decay to the dark U-boson in heavy-ion reactions, *Phys. Lett. B* **723**, pp. 388.

Yong, G. C. and Li, B. A. (2017). Interplay of short-range correlations and nuclear symmetry energy in hard-photon production from heavy-ion reactions at Fermi energies, *Phys. Rev. C* **96**, pp. 064614.

Yong, G. C., Gao, Y., Wei, G. F., Guo, Y. F., and Zuo, W. (2019). Determination of the density region of the symmetry energy probed by the π^-/π^+ ratio arXiv:1704.05166.

Yong, G. C., Li, B. A., and Chen, L. W. (2007). The neutron/proton ratio of squeezed-out nucleons and the high density behavior of the nuclear symmetry energy, *Phys. Lett. B* **650**, pp. 344.

Yong, G. C., Li, B. A., and Chen, L. W. (2008). Neutron–proton bremsstrahlung from intermediate energy heavy-ion reactions as a probe of the nuclear symmetry energy? *Phys. Lett. B* **661**, pp. 82.

Yong, G. C., Li, B. A., Chen, L. W., and Zhang, X. C. (2009). Triton–^{3}He relative and differential flows as probes of the nuclear symmetry energy at supra-saturation densities, *Phys. Rev. C* **80**, pp. 044608.

Yong, G. C., Li, B. A., Chen, L. W., and Zuo, W. (2006). Single and double ratios in heavy-ion reactions as probes of the high-density behavior of the nuclear symmetry energy, *Phys. Rev. C* **73**, pp. 034603.

Yong, G. C., Zuo, W., and Zhang, X. C. (2011). Subthreshold pion production within a transport description of central Au + Au collisions, *Phys. Lett. B* **705**, pp. 240.

Yong, G.C. (2017). Modeling pion production in heavy-ion collisions at intermediate energies, *Phys. Rev. C* **96**, pp. 044605.

Zhang, F. and Yong, G. C. (2016). Effects of the high-momentum tail of nucleon momentum distribution on the isospin-sensitive observables, *Eur. Phys. J. A* **52**, pp. 350.

Zhang, N. B., Cai, B. J., Li, B. A., and Newton, W. G. (2017). How tightly is the nuclear symmetry energy constrained by a unitary Fermi gas? *J. Xu. Nucl. Sci. Tech.* **28**, pp. 181.

Zhang, N. B. and Li, B. A. (2019). Extracting nuclear symmetry energies at high densities from observations of neutron stars and gravitational waves, *Eur. Phys. J. A* **55**, pp. 39.

Chapter 9

How Isospin Effects Influence Transverse In-Plane Flow and Its Disappearance?

Sakshi Gautam

Department of Physics, Panjab University, Chandigarh 160014, India

sakshigautam@pu.ac.in

9.1 Introduction

The pioneering experiments at the University of Manchester between 1908 and 1913 marked the beginning of *Nuclear Physics*. Since then, we have witnessed tremendous growth in the field and our understanding of nucleus and its properties, nature of interactions between nucleons, etc. has become far clear. These continuous efforts still leave a few questions unanswered. How nuclear matter behaves at extreme thermodynamical conditions of high temperatures and pressures? How to describe various cosmological phenomena of the formation of neutron stars, mechanism of supernova explosions, etc.? In order to answer these questions, heavy-ion collisions (HIC) at intermediate energies are of great importance as one can study nuclear equation of state and related properties using these as potential candidates in terrestrial laboratories. In the recent past, the nuclear community has witnessed plenty of research on various

Multifragmentation in Heavy-Ion Reactions: Theory and Experiments
Edited by Rajeev K. Puri, Yu-Gang Ma, and Arun Sharma

ISBN 978-981-4968-69-0 (Hardcover), 978-1-003-38513-4 (eBook)
www.jennystanford.com

phenomenon observed in these HIC, such as nuclear fragmentation, collective flow, nuclear stopping, particle production, and so on. The field has been quite active for a while now, and our understanding of the dynamics of such collisions gets updated every now and then. With this in mind, we will focus ourselves on the studies of collective flow and its dynamics and will present some recent developments in the field of HIC established using collective flow as an observable.

During the initial evolution of the reaction, compression of colliding nuclei leads to the creation of a supra-saturation density region. At the later stages of expansion, this pressure gradient influences the velocity of nucleons and they tend to escape the higher-pressure region. This pressure-dependent correlation between the positions and momenta of nucleons constitutes flow. There are various types of collective flow, namely, radial flow, directed flow, elliptic flow, triangular flow, and higher-order anisotropic flows. The transverse in-plane flow (or directed flow) is a preferential emission of nucleons within, and to a particular side of the reaction plane [Schied (1974)]. Note that the reaction plane is defined by the beam axis and a line joining the centers of the two colliding nuclei. The transverse in-plane flow has been reported to be highly sensitive to the nuclear equation of state, in-medium effective interactions [Westfall (1993), Klakow (1993), Insolia (1994)] as well as various entrance channel parameters such as the total colliding mass [Ogilvie (1989), Andronic (2003)], centrality of a reaction [Pan (1993), Bertsch (1987)], beam energy, etc. [Zhang (1990), B. Hong (2000)]. Figure 9.1 illustrates the preferential emission of the nucleons to the positive scattering angles in the reaction plane giving rise to a positive directed flow.

The strength of the directed flow is measured by the slope of the transverse momentum of particles as a function of the rapidity, i.e.,

$$F = \frac{d < p_x/A >}{dY}|_{Y_{\text{c.m.}}}, \tag{9.1}$$

where $Y_{\text{c.m.}}$ is the rapidity in the center-of-mass frame.

A more integrated quantity called *directed transverse momentum* $< p_x^{\text{dir}} >$ is also proposed to calculate the transverse momentum and

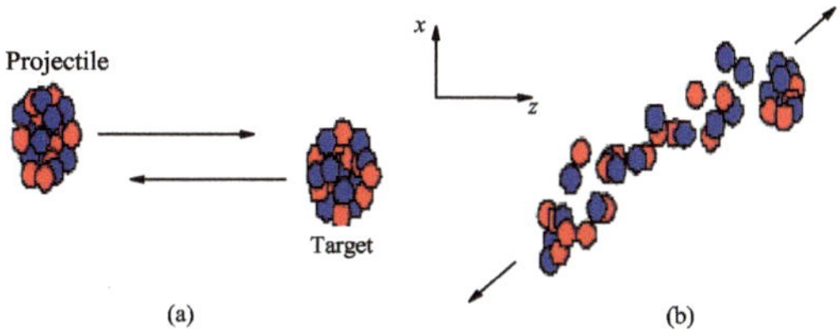

Figure 9.1 A schematic representation of the positive directed flow in the reaction plane. Two nuclei incident, collide and nucleons scattered to positive scattering angles because of the repulsive interactions, causing the flow to become positive. The figure is taken from Ref. [Majestro (2000)].

is defined as [Lehmann (1996), Sood (2006)]

$$<p_x^{\mathrm{dir}}> = \frac{1}{A}\sum_{i=1}^{A} sign\{\mathrm{Y}(i)\}p_x(i), \tag{9.2}$$

where $\mathrm{Y}(i)$ and p_x (i) are, respectively, the rapidity and the momentum of the ith particle.

The dependence of the transverse in-plane flow on the above mentioned parameters has revealed much interesting physics, especially the beam energy dependence, which has also led to its disappearance. The dominance of an attractive mean field at lower beam energies prompts the scattering of particles into negative deflection angles resulting in a negative flow. However, frequent nucleon–nucleon collisions and repulsive mean field dominate the dynamics at higher beam energies and thus results in the emission of particles into positive deflection angles yielding positive flow. While going through the incident energies, transverse in-plane flow disappears at a particular incident energy termed the *balance energy* (E_{bal}) or *energy of vanishing flow* (EVF) [Krofcheck (1989)]. The EVF has been studied experimentally as well as theoretically over a wide range of mass, varying from ^{12}C+^{12}C to ^{238}U+^{238}U and found to be strongly dependent on the total colliding mass [Mota (1992), Magestro (2000)] and centrality of the reaction [Magestro (2000), Lukasik (2005)].

At the same time, with the advancements of upcoming and existing radioactive-ion beam facilities around the world, the interest of the community has been shifted in studying the collisions of neutron- or proton-rich nuclei. Such studies shed light on the role of the isospin degree of freedom on the reaction dynamics. It is worth mentioning that the isospin dependence of in-medium effective interactions comes via both, nuclear equation of state and scattering cross section. The equation of state of isospin asymmetric nuclear matter has profound implications in governing the structure and evolution of many astrophysical objects such as neutron stars, supernova explosions, etc. In order to gain a deeper insight of the isospin dependence of cross sections and equation of state, both of which are unknown, it is desirable to describe the observables of HIC sensitive to both. In this direction, transverse in-plane flow and its disappearance have been reported to be highly sensitive to isospin effects. The isospin effects on the transverse flow and EVF were first reported by Li *et al.* [Li (1996)]. Later on, the measurements of flow and EVF for the reactions of ^{58}Ni+^{58}Ni and ^{58}Fe+^{58}Fe were studied at the MSU by Pak *et al.* [Pak (1997)], and they also demonstrated experimentally the isospin effects throughout the range of colliding geometries. The theoretical calculations using the isospin-dependent Boltzmann–Uehling–Uhlenbeck (IBUU) model were also confronted with measurements. The calculations under-predicted the experimentally measured EVF. Chen *et al.* [Chen (1998)] studied isospin effects on the transverse flow and EVF for the same colliding pairs using the isospin-dependent quantum molecular dynamics (IQMD) model, which is an improved version of the original QMD model [Aichelin (1991)]. Their calculations were also different from the data at all colliding geometries. The authors put forward the reason for the large deviation to be a low saturation density in the initialized nuclei (about 0.12 fm^{-3}) compared to the normal saturation density of 0.16 fm^{-3} and to the fact that the mean field due to the isospin-independent part of EOS would be more attractive at low density. However, it has been shown in Ref. [Khoa (1992)] that the mean field potential is rather the same both at ρ/ρ_0 = 1 and 0.75 for the equations of state used in Ref. [Chen (1998)]. Only at values larger than the normal nuclear matter density ρ_0, the mean field potential begins to differ. Moreover in Ref. [Hartnack (1998)] also, it has been shown that significant differences in the transverse flow values due to the different initial densities occur only at high incident energies. These differences vanish in the EVF domain. Another study by Scalone *et al.* [Scalone (1999)] also demonstrates the isospin effects on the transverse flow. Their calculations indicate toward different neutron and proton flows (see [Baran (2005), Li (2008)] also). Their results of EVF were in good agreement with the data at $\hat{b}$ = 0.45. It is worth mentioning that the isospin dependence of transverse flow has been explained as the competition among various reaction mechanisms, such as nucleon–nucleon collisions, symmetry energy, surface property of colliding nuclei, and Coulomb force. However, the relative importance among these mechanisms is not yet clear.

In addition, the investigation of system size effects on various phenomenon observed in HIC has also attracted a lot of attention. For example, in the low-energy regime where phenomena like fusion, fission, cluster radioactivity, as well as formation of super heavy nuclei, etc. [Puri (1992)] take place, one has tried to scale the fusion barrier heights as well as positions in terms of the masses of colliding nuclei [Jiang (2006)]. Similarly, the system size effects have been reported in other phenomena like collective flow, particle production, multifragmentation, density, temperature, and so on. For instance, in Ref. [Sturm (2001)], the power law scaling $\propto A_{tot}^{\tau}$ (where A_{tot} is the sum of projectile and target mass) of pion/kaon production with mass of the colliding system has been reported. A similar power law behavior for the system size dependence has been reported for fragment yields as well. Additionally, the transverse in-plane flow and its disappearance (at EVF) has also been found to be strongly dependent on the total colliding mass. Westfall *et al.* [Westfall (1993), Klakow (1993)] presented the experimental data for the EVF for the collisions of $^{12}C+^{12}C$, $^{20}Ne+^{27}Al$, $^{40}Ar+^{41}Sc$, $^{86}Kr+^{93}Nb$ and showed that the energy of vanishing flow scaled as $A_{tot}^{-1/3}$. This $A_{tot}^{-1/3}$ dependence was interpreted as a result of the competition between mean field and the nucleon–nucleon scattering. Comparison of their measurements with BUU calculations using soft and hard equations of state yields power law parameters to be 0.32 and 0.28, respectively. A year later, Zhou *et al.* [Zhou (1994)] studied the role of the momentum dependence of mean field on EVF and its mass dependence. Their study indicated a higher role of momentum-dependent interactions in the EVF of lighter systems compared to heavier ones. Nevertheless, the authors argued that the effect of nucleon–nucleon collisions is predominant, and the scaling value ($A_{tot}^{-1/3}$) of mass dependence was preserved, suggesting that MDI effect (in lighter systems) was balanced by Coulomb force (in heavier systems). Later on, Sood and Puri studied the mass dependence of the EVF for nearly symmetric systems from ^{20}Ne+^{27}Al to ^{197}Au+^{197}Au using the QMD model and found power law parameter to be close to 0.4 [Sood (2006)]. Also, when the study was restricted to heavier systems only, $1/\sqrt{A_{tot}}$ dependence of EVF is observed, which was also close to the experimental value of 0.53 [Sood (2004)] of the power law parameter. Thus, we see that in all studies mentioned above, a power law dependence of the EVF with the system mass has been reported. Thus, we see that system size plays an important role in the dynamics of heavy-ion collisions.

Therefore, from the above literature survey, we observed that none of the theoretical approaches could successfully reproduce the measured EVFs for the reactions of ^{58}Ni+^{58}Ni and ^{58}Fe+^{58}Fe. At the same time, the relative role of various factors governing the isospin effects is not clear. With this in mind, here in this chapter, we shall attempt:

(1) To see whether we can reproduce all the measured EVFs for the reactions of ^{58}Ni+^{58}Ni and ^{58}Fe+^{58}Fe (used to demonstrate isospin effect by [Pak (1997)] and [Chen (1998)]) using isospin-dependent quantum molecular dynamics model;

(2) To explain in part why the calculations of Chen *et al.* [Chen (1998)] using the IQMD model show a large deviation from the measured EVF at all colliding geometries;

(3) To study the role of isospin degree of freedom on the EVF throughout the mass and colliding geometry range and further understanding various reaction mechanisms responsible for isospin effects. For the present study, we use the IQMD model, a brief review of which is presented below and the details can be found in [Hartnack (1998)].

9.2 The Model

The IQMD model is an extension of the QMD model [Aichelin (1991)], which treats different charge states of nucleons, deltas, and pions explicitly, as inherited from the Vlasov–Uehling–Uhlenbeck (VUU) model. The IQMD model has been used successfully for the analysis of a large number of observables from low to relativistic energies. The isospin degree of freedom enters into the calculations via symmetry potential, cross sections, and Coulomb interaction.

In this model, baryons are represented by Gaussian-shaped density distributions

$$f_i(\vec{r},\vec{p},t) = \frac{1}{\pi^2\hbar^2}\exp(-[\vec{r}-\vec{r}_i(t)]^2\frac{1}{2L}) \times \exp(-[\vec{p}-\vec{p}_i(t)]^2\frac{2L}{\hbar^2}) \tag{9.3}$$

Nucleons are initialized in a sphere with radius R = 1.12 $A^{1/3}$ fm, in accordance with the liquid-drop model. Each nucleon occupies a volume of h^3, so that phase space is uniformly filled. The initial momenta are randomly chosen between 0 and Fermi-momentum ($\vec{p}_F$). The nucleons of the target and projectile interact by two- and three-body Skyrme forces, Yukawa potential, Coulomb interactions, and momentum- dependent interactions. A symmetry potential between protons and neutrons corresponding to the Bethe–Weizsacker mass formula has been included. The hadrons propagate using the Hamilton equations of motion:

$$\frac{d\vec{r}_i}{dt} = \frac{d\langle H\rangle}{d\vec{p}_i}; \qquad \frac{d\vec{p}_i}{dt} = -\frac{d\langle H\rangle}{d\vec{r}_i} \tag{9.4}$$

with

$$\langle H\rangle = \langle T\rangle + \langle V\rangle = \sum_i \frac{p_i^2}{2m_i} + \sum_i \sum_{j>i} \int f_i(\vec{r},\vec{p},t)V^{ij}(\vec{r}\,',\vec{r}) \times f_j(\vec{r}\,',\vec{p}\,',t)d\vec{r}\,d\vec{r}\,'\,d\vec{p}\,d\vec{p}\,'. \tag{9.5}$$

The baryon potential V^{ij}, in the above relation, reads as

$$\begin{aligned} V^{ij}(\vec{r}\,' - \vec{r}) &= V^{ij}_{Skyrme} + V^{ij}_{Yukawa} + V^{ij}_{Coul} + V^{ij}_{mdi} + V^{ij}_{sym} \\ &= [t_1\delta(\vec{r}\,' - \vec{r}) + t_2\delta(\vec{r}\,' - \vec{r})\rho^{\gamma-1}(\frac{\vec{r}\,' + \vec{r}}{2})] \\ &\quad + t_3\frac{\exp(|(\vec{r}\,' - \vec{r})|/\mu)}{(|(\vec{r}\,' - \vec{r})|/\mu)} + \frac{Z_iZ_je^2}{|(\vec{r}\,' - \vec{r})|} \\ &\quad + t_4\ln^2[t_5(\vec{p}\,' - \vec{p})^2 + 1]\delta(\vec{r}\,' - \vec{r}) \\ &\quad + t_6\frac{1}{\varrho_0}T_{3i}T_{3j}\delta(\vec{r_i}\,' - \vec{r_j}). \end{aligned} \tag{9.6}$$

Here t_6 = 4C with C = 32 MeV and Z_i and Z_j denote the charges of the ith and j^{th} baryon, and T_{3i} and T_{3j} are their respective T_3 components (i.e., 1/2 for protons and $-1/2$ for neutrons). The parameters μ and $t_1,\ldots,t_5$ are adjusted to the real part of the nucleonic optical potential. For the density dependence of the nucleon optical potential, standard Skyrme-type parametrization is employed. The momentum dependence V^{ij}_{mdi} of the nucleon–nucleon interactions, which may optionally be used in IQMD, is fitted to the experimental data in the real part of the nucleon optical potential. We also use the standard energy-dependent free nucleon–nucleon cross section σ^{free}_{nn} as well as the cross section reduced by 20%, i.e., σ = 0.8 σ^{free}_{nn}. The details about the elastic and inelastic cross sections for proton–proton and proton–neutron collisions can be found in [Hartnack (1998), Cugnon (1981)]. The cross sections for neutron–neutron collisions are assumed to be equal to the proton–proton cross sections. Two particles collide if their minimum distance d fulfills

$$d \leq d_0 = \sqrt{\frac{\sigma_{tot}}{\pi}}, \sigma_{tot} = \sigma(\sqrt{s}, type), \tag{9.7}$$

where "type" denotes the ingoing collision partners (N-N...). Explicit Pauli blocking is also included; i.e., Pauli blocking of the neutrons and protons is treated separately. We assume that each nucleon occupies a sphere in the coordinate and momentum space. This trick yields the same Pauli blocking ratio as an exact calculation of the overlap of the Gaussians will yield. We calculate the fractions P_1 and P_2 of the final phase space for each of the two scattering partners that are already occupied by other nucleons with the same isospin as that of the scattered ones. The collision is blocked with the probability

$$P_{block} = 1 - [1 - min(P_1, 1)][1 - min(P_2, 1)], \qquad (9.8)$$

and, correspondingly, is allowed with the probability $1 - P_{block}$. For a nucleus in its ground state, we obtain an averaged blocking probability $\langle P_{block} \rangle$ = 0.96. Whenever an attempted collision is blocked, the scattering partners maintain the original momenta prior to scattering.

9.3 Results and Discussion

For the first part of the present analysis, we simulated several thousands of events for the reactions of ^{58}Ni+^{58}Ni and ^{58}Fe+^{58}Fe in the incident energy range from 50 to 150 MeV/nucleon. The impact parameters are guided by [Pak (1997)]. We also use isospin-dependent nucleon–nucleon scattering cross section (labeled as σ_{NN} and is same as $\sigma(\sqrt{s})$) as well as σ_{NN} reduced by 20%. We use a soft EOS along with momentum-dependent interactions labeled as SMD. The reactions are followed till the transverse flow saturates. The saturation time is around 150 fm/c for the reactions in the present study.

9.3.1 Time Evolution of Directed Transverse Flow

As discussed in the introduction, there are several methods reported in the literature to define the transverse in-plane flow. In most of the studies, the EVF is extracted from $< p_x/A >$ plots, where one plots $< p_x/A >$ as a function of reduced rapidity $Y_{c.m.}/Y_{beam}$. Using a linear fit to the slope, one can find the so-called reduced flow "F." Naturally, the energy at which the reduced flow passes through zero is called the EVF. Alternatively, one can also use a more integrated quantity "*directed transverse momentum* $< p_x^{dir} >$." However, $< p_x^{dir} >$ presents an easier way to measure the in-plane flow rather than complicated functions such as $< p_x/A >$ plots. It has been shown by [Sood (2006)]

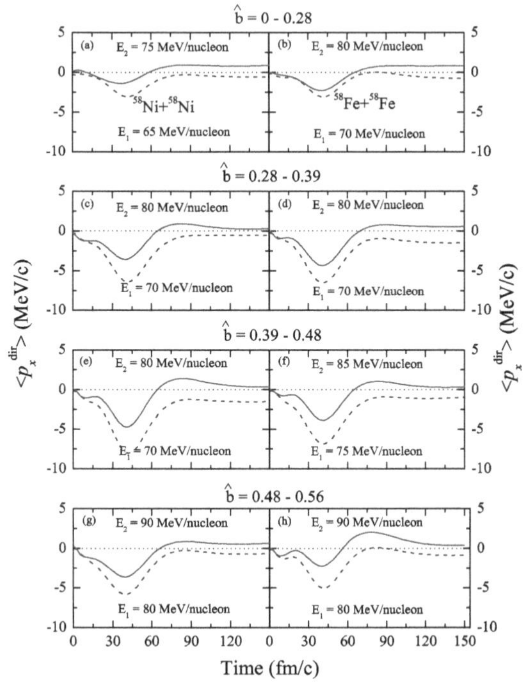

Figure 9.2 The time evolution of $<p_x^{\mathrm{dir}}>$ for the reactions of ^{58}Ni+^{58}Ni (left panels) and ^{58}Fe+^{58}Fe (right) at different colliding geometries. The shown results are for the incident energies around the energy of vanishing flow. Various lines are explained in the text.

that the disappearance of flow occurs at the same incident energy in both the cases showing the equivalence between $<p_x/A>$ plots and $<p_x^{\mathrm{dir}}>$ as far as the EVF is concerned.

In Fig. 9.2, we display the time evolution of the $<p_x^{\mathrm{dir}}>$ for ^{58}Ni+^{58}Ni (left panels) and ^{58}Fe+^{58}Fe (right panels) at two incident energies between which the $<p_x^{\mathrm{dir}}>$ changes sign at all the colliding geometries $\hat{b}$ (where $\hat{b}$ = b/b$_{max}$, b$_{max}$ = R$_P$ (projectile radius)+R$_T$ (target radius)) = 0–0.28 (a, b), 0.28–0.39 (c, d), 0.39–0.48 (e, f), and 0.48–0.56 (g, h), which are guided by [Pak (1997)]. The solid (dashed) lines represent the $<p_x^{\mathrm{dir}}>$ at higher (lower) incident energies. From the figure, we see that the $<p_x^{\mathrm{dir}}>$ is negative at the start of the reaction because of the dominance of the attractive mean field interactions during the initial phase of the reaction. These interactions remain attractive or may turn repulsive depending on the incident energies. The $<p_x^{\rm dir}>$ remains negative for the lower incident energies but turns positive at higher beam energies. We also see that the $<p_x^{\rm dir}>$ is more for the neutron-poor colliding pair of ^{58}Ni+^{58}Ni compared to the neutron-rich ^{58}Fe+^{58}Fe pair at all colliding geometries. This is because of the fact that (1) enhanced Coulomb repulsion in ^{58}Ni+^{58}Ni will throw the nucleons to positive scattering angles resulting in higher flow and (2) the cross section for neutron–proton collisions is three times as that of proton–proton (neutron–neutron) collisions [Hartnack (1998)] at the incident energies around the energies of vanishing flow. Therefore, there will be less binary collisions in the neutron-rich colliding pair, again resulting to reduced directed flow in ^{58}Fe+^{58}Fe. We also see that $<p_x^{\rm dir}>$ decreases when we move from the central to peripheral collisions. This is because of the reduction in the binary nucleon–nucleon collisions at peripheral geometries that yield lesser flow.

9.3.2 Energy of the Vanishing Flow as a Function of Impact Parameter

The $<p_x^{\rm dir}>$ changes sign between the two incident energies as displayed in Fig. 9.2. The EVF is calculated by a straight line fitting of the $<p_x^{\rm dir}>$ between these incident energies. In Fig. 9.3a, we display the EVF as a function of the reduced impact parameter $\hat{b}$ for the reactions of ^{58}Ni+^{58}Ni (solid symbols) and ^{58}Fe+^{58}Fe (open symbols). The stars in the figure represent the experimental data of Pak *et al.* [Pak (1997)], whereas diamonds correspond to our theoretical calculations. The experimental results and our theoretical calculations are plotted at the upper limit of each impact parameter bin. The squares (circles) represent the IQMD (IBUU) calculations of Ref. [Chen (1998)] ([Pak (1997)]). The pentagons represent the theoretical calculations of Ref. [Scalone (1999)] using the Boltzmann–Nordheim–Vlasov (BNV) model for $\hat{b}$ = 0.45. The lines are only to guide the eye. Our results of the EVF and experimental data for the reactions of ^{58}Fe+^{58}Fe have been slightly

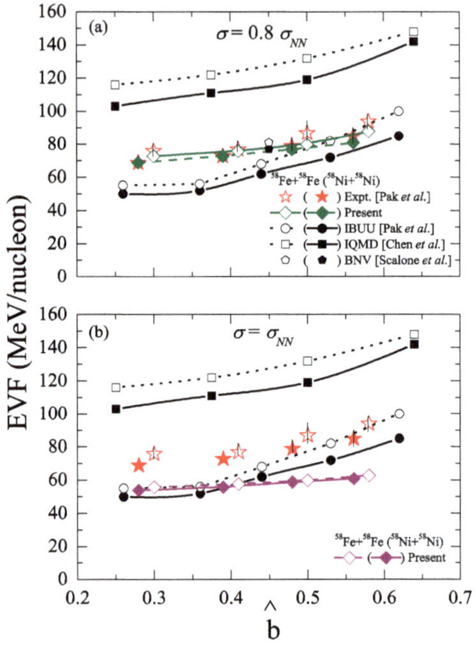

Figure 9.3 (a) The EVF as a function of reduced impact parameter with reduced nucleon–nucleon cross section. (b) Same as (a), but with full nucleon–nucleon cross section. Various symbols are explained in the text. Lines are only to guide the eye. Figure (a) extracted from S. Gautam, R. Chugh, A. D. Sood, R. K. Puri, C. Hartnack, and J. Aichelin, J. Phys. G: Nucl. Part. Phys. 37, 085102 (2010); doi: 10.1088/0954-3899/37/8/085102 @IOP Publishing. Reproduced with permission. All rights reserved.

offset in the horizontal direction for clarity. The vertical lines on the data points represent statistical errors. The statistical error bars on the theoretical points of [Pak (1997)] and [Chen (1998)] are not displayed, again for clarity. For the calculations of EVF, we use the standard energy-dependent *NN* cross section reduced by 20%. It is worth mentioning here that the choice of the reduced cross section has also been motivated by many previous studies [Andronic (2003), Zhang (1990), Daffin (1990)]. Daffin and Bauer [Daffin (1990)] have suggested the factor of 0.2–0.3 for the density-dependent reduction in the in-medium cross section. Their theoretical results (not shown here) were much closer to the data when using the in-medium reduction in the scattering cross section. From the figure, we see that our calculations (with reduced nucleon–nucleon cross section) [Gautam (2010)] are in good agreement with the data and the calculations of Scalone *et al.* [Scalone (1999)]. The results of IBUU calculations by [Pak (1997)] under-predict the data, whereas the IQMD calculations of [Chen (1998)] over-predict the data consistently.

To see the effect of the reduction in nucleon–nucleon cross section, we also perform the calculations with full cross section. The results are displayed in Fig. 9.3b. From the figure, we see that the EVF reduces by about 30% at all colliding geometries for both colliding partners. Thus, we can conclude that in-medium effects in lowering the nucleon–nucleon cross section play an important role in reaction dynamics and can have drastic effect on the observables.

9.3.3 Percentage Difference of the Energy of Vanishing Flow

In Fig. 9.4, we show the percentage deviation $\triangle$EVF (%) of the calculated EVF over experimental data with $\triangle \text{EVF}\ (\%) = \frac{\text{EVF}^{theo} - \text{EVF}^{expt}}{\text{EVF}^{expt}} \times 100$. The symbols have the same meaning as in Fig. 9.2. For IBUU calculations, the percentage deviation for central collisions is about 28%, whereas it is about 10% for the peripheral collisions. The average deviation is about 19% over all colliding geometries. On the other hand, for the IQMD model calculations of [Chen (1998)], we see that the percentage deviation $\triangle$EVF (%) is about 54% at all geometries, i.e., the calculated EVFs are consistently higher compared to the data. Note that for the present calculations, the percentage deviation is only about 3%. So, for the first time, we are able to reproduce the measured EVF for ^{58}Ni+^{58}Ni and ^{58}Fe+^{58}Fe at all colliding geometries.

9.3.4 Energy of Vanishing Flow and Interaction Range

As mentioned by [Hartnack (1998)], the width of the Gaussian in IQMD depends on the nuclear mass so as to achieve the maximum

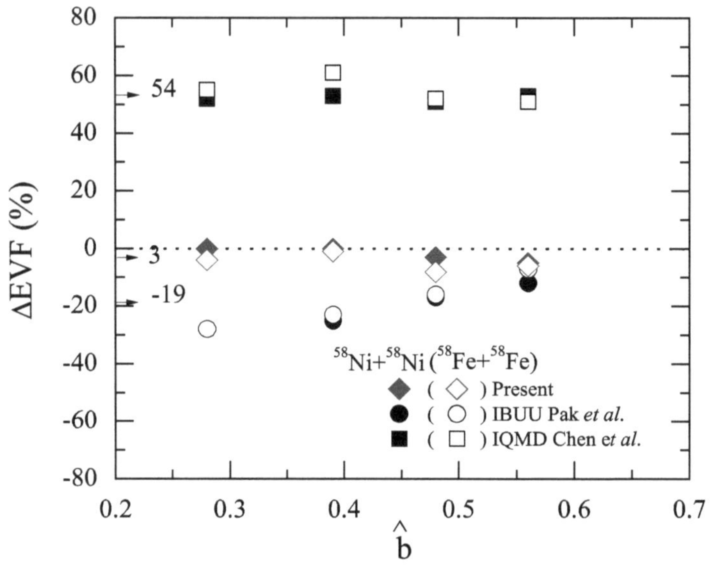

Figure 9.4 The percentage deviation of the EVF values for different calculations over the experimentally measured EVF as a function of the reduced impact parameter. Figure extracted from S. Gautam, R. Chugh, A. D. Sood, R. K. Puri, C. Hartnack, and J. Aichelin, J. Phys. G: Nucl. Part. Phys. 37, 085102 (2010); doi: 10.1088/0954-3899/37/8/085102 @IOP Publishing.

stability of a nucleus. It has also been shown by [Hartnack (1998), Westfall (1993), Klakow (1993)] that the collective flow and its disappearance depend strongly on the interaction range. The higher the interaction range, the smaller the collective flow (as higher interaction range will smear out the density profile and thus reduce the flow) and, therefore, the larger the corresponding EVF. In the IQMD calculations of [Chen (1998)], the interaction range was taken to be 2 fm^2, which in part could have led to reduced flow and corresponding enhancement in the EVF at all colliding geometries. For our calculations, we use the interaction range 60% of L_{Au} (where L_{Au} = 2.16 fm^2). As per Fig. 9.3, a good agreement with the experimental data was achieved at all colliding geometries. Note that the treatment of various potential terms such as Yukawa, Coulomb, and MDI is quite similar in our IQMD model and the IQMD model

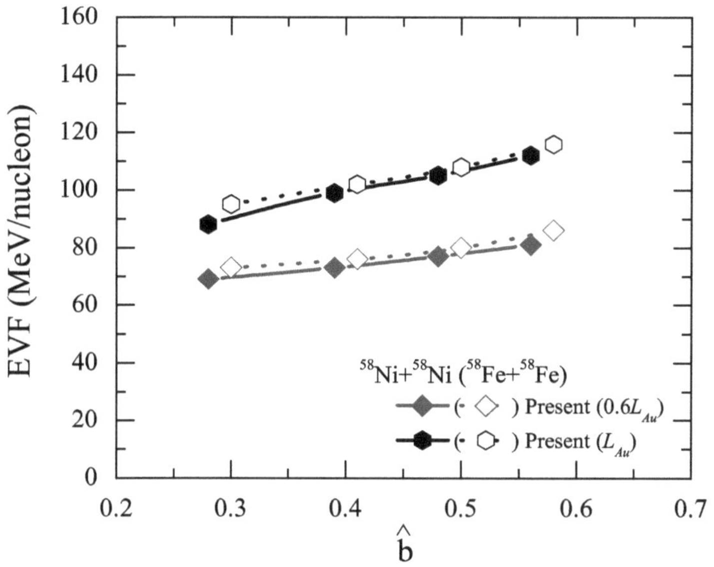

Figure 9.5 The EVF as a function of the reduced impact parameter for different Gaussian widths. The solid (open) symbols represent the reactions of ^{58}Ni+^{58}Ni (^{58}Fe+^{58}Fe). Hexagons (diamonds) correspond to L_{Au} ($0.6L_{Au}$). The lines are only to guide the eye. Figure extracted from S. Gautam, R. Chugh, A. D. Sood, R. K. Puri, C. Hartnack, and J. Aichelin, J. Phys. G: Nucl. Part. Phys. 37, 085102 (2010); doi: 10.1088/0954-3899/37/8/085102 @IOP Publishing. Reproduced with permission. All rights reserved.

of Chen *et al.* [Chen (1998)]. The range of Yukawa force, however, is different in both versions; the present range is 0.4 fm compared to 1.2 fm of Chen *et al.* However, it has been shown by [Hartnack (1998)] that the different ranges of Yukawa force does not alter the collective flow significantly. The treatment of the asymmetry term is also similar in both the models with the strength of the symmetry energy equal to 32 MeV. To pen down the cause of different energy of vanishing flow in both models, we performed calculations with different interaction ranges (i.e., full range, L_{Au}). A huge enhancement is found in the calculated EVF of ^{58}Ni+^{58}Ni (solid hexagons) and ^{58}Fe+^{58}Fe (open hexagons), thus making our calculations close to that of [Chen (1998)] as shown in Fig. 9.5. Thus, the EVF values for the two different choices of the Gaussian width are quite different.

This also indicates that the choice of Gaussian width affects the collective flow and EVF significantly, which is in agreement with [Westfall (1993), Klakow (1993), Hartnack (1998)].

For the second part of the analysis, in order to study the role of system size and colliding geometry on the isospin effects of EVF, we simulated the reactions of isotopic as well as isobaric colliding pairs.

9.3.5 Mass Dependence Analysis: Collisions of Isotopic Pairs

Further, we simulate the isotopic reactions of ^{26}Mg+^{26}Mg, ^{65}Zn+^{65}Zn, ^{91}Mo+^{91}Mo, ^{117}Xe+^{117}Xe, and ^{164}Os+^{164}Os having N/Z = 1.16 and reactions ^{28}Mg+^{28}Mg, ^{70}Zn+^{70}Zn, ^{98}Mo+^{98}Mo, ^{126}Xe+^{126}Xe, and ^{177}Os+^{177}Os having N/Z = 1.33, respectively, at various colliding geometries from central to peripheral collisions of b/b_{max}= 0.15–0.25, 0.35–0.45, 0.55–0.65, and 0.75–0.85. The N/Z for a given pair is varied by adding the neutron content only keeping the charge fixed, so that the effect of the Coulomb potential is same for a given mass pair. However, this will lead to the increase in mass for systems with higher neutron content. In Fig. 9.6, we display the EVF as a function of the total mass of the colliding system for two sets of isotopic systems with different neutron content. Closed (open) circles represent systems with less (more) neutron content. Lines are the power law fit $\propto A^{\tau}$. As expected, both isotopic series follow a power law behavior $\propto A^{\tau}$, at all colliding geometries. Note that the values of the power law parameter τ are -0.44 ± 0.01 for systems with less neutron content (labeled as $\tau^{a}_{1.16}$) and -0.42 ± 0.01 for systems with more neutron content (labeled as $\tau^{a}_{1.33}$) for semi-central collisions at 0.35–0.45. Note that these values of $\tau^{a}_{1.16}$ and $\tau^{a}_{1.33}$ are very close to the previous values of τ by [Magestro (2000)]. Interestingly, isospin effects are not visible for any mass system. As per the literature [Li (1996), Pak (1997)], the neutron-rich system has a higher EVF due to the isospin dependence of the scattering cross section. However, in our case, system size effects seem to dominate the isospin effects throughout the mass range as neutron-rich colliding pairs have lower EVF. Note that the Coulomb forces play the same role for a given isotopic pair.

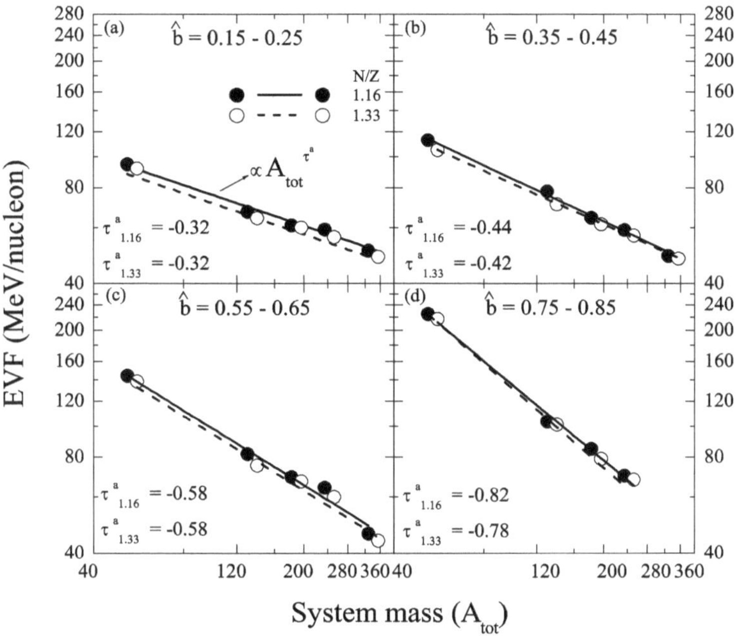

Figure 9.6 The EVF as a function of the combined reacting mass (A_{tot}) at different colliding geometries for isotopic pairs having N/Z ratio equal to 1.16 and 1.33. Various symbols are explained in the text. Lines are the power law fits $\propto A_{tot}^{\tau^a}$.

9.3.6 Collisions of Isobaric Pairs

Next, we carried out the study for isobaric pairs throughout the mass range. In particular, we simulate the reactions of $^{24}Mg+^{24}Mg$, $^{58}Cu+^{58}Cu$, $^{72}Kr+^{72}Kr$, $^{96}Cd+^{96}Cd$, $^{120}Nd+^{120}Nd$, and $^{135}Ho+^{135}Ho$, having N/Z = 1.0 and reactions $^{24}Ne+^{24}Ne$, $^{58}Cr+^{58}Cr$, $^{72}Zn+^{72}Zn$, $^{96}Zr+^{96}Zr$, $^{120}Sn+^{120}Sn$, and $^{135}Ba+^{135}Ba$ having N/Z = 1.4, respectively, in the whole range of colliding geometry as done for the isotopic pairs. Here N/Z is changed by keeping the total reacting mass fixed. First, we display the results for the semi-central collisions ($\hat{b}$ = 0.35–0.45). In Fig. 9.7a, we display the EVF as a function of the total system mass for two sets of isobars. The experimental data (stars) and our theoretical calculations (pentagons) (both displaced

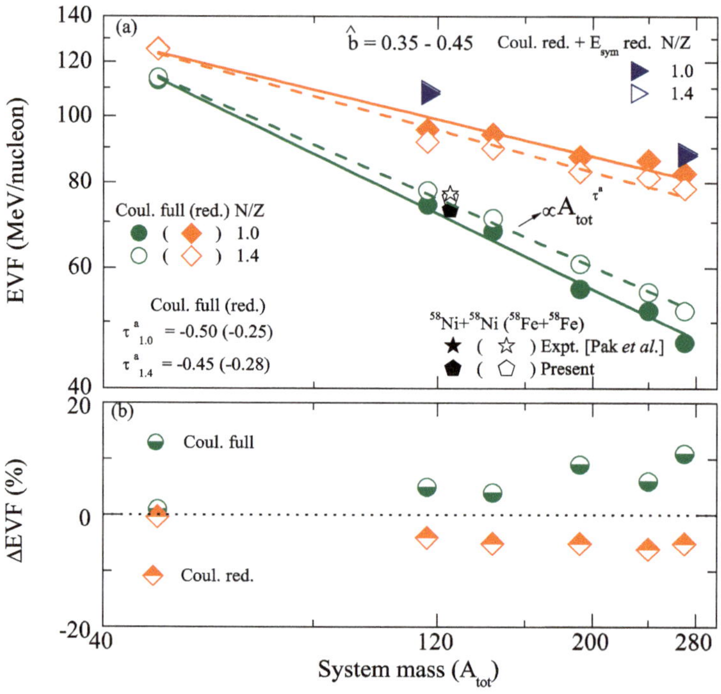

Figure 9.7 (a) The EVF as a function of combined reacting mass at $\hat{b}$ = 0.35–0.45. (b) The percentage difference $\triangle$EVF (%) as a function of the combined reacting mass. The solid (open) symbols are for N/Z = 1.0 (1.4). Lines are the power law fits $\propto A_{tot}^{\tau^a}$. Various symbols and lines are explained in the text. Reprinted with permission from Sakshi Gautam and Aman D. Sood, Physical Review C, 82, 014604, 2011. Copyright (2019) from the American Physical Society.

horizontally for clarity) for the reactions of ^{58}Ni+^{58}Ni (solid symbols) and ^{58}Fe+^{58}Fe (open) in the present range of colliding geometry ($\hat{b}$ = 0.35–0.45) is also displayed. The solid and open circles represent the EVF for systems with less and more neutron content, respectively. Lines are the power law fits $\propto A_{tot}^{\tau^a}$. Interestingly, throughout the mass range, the neutron-rich system has a higher EVF. The calculated EVF values fall on a line that is a fit of power law nature ($\propto A_{tot}^{\tau^a}$), where τ^a = -0.45 and -0.50 for N/Z = 1.4 and 1.0 (labeled by $\tau^a_{1.4}$ and $\tau^a_{1.0}$), respectively. The subscripts to the power law parameter represent the N/Z ratio. The different values of τ^a for two curves can be attributed to the larger role of the Coulomb force in the case of systems with more proton (less neutron) content. Our value of $\tau^a_{1.4}$ is equal/close to the value -0.46/-0.41 by [Magestro (2000)], both of which show deviation from the standard value $\simeq$ -1/3 [Mota (1992), Westfall (1993), Klakow (1993)] where analysis was done for the lighter mass nuclei only ($\leq$ 200). However, when heavier systems like ^{139}La+^{139}La and ^{197}Au+^{197}Au were included, τ^a increased to -0.45 [Magestro (2000)], suggesting the dominance of Coulomb repulsion. A further analysis by [Sood (2004)] showed that when only heavier nuclei were taken into account, the value of τ^a increased to -0.53, which is very close to our value of $\tau^a_{1.0}$ (-0.50) in the present case. Although the mass ranges in the present study and by [Sood (2004)] differ substantially, the present power law parameter $\tau^a_{1.0}$ is very close to the power law parameter as obtained in Ref. [Sood (2004)], which could be because of higher Coulomb repulsion for neutron-deficient systems. This indicates that the difference in the EVF for a given pair of isobaric systems may be dominantly due to the Coulomb potential, which is further supported by the fact that since both nuclear symmetry energy and nucleon–nucleon cross section add to the repulsive interactions, so both lead to the reduction in EVF. In systems with more neutron content, the role of symmetry energy could be larger (which will enhance the flow), whereas effects due to isospin-dependent cross section could play a dominant role in systems with less neutron content (more proton content). Therefore, there is a possibility that the combined effect of symmetry energy and cross section could be approximately the same for two isobaric systems with different neutron and proton content.

Further, to demonstrate the dominance of the Coulomb potential, the Coulomb potential is reduced by a factor of 100 and corresponding EVF values are displayed by solid and open (orange) diamonds representing systems with less and more neutron content, respectively, in Fig. 9.7a. From the figure, one can clearly notice the dominance of the Coulomb repulsion in both mass dependence and isospin effects. The values of $\tau^a_{1.4}$ and $\tau^a_{1.0}$ are now, respectively, -0.28 and -0.25. As expected, there is a large enhancement in the EVF for medium (e.g., ^{58}Cu+^{58}Cu) and heavy (^{120}Nd+^{120}Nd) mass systems, thus, reducing the value of both $\tau^a_{1.0}$ and $\tau^a_{1.4}$. The effect is small in lighter masses (^{24}Mg+^{24}Mg). One also observes that isospin effects on the EVF are reversed when we reduce the strength of the Coulomb interaction as neutron-rich systems have lower EVF. As in the absence of Coulomb interactions, repulsive symmetry energy will be more dominant in neutron-rich systems, which will in turn enhance flow and lead to less EVF for neutron-rich systems and hence an opposite trend for $\tau^a_{1.4}$ and $\tau^a_{1.0}$ for two different cases (Coulomb full and reduced). To further check this point, we reduce the strength of both symmetry energy and Coulomb potential and calculate EVF for isobaric pairs with combined mass of the system equal to 116 and 270 (shown by blue triangles in Fig. 9.7a). We now found the same EVF values for both the isobaric systems. This detailed analysis clearly points toward the dominance of Coulomb repulsion over symmetry energy for medium as well as heavy mass systems, whereas their impact is small in lighter masses. In Fig. 9.7b, we display the percentage difference $\triangle$EVF (%) between the systems of isobaric pairs as a function of the combined mass of system where $\triangle \text{EVF}\ (\%) = \frac{\text{EVF}^{1.4} - \text{EVF}^{1.0}}{\text{EVF}^{1.4}} \times 100$. The half-filled (green) circles are for full Coulomb and the half-filled (orange) diamonds are for reduced Coulomb. The negative (positive) values of $\triangle$EVF (%) shows that the EVF$^{1.0}$ is more (less) than EVF$^{1.4}$. From Fig. 9.7b (circles), we see that the percentage difference between the two masses of a given pair is larger for heavier masses compared to lighter ones. In lighter masses, the magnitude of the Coulomb repulsion is small, so there is small difference, whereas in the heavier masses, due to the large magnitude of Coulomb repulsion, there is a large difference in the EVF for a given pair of isobars. However, this trend is not visible when we reduce the Coulomb (diamonds). The values of $\triangle$EVF (%) are almost constant for medium and heavy masses.

9.3.7 Impact Parameter Dependence of Isospin Effects in Isobaric Pairs as an Example

The role of isospin degree of freedom in EVF is found to be more pronounced at peripheral collisions for the reactions of ^{58}Ni+^{58}Ni and ^{58}Fe+^{58}Fe [Li (1996), Pak (1997), Gautam (2010)]. So, as the next step, we studied the system size dependence of the EVF for isobaric pairs (having N/Z = 1.0 and 1.4) for all impact parameter bins.

Figure 9.8 displays the mass dependence of the EVF for four impact parameter bins. The solid (open) green circles indicate EVF

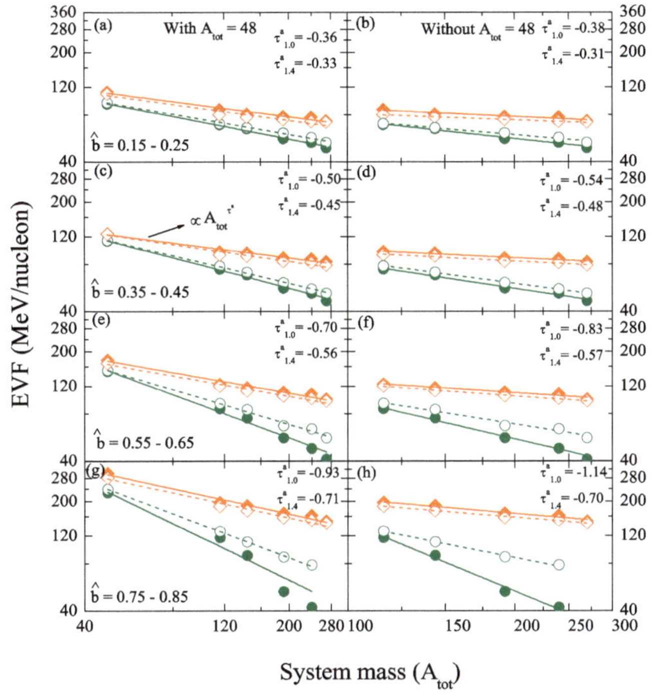

Figure 9.8 The left (right) panels display the EVF as a function of combined reacting mass at different colliding geometries with (without) A_{tot} = 48. The solid (open) symbols are for the systems having N/Z = 1.0 (1.4). The circles (diamonds) are for the calculations with full (reduced) Coulomb interactions. Lines are the power law fits $\propto A_{tot}^{\tau^a}$. Calculated τ^a values for full Coulomb interactions are displayed in the figure. The detailed values of τ^a are given in Tables 9.1 and 9.2. Reprinted with permission from Sakshi Gautam, Aman D. Sood, Rajeev K. Puri, and J. Aichelin, Physical Review C, 83, 014603, 2011. Copyright (2019) from the American Physical Society.

for systems with lower (higher) neutron content. Lines are the power law fits $\propto A_{tot}^{\tau^a}$. Left (right) panels are for the mass dependence of EVF when we include (exclude) A_{tot} = 48. First, we discuss the left panels for all the four bins. The energy of vanishing flow follows a power law behavior $\propto A_{tot}^{\tau^a}$ for both N/Z = 1.0 and 1.4 (τ^a being labeled as $\tau^a_{1.0}$ and

Table 9.1 The values of τ_1^a and $\tau_{1.4}^a$ for the whole range of colliding geometry for calculations with Coulomb potential with and without A_{tot} = 48. Reprinted with permission from Sakshi Gautam, Aman D. Sood, Rajeev K. Puri, and J. Aichelin, Physical Review C, 83, 014603, 2011]. Copyright (2019) from the American Physical Society.

$\hat{b}$	$\tau_{1.0}^a$ With A_{tot}=48	$\tau_{1.0}^a$ Without A_{tot}=48	$\tau_{1.4}^a$ With A_{tot}=48	$\tau_{1.4}^a$ Without A_{tot}=48
0.15–0.25	-0.36	-0.38	-0.33	-0.31
0.35–0.45	-0.50	-0.54	-0.45	-0.48
0.55–0.65	-0.70	-0.83	-0.56	-0.57
0.75–0.85	-0.93	-1.14	-0.70	-0.70

$\tau_{1.4}^a$ for systems having N/Z = 1.0 and 1.4, respectively) at all colliding geometries. The values of τ^a are listed in Table 9.1.

There are small deviations from the power law behavior for heavy mass systems with N/Z = 1.0 at peripheral colliding geometry. Isospin effects are clearly visible for all four impact parameter bins as neutron-rich system has higher EVF throughout the mass range in agreement with the previous studies [Li (1996), Pak (1997), Gautam (2010)]. The magnitude of the isospin effects increases with an increase in the mass of the system at all colliding geometries. The effect is much more pronounced at larger colliding geometries. One can see that the difference between $\tau_{1.0}^a$ and $\tau_{1.4}^a$ increases as we move from central to peripheral collisions (green circles). To demonstrate the role of Coulomb potential with impact parameter, we again reduce the Coulomb potential by a factor of 100 (as done earlier for $\hat{b}$ = 0.35–0.45) and calculate the EVF for all systems at all colliding geometries. The solid (open) diamonds represent EVF calculated with reduced Coulomb for systems with lower (higher) neutron content. Lines are power law fits $\propto A_{tot}^{\tau^a}$. The values of τ^a are listed in Table 9.2.

From the figure, we observe the following:

(1) The magnitude of isospin effects (difference in EVF for a given pair) is now nearly the same throughout the mass range indicating that the effect of symmetry energy is uniform throughout the mass range for all colliding geometries. This is supported by [Sood (2004)] where Sood and Puri studied the average density as a function of

Table 9.2 Same as Table 9.1 but for calculations with Coulomb potential reduced by a factor of 100. Reprinted with permission from Sakshi Gautam, Aman D. Sood, Rajeev K. Puri, and J. Aichelin, Physical Review C, 83, 014603, 2011. Copyright (2019) from the American Physical Society.

	$\tau^a_{1.0}$		$\tau^a_{1.4}$	
$\hat{b}$	With A_{tot}=48	Without A_{tot}=48	With A_{tot}=48	Without A_{tot}=48
0.15–0.25	-0.25	-0.17	-0.26	-0.14
0.35–0.45	-0.25	-0.17	-0.28	-0.19
0.55–0.65	-0.33	-0.24	-0.33	-0.26
0.75–0.85	-0.41	-0.31	-0.41	-0.29

the mass of the system at energies equal to EVF for each given mass. There they found that although both EVF and average density follow a power law behavior $\propto A_{tot}^{\tau^a}$, EVF decreases more sharply with the combined mass of the system (with τ^a = -0.42), whereas the average density (calculated at an incident energy equal to EVF) is almost independent of the mass of the system with τ^a = -0.05. It is worth mentioning here that the trend will be different at fixed incident energy, in which case density increases with increase in the mass of the system [Ogilvie (1989), Khoa (1992)]. We also note that the power law with reduced Coulomb interaction is now absolute for N/Z = 1.0 (solid diamonds in Fig. 9.8g). This indicates that the deviations from the power law behavior for heavy mass systems (with N/Z = 1.0) are due to the dominance of the Coulomb repulsion.

(2) One can also see that the enhancement in EVF (by reducing Coulomb) is more in heavier systems compared to lighter systems at all colliding geometries. The effect is more pronounced for peripheral collisions.

(3) With reduced Coulomb calculations, neutron-rich systems have lower EVF compared to neutron-deficient systems, which is a reverse of the default behavior. This (as explained earlier also) is due to the dominance of repulsive symmetry energy in neutron-rich systems. Similar conclusions can also be drawn qualitatively when we exclude the lighter mass of A_{tot} = 48 (Fig. 9.8 right panels), except that now the values of τ^a are different.

9.3.8 Role of Coulomb Interaction

The above detailed analysis pointed toward the dominance of Coulomb potential in isospin effects on the EVF. To further strength our point, in Figs. 9.9a–c, we display the EVF as a function of the reduced impact parameter $\hat{b}$ for masses 116, 192, and 240, respectively, using both the full and reduced Coulomb strengths. Symbols have the same meaning as in Fig. 9.8. For full Coulomb strength (green circles), for all the masses at all colliding geometries, a system with higher N/Z ratio has a larger EVF in agreement with previous studies [Klakow (1993), Magestro (2000), Li (1996), Pak (1997)]. Moreover, the difference between EVF for a given mass pair increases with an increase in the colliding geometry. This is more clearly visible in heavier masses. Also for N/Z = 1.4, the EVF increases with an increase in the impact parameter in agreement with [Magestro (2000)]. This is due to the decreased participant zone in peripheral collisions that decreases the amount of repulsive nucleon–nucleon collisions and, therefore, higher energy is required to overcome the attractive nuclear force. This effect is less pronounced in heavier systems since even at peripheral geometry, a significant number of nucleon–nucleon collisions will occur. Moreover, the effect of a stronger Coulomb repulsion in heavier systems will increase with colliding geometry since it will push a greater number of nucleons in the transverse direction away from the participant zone. However, for N/Z = 1.0, an increase in the EVF with the impact parameter is true only for a lighter mass system, such as A_{tot} = 116. For heavier masses, the EVF, in fact, begins to decrease with an increase in the impact parameter in contrast to the previous studies [Magestro (2000), Li (1996), Pak (1997), Gautam (2010)].

On the other hand, the calculations with reduced Coulomb potential depicts:

(1) Neutron-rich systems have a decreased EVF compared to neutron-deficient systems for all centralities. This clearly shows the dominance of the Coulomb repulsion over symmetry energy in isospin effects throughout the mass range at all colliding geometries.

(2) The difference between the EVF for systems with different N/Z ratios remains almost constant as a function of the colliding

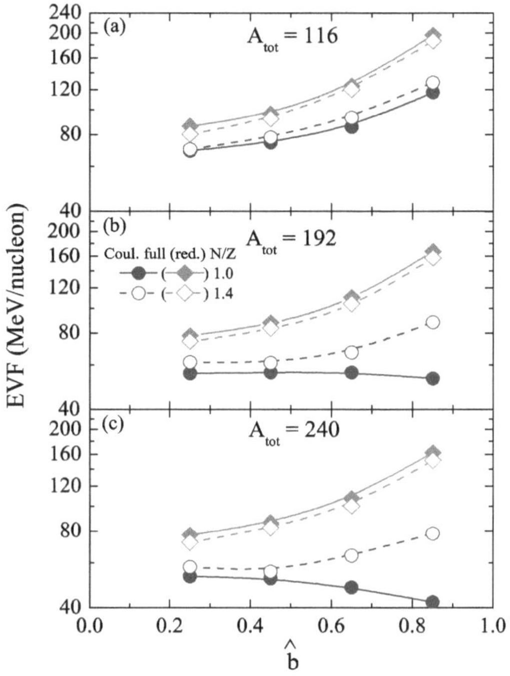

Figure 9.9 The EVF as a function of the impact parameter for different system masses with A_{tot} = 116 (top panel), 192 (middle), and 240 (bottom). Various symbols have the same meaning as used in Fig. 9.8. Lines are only to guide the eye. Reprinted with permission from Sakshi Gautam, Aman D. Sood, Rajeev K. Puri and J. Aichelin, Physical Review C, 83, 014603, 2011, Copyright (2019) from the American Physical Society.

geometry indicating that the effect of symmetry energy is uniform throughout the range of $\hat{b}$ as well.

(3) In heavier systems, at peripheral colliding geometry, the increase in the EVF is more in systems with N/Z =1.0 compared to N/Z = 1.4, which shows the much dominant role of Coulomb repulsion at high colliding geometry [Gautam (2011)].

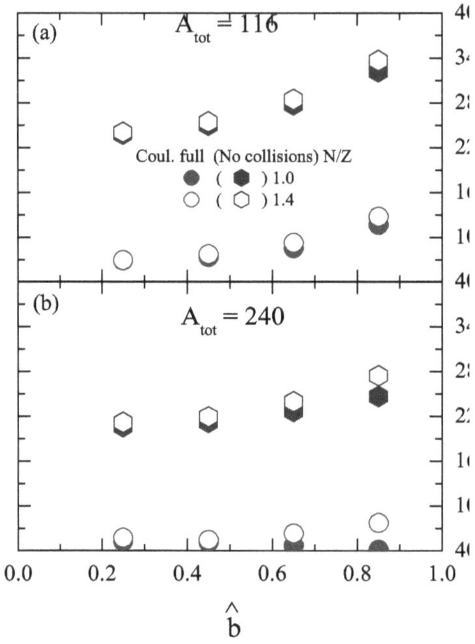

Figure 9.10 The EVF for the reactions having A_{tot} = 116 (upper panel) and 240 (lower panel) without nucleon–nucleon binary collisions (Hexagons). Reprinted with permission from Sakshi Gautam, Aman D. Sood, Rajeev K. Puri, and J. Aichelin, Physical Review C, 83, 014603, 2011, Copyright (2019) from the American Physical Society.

9.3.9 Relative Role of Coulomb Potential and Nucleon–Nucleon Cross Section

Next, we also see the relative contribution of Coulomb repulsion and nucleon–nucleon cross section in lighter and heavier systems. For this, we switch off the collision term and calculate the EVF for A_{tot} = 116 and 240 at all the four impact parameter bins. The results are displayed in Fig. 9.10. The hexagons represent the calculations without collisions. The other symbols have the same meaning as in Fig. 9.8. We find that at a given impact parameter, the EVF increases by a large magnitude for both systems, which shows the importance of collisions. The increase is of the same order for both the masses at central as well as peripheral collisions, indicating the same role of cross section for both lighter and heavier masses as we have expected in the discussion of Fig. 9.9. This is supported by [Ogilvie (1989), Sood (2004)], which shows that since the mean field is independent of the mass of the system, one needs the same amount of collisions to counter balance the mean field in both lighter and heavier masses. Moreover, the same order of increase in the EVF at central and peripheral colliding geometries (when we switch off the collisions) indicates the importance of collisions at peripheral colliding geometries as well. The effect that EVF decreases with an increase in the impact parameter (due to the dominance of Coulomb) for heavy mass systems with N/Z = 1.0 (Fig. 9.9, lower panel) does not appear here for lighter and heavier masses. Therefore, in Fig. 9.9, the reduced Coulomb strength allows one to examine the balance of nucleon–nucleon collisions and mean field, while in Fig. 9.10, the removal of nucleon–nucleon collisions allows one to examine the balance of Coulomb repulsion and mean field.

In Fig. 9.11, we display the percentage difference in the EVF with respect to that without Coulomb potential and without *NN* collisions. We calculate ΔEVF (%) in the EVF when we reduced the Coulomb potential ($\Delta\text{EVF}\ (\%) = \frac{\text{EVF}^{Coul.\ red.} - \text{EVF}}{\text{EVF}^{Coul.\ red.}} \times 100$) (up triangles) and without nucleon–nucleon collisions ($\Delta\text{EVF}\ (\%) = \frac{\text{EVF}^{No\ colls.} - \text{EVF}}{\text{EVF}^{No\ colls.}} \times$ 100 100) (down triangles) for two colliding pairs of masses A_{tot} = 116 (upper panel) and 240 (lower panel) at the four impact parameter bins. The solid (open) symbols represent the systems with less (more) neutron content. Superscripts represent the EVF with reduced Coulomb potential and without nucleon–nucleon collisions. From the figure, we find that the percentage change in the EVF is more when we exclude *NN* collisions from the calculations compared to when we reduce the Coulomb potential (compare down and up triangles) for both the systems at central collisions. This indicates the importance of the binary *NN* collisions compared to Coulomb repulsion at central colliding geometries. However, the percentage change is almost same for both A_{tot} = 116 & 240 colliding pairs. On the other hand, for peripheral collisions, the percentage change in the EVF with reduced Coulomb and without *NN* collisions is comparable for both masses, indicating the importance of both Coulomb potential

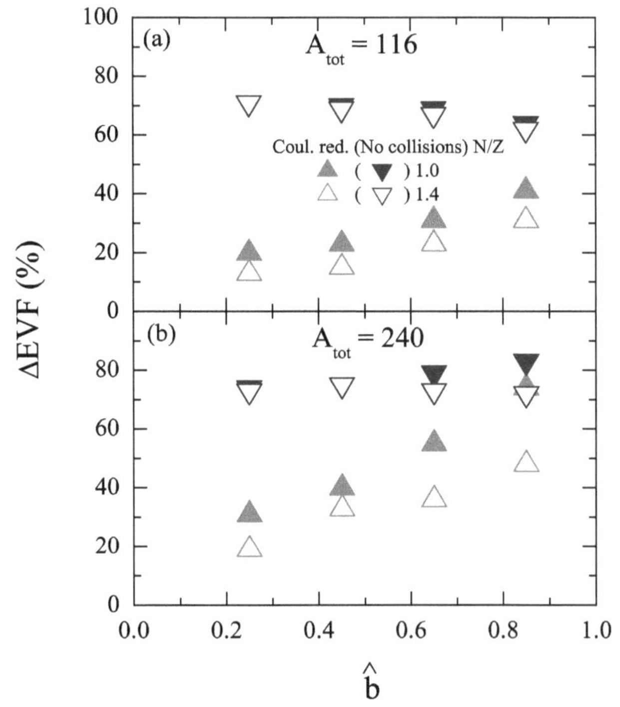

Figure 9.11 The percentage difference $\triangle$EVF (%) for reduced Coulomb interactions and without nucleon–nucleon binary collisions for the reactions having A_{tot} = 116 (upper panel) and 240 (lower). Various symbols are explained in the text.

and nucleon–nucleon collisions at peripheral collisions for heavier masses.

As we have seen in Fig. 9.10 that EVF is much higher than the actual EVF when we switch off the cross section, so to explore whether Coulomb repulsion affects the collisions in the EVF domain, we display in Fig. 9.12 the collision rate $< \frac{dN_{coll}}{dt} >$ for A_{tot} = 48 (upper panel), 116 (middle panel), and 240 (bottom panel) at an incident energy of 50 MeV/nucleon. The solid (dashed) lines represent Coulomb full (reduced) calculations. Higher (lower) peaks represent results for central (peripheral) impact parameter. We find that the Coulomb potential decreases the collision rate in

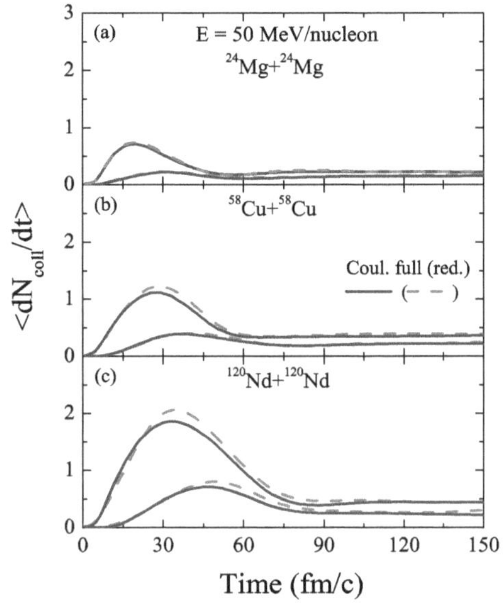

Figure 9.12 The time evolution of the collision rate ($<\frac{dN_{coll}}{dt}>$) for various reacting masses at an incident energy of 50 MeV/nucleon for central ($\hat{b}$ = 0.15–0.25) and peripheral ($\hat{b}$ = 0.75–0.85) colliding geometries. Various lines are explained in the text. Reprinted with permission from Sakshi Gautam, Aman D. Sood, Rajeev K. Puri, and J. Aichelin, Physical Review C, 83, 014603, 2011, Copyright (2019) from the American Physical Society.

medium and heavier mass systems for central collisions, whereas for peripheral collisions, the effect of Coulomb potential on collisions is significant only for the heavier masses (lower panel). Comparing top and bottom peaks within a panel, we see that there is still a significant number of collisions at peripheral colliding geometries.

9.4 Summary

We have studied the role of the isospin degree of freedom on the energy of vanishing flow using the reactions of ^{58}Ni+^{58}Ni and ^{58}Fe+^{58}Fe, for which the experimental data is also available. Our calculations have reproduced, for the first time, all the measured

EVF for these colliding pairs throughout the colliding geometry. Our study has also indicated that the EVF for the neutron-rich $^{58}Fe+^{58}Fe$ pair is higher than that for the proton-rich $^{58}Ni+^{58}Ni$ pair. We have also investigated the details of the under- and over-prediction of the EVF in earlier attempts by Pak *et al.* and Chen *et al.* using the IBUU and IQMD models, respectively. Our investigations revealed that the interaction range has a significant role to play in the energy of vanishing flow. Further, we extended the study of isospin effects on the mass dependence of EVF for isotopic as well as isobaric pairs. The study has indicated the dominance of system size effects over isospin effects on the energy of vanishing flow for isotopic colliding pairs throughout the range of colliding geometry. On the other hand, for isobaric pairs, we have found a relative competition between Coulomb and symmetry potential, where the dominance of Coulomb potential is observed in isospin effects on EVF. We also found that the role of the isospin degree of freedom in system size effects on the energy of vanishing flow is much more pronounced at peripheral collisions. The study has also indicated that the effect of the symmetry energy and nucleon–nucleon cross section is uniform throughout the mass range and the range of colliding geometry in isospin effects for isobaric colliding pairs. We have also presented the counterbalancing of nucleon–nucleon collisions and mean field by reducing the Coulomb potential and counterbalancing of Coulomb and mean field by removing the nucleon–nucleon collisions. Our study has also indicated the importance of nucleon–nucleon collisions over Coulomb repulsion in central colliding geometries, whereas showed the dominance of Coulomb potential for peripheral collisions.

Acknowledgment

The authors are thankful to Prof. Rajeev K. Puri for giving access to various computer programs used for the present study.

References

Aichelin, J. (1991). "Quantum" molecular dynamics: A dynamical microscopic n-body approach to investigate the fragment formation and the nuclear equation of state in heavy-ion collisions, *Phys. Rep.* **202**, pp. 233–360.

Andronic, A. *et al.* (2003). Directed flow in Au+Au, Xe+CsI, and Ni+Ni collisions and the nuclear equation of state, *Phys. Rev. C* **67**, pp. 034907.

Baran, V., Colonna, M., Greco, V., and Di Toro, M. (2005). Reaction dynamics with exotic nuclei, *Phys. Rep.* **410**, 335–466.

Bertsch, G. F., Lynch, W. G., and Tsang, M. B. (1987). Transverse momentum distributions in intermediate-energy heavy-ion collisions, *Phys. Lett. B* **189**, pp. 384–387.

Chen, L. W., Zhang, F. S., and Jin, G. M. (1998). Analysis of isospin dependence of nuclear collective flow in an isospin-dependent quantum molecular dynamics model, *Phys. Rev. C* **58**, pp. 2283–2291.

Cugnon, J., Mizutani, T., and Vandermeulen, J. (1981). Equilibration in nuclear collisions. A Monte Carlo calculation, *Nucl. Phys. A* **352**, pp. 505–534.

Daffin, F. and Bauer, W. (1998). Effects of isospin asymmetry and in-medium corrections on balance energy, arxiv: nucl-th/9809024v1.

de la Mota, V., Sebille, F., Farine, M., Remaud, B., and Schuck, P. (1992). Analysis of transverse momentum collective motion in heavy-ion collisions below 100 MeV/nucleon, *Phys. Rev. C* **46**, pp. 677–686.

Gautam, S. and Sood, A. D. (2010). Isospin effects on the mass dependence of the balance energy, *Phys. Rev. C* **82**, pp. 014604.

Gautam, S. *et al.* (2010). Isospin effects on the energy of vanishing flow in heavy-ion collisions, *J. Phys. G: Nucl. Part. Phys.* **37**, pp. 085102.

Gautam, S., Sood, A. D., Puri, R. K., and Aichelin, J. (2011). Isospin effects in the disappearance of flow as a function of colliding geometry, *Phys. Rev. C* **83**, pp. 014603.

Hartnack, C. *et al.* (1998). Modelling the many body dynamics of heavy-ion collisions: Present status and future perspective, *Eur. Phys. J. A* **1**, pp. 151–169.

Hong, B. *et al.* (2000). Proton and deutron rapidity distributions and nuclear stopping in ^{96}Ru (^{96}Zr)+^{96}Ru (^{96}Zr) collisions at 400 AMeV, *Phys. Rev. C* **66**, pp. 034901.

Insolia, A., Lombardo, U., Sandulescu, N. G., and Bonasera, A. (1994). Nuclear dynamics for heavy-ion collisions with a momentum dependent potential, *Phys. Lett. B* **334**, pp. 12–17.

Jiang, C. L., Back, B. B., Esbensen, H., Janssens, R. V. F., and Rehm, K. E. (2006). Systematics of heavy-ion fusion hindrance at extreme sub-barrier energies, *Phys. Rev. C* **73**, 014613.

Khoa, D. T. *et al.* (1991). Photon production in heavy-ion collisions and nuclear equation of state, *Nucl. Phys. A* **529**, pp. 363–386.

Klakow, D., Welke, G., and Bauer, W. (1993). Nuclear flow excitation function, *Phys. Rev. C* **48**, pp. 1982–1987.

Krofcheck, D. *et al.* (1989). Disappearance of flow in heavy-ion collisions, *Phys. Rev. Lett.* **63**, pp. 2028–2031.

Lehmann, E., Faessler, A., Zipprich, J., Puri, R. K., and Huang, S. W. (1996). Study of in-medium effects on the disappearance of the sidewards flow in heavy-ion collisions, *Z. Phys. A* **355**, pp. 55–60.

Li, B. A., Chen, L. W., and Ko, C. M. (2008). Recent progress and new challenges in isospin physics with heavy-ion collisions, *Phys. Rep.* **464**, pp. 113–281.

Li, B. A., Ren, Z., Ko, C. M., and Yennello, S. J. (1996). Isospin dependence of collective flow in heavy-ion collisions at intermediate energies, *Phys. Rev. Lett.* **76**, pp. 4492–4495.

Łukasik, J. *et al.* (2005). Directed and elliptic flow in ^{197}Au+^{197}Au at intermediate energies, *Phys. Lett. B* **608**, pp. 223–230.

Majestro, D. J. (2000). Ph.D. Thesis, Michigan State University, USA.

Majestro, D. J. *et al.* (2000). Disappearance of transverse flow in Au+Au collisions, *Phys. Rev. C* **61**, pp. 021602 (R).

Ogilvie, C. A. *et al.* (1989). Transverse collective motion in intermediate-energy heavy-ion collisions, *Phys. Rev. C* **40**, pp. 2592–2599.

Pak, R. *et al.* (1997). Isospin dependence of the balance energy, *Phys. Rev. Lett.* **78**, pp. 1026–1029.

Pan, Q. and Danielewicz, P. (1993). From sideward flow to nuclear compressibility, *Phys. Rev. Lett.* **70**, pp. 2062–2065.

Puri, R. K. and Gupta, R. K. (1992). Alpha cluster transfer process in colliding S-D shell nuclei using the energy density formalism, *J. Phys. G: Nucl. Part. Phys.* **18**, pp. 903–915.

Scalone, L., Colonna, M., and Di Toro, M. (1999). Transverse flows in charge symmetric collisions, *Phys. Lett. B* **461**, pp. 9–14.

Scheid, W., Müller, H., and Greiner, W. (1974). Nuclear shock waves in heavy-ion collisions, *Phys. Rev. Lett.* **32**, pp. 741–745.

Sood, A. D. and Puri, R. K. (2004). Mass dependence of disappearance of transverse in-plane flow, *Phys. Rev. C* **69**, pp. 054612.

Sood, A. D. and Puri, R. K. (2004). Nuclear dynamics at the balance energy, *Phys. Rev C* **70**, pp. 034611.

Sood, A. D. and Puri, R. K. (2006). Influence of momentum-dependent interactions on balance energy and mass dependence, *Eur. Phys. J. A* **30**, pp. 571–577.

Sood, A. D., Puri, R. K., and Aichelin, J. (2004). Study of balance energy in central collisions for heavier nuclei, *Phys. Lett. B* **594**, pp. 260–264.

Sturm, C. *et al.* (2001). Evidence for a soft nuclear equation of state from kaon production in heavy-ion collisions, *Phys. Rev. Lett.* **86**, pp. 39–42.

Westfall, G. D. *et al.* (1993). Mass dependence of disappearance of flow in nuclear collisions, *Phys. Rev. Lett.* **71**, pp. 1986–1989.

Bansal, R., Gautam, S., Puri, R. K., and Aichelin, J. (2013). Role of structural effects on the collective transverse flow and the energy of vanishing flow in nuclear collisions, *Phys. Rev. C* **87**, 061602 (R).

Zhang, W. M. *et al.* (1990). Onset of flow of charged fragments in Au+Au collisions, *Phys. Rev. C* **42**, R491–R494.

Zhou, H., Li, Z., and Zhuo, Y. (1994). Momentum-dependent Vlasov–Uehling–Uhlenbeck calculation of mass dependence of the flow disappearance in heavy-ion collisions, *Phys. Rev. C* **50**, R2664–R2667.

Chapter 10

Exploring the Role of Structure Effects on Nucleon–Nucleon Collisions at Intermediate Energy

Sakshi Gautam[a] and Rajni Bansal[b]

[a]*Department of Physics, Panjab University, Chandigarh 160014, India*

[b]*Faculty of Engineering and Construction, Pembrokeshire College, Merlins Bridge, Haverfordwest, Pembrokeshire, SA61 1SZ, Wales, U. K.*

sakshigautam@pu.ac.in

10.1 Introduction

One of the most fundamental properties of an atomic nucleus is its size (i.e., the radius). The knowledge of the size of the nucleus (nuclear structure and its matter distribution) is of great importance for the low energy phenomenon like fusion, fission, cluster radioactivity, and formation of superheavy nuclei. Even in the framework of proximity potential, a suitable choice of radius parametrization is essential to reproduce the experimental data on the fusion barrier nicely [Dutt (2010), Myers (2000)]. The study of the nuclear charge radii gives us the useful information about the saturation properties of nuclear forces and the Coulomb energy

Multifragmentation in Heavy-Ion Reactions: Theory and Experiments
Edited by Rajeev K. Puri, Yu-Gang Ma, and Arun Sharma

ISBN 978-981-4968-69-0 (Hardcover), 978-1-003-38513-4 (eBook)
www.jennystanford.com

of nuclei and, therefore, becomes important for the nuclear mass formulae too [Royer (2008)]. Moreover, neutron-skin thickness, a measure of difference between the root mean square (rms) radius of neutron and proton, has also emerged as a sensitive probe to the symmetry energy as well as to the equation of state of asymmetric nuclear matter [Brown (2000), Dieperink (2003), Tsang (2012), Dai (2015)]. Therefore, finding the accurate value of radius has been a subject of keen investigation. The rms charge radius ($R = < r^2 > ^{1/2}$) of a nucleus can be measured via various experimental techniques. The most important methods being used for the measurements of the rms radii of the nuclei are elastic electron scattering, muonic atom X-ray method, K_αX-ray isotope shift (KIS), and optical isotope shift (OIS). The first two methods have been used only for stable isotopes, whereas the last two methods have been employed to measure the difference between the radii of different isotopes of the same element.

Since there are four techniques used to determine the nuclear radius, a more comprehensive picture of nuclear charge distribution can be found by combining the optical isotope shift results with those from the X-ray line and for the transition from the muoic atom together with the results of elastic electron scattering from nuclei. In this direction, the experimental data from the elastic scattering, muonic atom X-rays, K_α isotope shifts, and optical isotopic shifts are taken into account by Angeli for the 799 ground-state nuclear radii [Angeli (2004)]. Moreover, the possible correction and constrains between the data were applied to make the data system more consistent. On the other hand, on the theoretical front, the nuclear radii are given by the traditional liquid-drop model having the form $R = R_0 A^{'1/3}$, resulted from the saturation properties of the nuclear forces. The value of R_0 is assumed to be somewhere near 1.45 fm in the early 19th century. From the experimental work conducted by various collaborations [Hofstadter (1956)], it was observed that this value of R_0 yield is a much larger value for the radii of heavier systems than observed experimentally. Also, the results from the muonic experiments indicated that the lighter elements possessed smaller radii deduced from the same formula but with R_0 =1.20. The discrepancy between the values for the nuclear radii derived from mirror nuclei and μ-mesonic atoms having R_0 =1.45 and 1.2 fm, respectively, was also investigated by [Jancovici (1954)]. Now a large number of experiments carried out on the electron scattering and muonic atom X-rays provided the information about the rms radius of the nucleus ranging from proton to the Uranium. With the availability of the large amount of data on nuclear radii, various authors came with different parametrization of the radius, which can explain the experimental data reasonably well. It has also been reported that the traditional liquid-drop formula ($R = R_0 A^{\frac{1}{3}}$) to calculate nuclear charge radius is never obeyed in the case of isotopes and isotones [Elton (1967)]. Therefore, different parametrizations of radius to reproduce the experimental data on the rms of nuclear charge and its isotopic shifts (taking care of neutron excess) have been proposed in the literature time to time. In the literature, many well-established potentials used the radius of the form $aA_i^{1/3} + bA_i^{-1/3}$ [Dutt (2010), Blocki *et al.* (1977), Bass (1977), Christensen (1976), Reisdorf (1994), Winther (1995), Ngô (1980)]. Different values of a and b were chosen to give different parameterizations. The proximity-based potential also added a constant term in the above formula [Dutt (2010)]. Many attempts reported in the literature tried to incorporate isospin effects in the radius and modified the parametrization as: $\acute{a}A_i^{1/3} + \acute{b}A_i^{-2/3} - \acute{c}I$ where $I = \frac{A_i - 2Z_i}{A_i}$, where $\acute{a}$, $\acute{b}$, and $\acute{c}$ are again some constants [Pomorska (1994), Denisov (2002), Royer (2009)]. The various forms of nuclear radius proposed by different authors are given as follows. For example, the nuclear radius given by Blocki [Blocki *et al.* (1977)] reads as:

$$R_{Blocki} = 1.28A_i^{\frac{1}{3}} - 0.76 + 0.8A_i^{-\frac{1}{3}}\ fm \quad (i = 1, 2). \tag{10.1}$$

The form of nuclear radius given by Bass [Bass (1977)] reads as:

$$R_{Bass} = 1.16A_i^{\frac{1}{3}} - 1.39A_i^{-\frac{1}{3}}\ fm \quad (i = 1, 2). \tag{10.2}$$

The empirical expression for radius given by Christensen and Winther [Christensen (1976)] reads as:

$$R_{CW} = 1.233A_i^{\frac{1}{3}} - 0.978A_i^{-\frac{1}{3}}\ fm \quad (i = 1, 2), \tag{10.3}$$

The form of nuclear radius used by Brogila and Winther [Reisdorf (1994)] reads as:

$$R_{BW} = 1.233A_i^{\frac{1}{3}} - 0.98A_i^{-\frac{1}{3}}\ fm \quad (i = 1, 2), \tag{10.4}$$

The form of nuclear radius given by Winther [Winther (1995)] reads as:

$$R_{AW} = 1.20A_i^{\frac{1}{3}} - 0.09\ fm \quad (i = 1, 2), \tag{10.5}$$

The form of nuclear radius given by Ngô [Ngô (1980)] reads as:

$$R_{Ngo} = \frac{NR_{ni} + ZR_{pi}}{A_i}\ fm \quad (i = 1, 2), \tag{10.6}$$

where the equivalent sharp radii for proton and neutron are given as:

$$R_{pi} = r_{0_{pi}}A_i^{\frac{1}{3}}; \ \text{with} \ r_{0_{pi}} = 1.128\ fm \quad (i = 1, 2), \tag{10.7}$$

$$R_{ni} = r_{0_{ni}}A_i^{\frac{1}{3}}; \ \text{with} \ r_{0_{ni}} = 1.1375 + 1.875 \times 10^{-4}A_i\ fm \quad (i = 1, 2), \tag{10.8}$$

The form of radius given by Nerlo-Pomorska and Pomorski [Pomorska (1994)] reads as:

$$R_{Pomorska} = 1.240A_i^{\frac{1}{3}}\left\{1 + \frac{1.646}{A_i} - 0.191\left(\frac{A_i - 2Z_i}{A_i}\right)\right\}\ fm \quad (i = 1, 2), \tag{10.9}$$

for even–even nuclei with $8 \leq Z \leq 38$, and for nuclei with $Z \geq 38$, the above equation is modified as:

$$R_{Pomorska} = 1.256A_i^{\frac{1}{3}}\left\{1 - 0.202\left(\frac{A_i - 2Z_i}{A_i}\right)\right\}\ fm \quad (i = 1, 2) \tag{10.10}$$

The effective sharp radius used by Denisov [Denisov (2002)] reads as:

$$R_{Denisov} = R_{ip}\left(1 - \frac{3.413817}{R_{ip}^2}\right) + 1.284589\left\{\left(\frac{A_i - 2Z_i}{A_i}\right) - \left(\frac{0.4A_i}{A_i + 200}\right)\right\} fm \quad (i = 1, 2), \tag{10.11}$$

where R_{ip} is given by Eq (10.7). Royer [Royer (2009)] proposed different expressions for the nuclear charge radius with $N, Z \geq 8$ by adjusting the experimental charge radii for 2027 masses in the forms already existing in the literature and the recent one is

$$R_{Royer} = 1.2332A_i^{\frac{1}{3}}\left\{1 + \frac{2.348443}{A_i} - 0.151541\left(\frac{A_i - 2Z_i}{A_i}\right)\right\} fm \quad (i = 1, 2) \tag{10.12}$$

From the above discussion, it is evident that there is a large variation in the value of radii on experimental as well as theoretical fronts. This variation in radius is also observed in various theoretical dynamical models, which have been advocated to pin down various phenomena at intermediate energies [Hartnack (1998), Yariv (1979), Cugnon (1980), Aichelin (1991), Bauer (1988), Feldmeier (1997), Ono (1992), Abdel-Waged (2004), Ohtsuka (1987)].

Also most of the simulation models at intermediate energies use the liquid-drop formula for calculating the radius, i.e., $R = R_0 A_i^{1/3}$ with R_0 = 1.12, 1.142, and 1.18 being used in Isospin-dependent Quantum Molecular Dynamics (IQMD) [Hartnack (1998)] and Monte Carlo (MC) method [Cugnon (1980)], Inter Nuclear Cascade (INC) [Yariv (1979)], as well as Quantum Molecular Dynamics (QMD)[Aichelin (1991)], respectively. At the same time, adjustments were made while using models for ^{12}C nucleus. For example, R_0 = 1.3 was taken in the Monte Carlo method while considering the case of ^{12}C [Cugnon (1980)]. Also, the rms values used for ^{12}C nucleus within Fermionic Molecular Dynamics (FMD) [Feldmeier (1997)], Antisymmetrized Molecular Dynamics (AMD) [Ono (1992)], and nucleus–nucleus optical potential (op. pot.) [Ohtsuka (1987)] are 2.79 fm, 2.49 fm, and 2.355 fm, respectively. It is interesting to note that these various models have taken the radii of colliding nuclei in a very *adhoc* manner assuming that this will not yield any effect on the dynamics at all.

Though good progress has been made on the dynamical part at intermediate energies, clearly no serious attempts are reported on the inputs that enter the calculations via structural effects. One such attempt has been done by Yong *et al.* [Yong (2011)] where initialization effects (by using different parameterizations of the Skyrme forces) within the framework of the Isospin-dependent Bolzmann–Uehling–Uhlenbeck (BUU) model on the symmetry-energy-sensitive observables like free neutron to proton ratio (n/p), π^+/π^- ratio, and neutron to proton differential flow, F_x^{n-p}, have been studied. These observables were found to be quite sensitive to the initialization effects entering the calculations through Skyrme forces. In another study by Fang *et al.* [Fang (2011)], the role of the neutron-skin thickness on the reaction cross section and neutron or proton removal cross section was investigated. Recently, Li *et al.* [Li (2013)] studied the finite size effects of nuclei via Gaussian wave packet and surface energy term on the formation and the stability of the fragments in heavy-ion collisions at low and intermediate energies. Their study revealed that the sensitivity of finite size effects decreases with an increase in the incident energy. Similarly, the Gaussian width of the nucleons is found to affect the production of light-charged particles significantly by [Kaur (20014)]. Earlier, Hartnack *et al.* [Hartnack (1998)] have shown the sensitivity of directed transverse flow for different initializations. From the above discussion, it is evident that initialization effects play an important role on the dynamics of a reaction at intermediate energies. Since a large number of radius parameterizations are present in the literature, here we aim to investigate the initialization effects via nuclear radii on the dynamics of reactions at intermediate energies. In particular, we shall discuss directed flow and its disappearance (i.e., the energy of vanishing flow (EVF)) as an example to study the structural effects on the nuclear dynamics at intermediate energies [Bansal (2013)]. In this chapter, we shall attempt:

(1) To solve the long-standing problem of reproducing the energy of vanishing flow of the reaction of ^{12}C+ ^{12}C using the proper choice of nuclear radius.

(2) To study the role of the isospin dependence of nuclear radius on isospin effects in collective flow and fragmentation.

10.2 Directed Transverse Flow and Energy of Vanishing Flow (EVF)

There are several methods used in the literature to define the directed transverse flow. In most of the studies, one uses $\langle p_x/A \rangle$ plots where one plots $\langle p_x/A \rangle$ as a function of Y_{red}. Using a linear fit to the slope in mid-rapidity region, one can define the so-called reduced flow. Here we shall use a more integrated quantity "*directed transverse momentum* $\langle p_x^{dir} \rangle$," which is defined as

$$\langle p_x^{dir} \rangle = \frac{1}{A} \sum_{i=1}^{A} sign\{Y(i)\} p_x(i), \tag{10.13}$$

where $Y(i)$ is the rapidity and $p_x(i)$ is the transverse momentum of ith particle and A is the combined mass of the system. The rapidity $Y(i)$ is given by

$$Y(i) = \frac{1}{2} \ln \frac{E(i) + p_z(i)}{E(i) - p_z(i)}, \tag{10.14}$$

where $E(i)$ and $p_z(i)$ are, respectively, the total energy and the longitudinal momentum of the ith particle. In this definition, all rapidity bins are taken into account. Therefore, $\langle p_x^{dir} \rangle$ presents an easier way of measuring the in-plane flow than complicated functions such as $\langle p_x/A \rangle$ plots. It is worth mentioning that both the methods yield similar results, as far as the energy of vanishing flow is concerned [Sood (2004), Sood and Puri (2004), Lehmann (1996)]. The beam energy dependence of the directed transverse flow leads to its disappearance at a particular energy termed as the energy of vanishing flow. The energy of vanishing flow is the result of the counterbalancing of the attractive mean field (which is dominant at low incident energies) and the repulsive nucleon–nucleon scattering, which decides the fate of the reaction at higher incident energies. The energy of vanishing flow (representing the vanishing of flow) is of great significance because the experimentally determined energy of vanishing flow can be easily compared with various theoretical calculations as it is free from any experimental uncertainties. Experimentally, EVF is available for more than 16 systems ranging from $^{12}C + {}^{12}C$ to $^{197}Au + {}^{197}Au$ [Bansal (2014)]. It is evident from the literature that many theoretical attempts involving different models were made robust against the experimental energies of vanishing flow [Magestro *et al.* (2000), Magestro (2000), Kumar (2010), Westfall (1993), Klakow (1993), Sood (2006), Mota (1992), Ono (1993), Ono (1995), Sood (2004), Sood and Puri (2004)]. Interestingly, in most of the cases [Sood (2004), Sood and Puri (2004), Magestro (2000), Magestro *et al.* (2000), Kumar (2010)], one avoided to compare the reaction of $^{12}C+^{12}C$. This was due to the fact that one has tough time in reproducing the EVF for the reaction of $^{12}C+^{12}C$. One needed a variety of assumptions to evaluate the EVF in the $^{12}C+^{12}C$ reaction [Westfall (1993), Klakow (1993), Sood (2006), Mota (1992), Ono (1993), Ono (1995)]. For example, Westfall *et al.* [Westfall (1993)] expressed the need for a density-dependent parametrization of cross section to reproduce the EVF for the low density matter. For the rest of the systems, a constant reduction in the cross section was enough to reproduce the experimental energy of vanishing flow. Another study by Klakow *et al.* [Klakow (1993)] has shown that the proper choice of the surface thickness is necessary for calculating the energy of vanishing flow of the $^{12}C+^{12}C$ system using the BUU model. Sood *et al.* [Sood (2006)] felt the need for momentum-dependent interactions within the QMD model to include $^{12}C+^{12}C$ in the mainstream. Mota *et al.* [Mota (1992)] used the Landau Vlasov formulation (embedded with proper surface effects) to calculate the EVF of the $^{12}C+^{12}C$ reaction. Similar assumptions were also made in the AMD model [Ono (1993), Ono (1995)]. Unfortunately, none of these studies paid any serious attention to look carefully at the radii of colliding nuclei, where a very casual approach was used. We shall, however, show that this parameter can play a huge role in the case of lighter nuclei where all the above models were reported to have problems in reproducing experimental values [Westfall (1993), Sood (2006), Mota (1992), Klakow (1993), Ono (1993), Ono (1995)].

In Table 10.1, we have displayed the value of radii for the nuclei of ^{12}C and ^{197}Au using various parameterizations of radius defined earlier. From the table, we see that the range of the radius of ^{12}C using different parameterizations (discussed earlier) lies between 2.05 fm (by Bass [Dutt (2010)]) and 2.66 fm (by AW) [Dutt (2010)]). Note that the experimental value by Elton reads 2.3 fm [Elton (1961)]. At the same time, the radius used in different transport models at intermediate energies ranges between 2.15 fm (UrQMD) and 2.98 fm

Table 10.1 The radii of ^{12}C and ^{197}Au nuclei using different parametrization forms.

Parametrization	Radius (fm) for ^{12}C nucleus	Radius (fm) for ^{197}Au nucleus
Bass	2.05	6.51
Elton	2.3	—
Denisov	—	6.53
BW	2.39	7.01
CW	2.40	7.01
Blocki	2.52	6.83
Ngo	2.60	6.73
AW	2.66	6.89
Pomorska	—	7.02
Royer	—	7.05

(by MC). The standard liquid-drop formula $\propto 1.12A^{1/3}$ leads to radius = 2.56 fm. In contrast, the variation in the radius for the ^{197}Au nucleus is far less, lying between 6.5 fm (by Bass [Dutt (2010)]) and 7.05 fm (by Royer [Royer (2009)]), whereas the liquid-drop model gives 6.52 fm. From the above discussion, it is clear that the ^{12}C system should be dealt carefully. Note that the effect of different parameterizations will be drastic in the lighter nuclei compared to the heavier nuclei where surface contribution is negligible. Therefore, in this chapter, we shall next show that small differences in the initial density profile (by taking different radii) lead to 40–50% difference in the EVF for small systems such as $^{12}C+^{12}C$, whereas for large systems like $^{197}Au+^{197}Au$, the EVF is rather insensitive to the initial density profile.

10.3 Results and Discussion

For the first part, we simulated thousands of events for the reactions of $^{12}C+^{12}C$ and $^{197}Au+^{197}Au$ at the impact parameters of b_{red} = 0.4 and b = 0–4 fm (as guided by the experimental findings [Westfall (1993), Lukasik (2005)]), respectively, at incident energies between 40 MeV/nucleon and 200 AMeV at a step of 10 AMeV. We use here a soft momentum-dependent equation of state labeled as SMD with 20% reduced nucleon–nucleon scattering cross section ($\sigma = 0.8\sigma_{free}$). For this analysis, different radii were implemented in the IQMD model and the energy of vanishing flow was calculated in each case. For the calculations of EVF, a straight-line fitting of the $\langle p_x^{dir} \rangle$ between two incident energies is used. In addition, for the next part of the analysis, we simulated reactions of isobaric pairs (having the same liquid-drop radius) of ^{60}Ca+^{60}Ca and ^{60}Zn+^{60}Zn at an incident energy of 100 MeV/nucleon using the soft equation of state throughout the range of colliding geometry. To study the effect of radius and its isospin dependence, we used, apart from the liquid-drop radius, the recent parameterized form of radius proposed by Royer and Rousseau in 2009.

10.3.1 Role of Nuclear Radius on the Directed Transverse Flow

In Fig. 14.1, we display the time evolution of the directed transverse flow $\langle p_x^{dir} \rangle$ calculated using different radius parameterizations for the reactions of ^{12}C+^{12}C (upper panel) and ^{197}Au+^{197}Au (lower panel) at their corresponding experimental energies of vanishing flow, which reads 122 AMeV [Westfall (1993)] and 54 AMeV, respectively [Lukasik (2005)]. The dotted, dash-dotted, dash-dot-dotted, short-dashed, short-dotted, dashed, short-dash-dotted, and solid lines represent the calculations using the radius parametrization by Bass, Elton, BW, CW, Blocki, AW, Ngô, and IQMD, respectively. From the figure, we notice that the directed transverse flow decreases as the size of ^{12}C increases. We also see that the $\langle p_x^{dir} \rangle$ is negative (due to mean field) during the initial phase of the reaction for all parameterizations. As the reaction proceeds, binary nucleon–nucleon collisions contribute to the positive directed flow and, therefore, the directed transverse flow changes. Though the flow decreases with an increase in the radius for ^{12}C+^{12}C, it remains nearly same for the reaction of ^{197}Au+^{197}Au. The decrease in the flow with an increase in the radius is due to the decrease in the density gradient of the nuclear matter. Moreover, in lighter systems, the ratio of the surface diffuseness to radius is more compared to the heavier systems. Therefore, we expect this behavior of change in the density gradient to be quite significant. Also, in lighter colliding nuclei, due to the decrease in the density gradient, repulsive forces

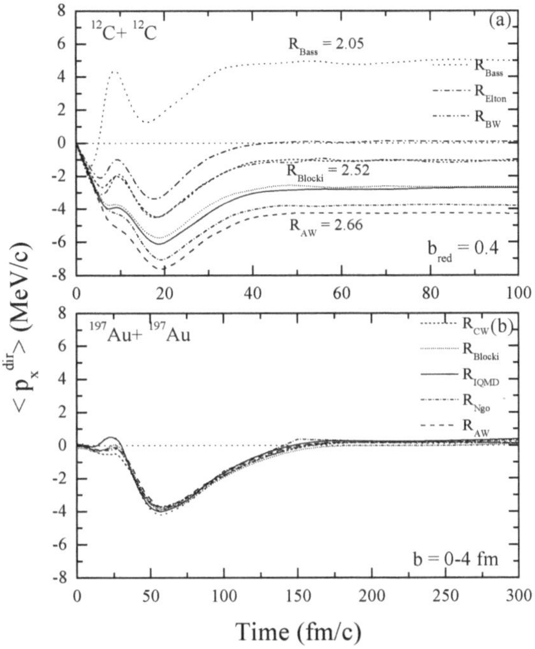

Figure 10.1 The time evolution of $\langle p_x^{dir} \rangle$ for the reactions of ^{12}C+^{12}C and ^{197}Au+^{197}Au using different parametrization of radius at their corresponding measured energies of vanishing flow. Various lines are explained in the text. Reprinted with permission from Rajni Bansal, Sakshi Gautam, Rajeev K. Puri, and J. Aichelin, Physical Review C, 87, 061602(R), 2013, Copyright (2019) from the American Physical Society.

($\propto (\frac{\rho}{\rho_0})^{\gamma}$) get weakened and decreases the momentum transfer in the transverse direction. Also, the number of the binary collisions decreases in lighter systems as the radius increases. Hence, all these factors collectively lead to the reduction in flow with an increase in radius for the reaction of ^{12}C+^{12}C.

10.3.2 The Energy of Vanishing Flow as a Function of Nucleus Radius

Further in Fig. 14.2(a), we display the energy of vanishing flow calculated by using different parameterizations of radius for the reaction of ^{12}C+^{12}C (upper panel). The solid squares represent the EVF calculated by varying the default radius systematically between

80% and 120%. This covers all the radius range (2.15 to 2.98 fm) used by all dynamics models at intermediate energies. The open circles represent the calculated energies of vanishing flow using different radius parameterizations and also include the measured one due to Elton [Elton (1961)] (as has been labeled). We see that the energy of vanishing flow increases with an increase in the radius. This is due to the decrease in the strength of repulsive forces with radius, thus diminishing the directed flow and, therefore, pushing the energy of vanishing flow toward the higher side. To see the effect of radius on a heavier system, we also display the energy of vanishing flow for the reaction of ^{197}Au+^{197}Au (lower panel) using different parameterizations of radius. In contrast to the ^{12}C case, we see that EVF remains almost constant, thus demonstrating an insignificant dependence of a heavier system on the radii of the colliding nuclei. The thick solid horizontal line represents the experimentally measured EVF for the reactions of ^{12}C+^{12}C and ^{197}Au+^{197}Au. From the figure, it is evident that the energy of vanishing flow is significantly affected by the radii of the colliding nuclei for the reaction of ^{12}C+^{12}C compared to that of ^{197}Au+^{197}Au. Different parameterizations yield energy of vanishing flow between 50–54 MeV/nucleon and 98–155 MeV/nucleon for the reactions of ^{197}Au+^{197}Au and ^{12}C+^{12}C, respectively. Note that an increase of 30% in radius results 58% change in the EVF for the ^{12}C+^{12}C system, whereas 8% change in the radius for the ^{197}Au+^{197}Au system results only 3% change in the energy of vanishing flow. It is evident that a lighter system like ^{12}C+^{12}C is very sensitive to the choice of radius, whereas the dynamics of the reactions involving heavy nuclei like ^{197}Au is less sensitive. From the above discussion, it is evident that the flow physics at intermediate energies is dominated entirely by the surface effect, whereas they are bulk dominated for heavy systems.

10.3.3 Percentage Deviation of the Energy of Vanishing Flow as a Function of Radius

In Fig. 14.3, we display the percentage deviation (ΔEVF (%)) of the calculated energies of vanishing flow from the experimental value as a function of radius for the reaction of ^{12}C+^{12}C (upper panel) and

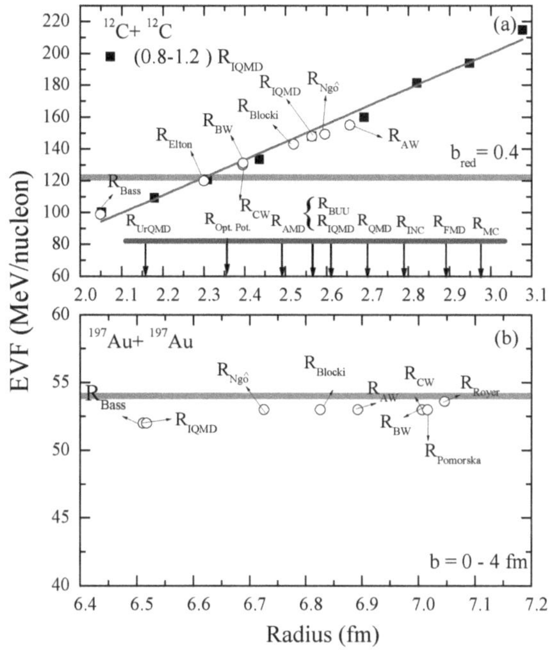

Figure 10.2 The energy of vanishing flow (EVF) for the reactions of $^{12}C+^{12}C$ (upper panel) $^{197}Au+^{197}Au$ (lower panel) as a function of the radii of the colliding nuclei. Various symbols are explained in the text. The thick solid horizontal line represents the measured EVF. The radius value of ^{12}C used in different transport models is shown by vertical arrows. The solid line is a linear fit to the EVF calculated by varying radius systematically in the IQMD model. Reprinted with permission from Rajni Bansal, Sakshi Gautam, Rajeev K. Puri, and J. Aichelin, Physical Review C, 87, 061602(R), 2013, Copyright (2019) from the American Physical Society.

$^{197}Au+^{197}Au$ (lower panel). Here, ΔEVF (%) is given by

$$\Delta \text{EVF} \, (\%) = \left(\frac{\text{EVF}^{theor} - \text{EVF}^{expt.}}{\text{EVF}^{expt.}} \right) \times 100. \tag{10.15}$$

The corresponding error bars are also displayed. The shaded area represents the region of percentage deviation in the energy of vanishing flow calculated using different radius parameterizations of ^{12}C nucleus. We see that when one uses the measured radius by Elton, the experimental energy of vanishing flow can be reproduced well. In the $^{197}Au+^{197}Au$ reaction, all parameterizations yield similar results.

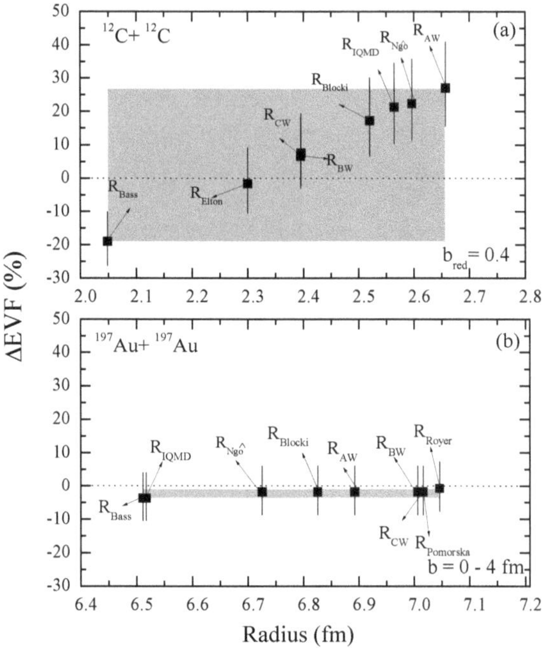

Figure 10.3 The percentage deviation ΔEVF (%) as a function of the radii of the colliding nuclei for the reactions of $^{12}C+^{12}C$ and $^{197}Au+^{197}Au$. The shaded area represents the region of percentage deviation of EVF calculated using different parameterizations of radius from the experimentally measured EVF. Reprinted with permission from Rajni Bansal, Sakshi Gautam, Rajeev K. Puri, and J. Aichelin, Physical Review C, 87, 061602(R), 2013, Copyright (2019) from the American Physical Society.

It is worth mentioning that the above calculations were performed by keeping the Fermi-momentum constant.

Further in Fig. 14.4, investigations are conducted by reducing the global Fermi momentum by 30% and by changing the Fermi momentum with radius that yields more stable density profile and choosing SMD as well as soft EOS with full and reduced cross sections. Though the absolute value of the flow (and hence EVF) changes, the above conclusions regarding the sensitivity of flow toward different choices of nuclear radius remain unaltered. In some cases, the sensitivity toward different choices of radius has even increased. We also performed the calculations using the QMD model, and the prediction remains unaltered. Therefore, here we can conclude that

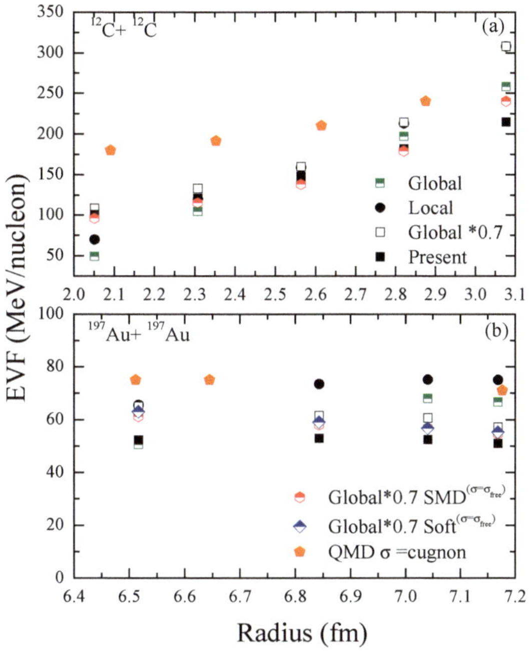

Figure 10.4 The energy of vanishing flow for the reactions of ^{12}C+^{12}C (upper panel) and ^{197}Au+^{197}Au (lower panel) as a function of the radii of the colliding nuclei. The half-shaded and open squares and solid circles represent the calculations performed by full global momentum and by reducing it by 30%, respectively. The solid circles correspond to full local Fermi-momentum calculations. The half-filled hexagons and diamonds represent the calculations for SMD and Soft EOS with full cross section, respectively, for a reduced global Fermi momentum by 30%. Also in all the above calculations, the Fermi momentum varies with radius. The pentagons represent the QMD calculation for the SMD EOS with Cugnon cross section. Whereas the squares represent our present calculations with constant Fermi momentum.

the radius parameter should be dealt with carefully while performing the calculations for lighter systems.

From the above investigation, it is clear that the radius of a colliding nucleus plays a huge role in the reaction dynamics of lighter nuclei. Also at intermediate energies, Fang [Fang (2011)] has shown that the neutron-skin thickness affects the reaction cross section, neutron or proton removal cross section as well as their ratios. Some of the proposed radius parameterizations have different radii for protons and neutrons and thus have explicit isospin dependence in them. It would be interesting to explore the effect of the isospin dependence of the radius on the reaction dynamics at intermediate energies. Among all different radii having explicit isospin dependence, for the present study, we use the recent one, which has modifications over the liquid-drop radius and also having an isospin dependence given by Royer and Rousseau. For this investigation, we simulated isobaric pairs of light colliding nuclei of ^{60}Ca+^{60}Ca and ^{60}Zn+^{60}Zn having different N/ Z ratio using a soft EOS (since the effect of radii is unaffected by the change in EOS as shown in Fig. 10.4) using the default IQMD radius and parameterization given by Royer and Rousseau.

In the next section, we will explore how an additional isospin dependence through nuclear structure effects may alter the isospin physics of nuclear reactions at intermediate energies.

10.3.4 Density Profile of the Nuclei using Different Radii

In Fig. 10.5, we display the density profiles of ^{60}Ca (solid lines) and ^{60}Zn (dashed) nuclei that are generated in IQMD using liquid-drop radius (labeled by R_{IQMD}) (thick lines) and Royer and Rousseau radius (labeled by R_{Royer}) (thin lines) at time t = 0 fm/c. From the figure, we see that the density profile is the same for both the nuclei using a liquid-drop radius initialization as it should be. On the other hand, the density profile gets extended for the ^{60}Zn nucleus compared to the ^{60}Ca nucleus when an isospin-dependent radius is used. This happens as the radius for ^{60}Zn gets larger compared to that for the ^{60}Ca nucleus. Thus, we see that with the isospin dependence of radius, a neutron-rich nucleus becomes compact compared to a neutron-deficient one and this may lead to a different reaction dynamics. The Royer and Rousseau radius, having isospin dependence, changes the radius of isospin symmetric nucleus (^{60}Zn, in the present case) also. Therefore, to see the role of isospin in radius explicitly, we also plot the density profile of the ^{60}Ca nucleus by initializing it with the same R_{Royer} as that of ^{60}Zn (i.e., without isospin dependence). We now see that the density profile of ^{60}Ca gets extended compared to that with the actual R_{Royer} for ^{60}Ca (compare dash-dotted and thin solid line) and also becomes close to that of ^{60}Zn (compare dash-dotted and dashed lines) because of the same radii of both. Therefore, by incorporating the isospin dependence of radius,

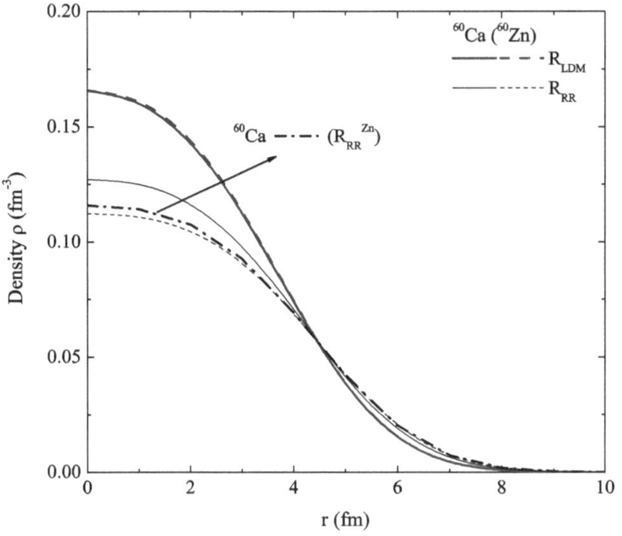

Figure 10.5 The density profile $\rho(r)$ for ^{60}Ca and ^{60}Zn nucleus initialized in IQMD with liquid-drop radius (thick lines) and Royer and Rousseau radius (thin lines) at time t = 0 fm/c. The dash-dotted line represents the case of the ^{60}Ca nucleus initialized with the same Royer and Rousseau radius without isospin dependence. Reprinted with permission from Sakshi Gautam, Physical Review C, 88, 057603, 2013, Copyright (2019) from the American Physical Society.

the neutron-rich nucleus becomes compact and thus can affect the dynamics.

10.3.5 Isospin Radius: Influence on Transverse Flow

In Fig. 10.6, we display the impact parameter dependence of $<p_x^{\mathrm{dir}}>$ for the reactions of ^{60}Ca+^{60}Ca (solid symbols) and ^{60}Zn+^{60}Zn (open symbols) using the default liquid-drop radius as well as with Royer and Rousseau parametrization. From the figure, we see that the transverse flow increases for semi-central reactions and then again decreases (see triangles). We find that the transverse flow is higher for the reaction of ^{60}Zn+^{60}Zn compared to that of

^{60}Ca+^{60}Ca with liquid-drop radius (see triangles). This is because of the dominance of Coulomb potential in isospin effects compared to the symmetry energy and nucleon–nucleon scattering cross section, as has been predicted by [Gautam (2010), Gautam (2011)], where Coulomb potential dominates in the isospin effects for isobaric colliding pairs. To be sure of this, we also calculated the transverse flow by neglecting the Coulomb potential, symmetry potential, and with isospin-independent nucleon–nucleon scattering cross section (see diamonds). We noticed that the transverse flow decreases at all the colliding geometries due to the weakening of repulsive forces (because of the absence of Coulomb and symmetry potentials). Also, the transverse flow now becomes almost equal in both the reactions, thus signifying the dominance of Coulomb potential in the isospin degree of freedom. It is worth mentioning that to explicitly see the dominance of Coulomb repulsion in isospin effects, only the Coulomb potential should be turned off (and not the symmetry potential and with isospin dependence of nucleon–nucleon scattering cross section as done by [Gautam (2010), Gautam (2011)]). If we had switched off only the Coulomb potential, the behavior of isospin effects in the transverse flow for the reactions of ^{60}Ca+^{60}Ca and ^{60}Zn+^{60}Zn would have reversed with significant difference between the two. Since we have neglected all the isospin effects, the difference between the two systems vanishes (because of the counteracting symmetry potential and scattering cross section). Similar dominance of Coulomb potential in the fusion cross section has been found for the fusion reactions of isotonic pairs [Dhiman (2006)]. Next, to see the role of isospin dependence of the radius, we initialized the nucleus of ^{60}Ca and ^{60}Zn by Royer and Rousseau parameterization and then calculated the transverse flow (circles). We find that for semi-central reactions, the above-mentioned trend gets reversed, and the flow now becomes higher for the ^{60}Ca+^{60}Ca reaction. This is due to the fact that the R_{Royer} radius is more for the ^{60}Zn nucleus compared to the ^{60}Ca radius. Due to this enhanced radius, the density profile gets extended, which in turn will lead to weak repulsive forces and thus transverse flow decreases. On the other hand, the flow behavior remains the same at peripheral collisions and the radius effect does not come into picture. This is because of the less overlapping region in peripheral collisions, so density achieved will be less and, therefore, density-dependent interactions would not be dominating and Coulomb potential still dominates. To further strengthen our point, we increase the normal liquid-drop radius in IQMD by 30% and 60% and calculate the flow for the reactions of $^{60}Ca+^{60}Ca$. Again, we found that on increasing the radius, flow decreases (results not shown here). Similarly, when all isospin effects are neglected (by switching off Coulomb and symmetry potentials and with isospin-independent scattering cross section) with R_{Royer} parameterization, the transverse flow is higher for the reactions of $^{60}Ca+^{60}Ca$ (solid squares) compared to the reactions of $^{60}Zn+^{60}Zn$ at semi-central reactions (which otherwise was same for both with the liquid-drop radius, see diamonds). Thus, we see that the isospin dependence of radius has a significant effect on transverse flow. As we have mentioned earlier, the size of the ^{60}Zn nucleus gets modified with R_{Royer} (because of modifications over the liquid-drop radius). Therefore, to see explicitly the role of the isospin-dependent radius, we calculate the flow for the reactions of $^{60}Ca+^{60}Ca$ by initializing the nucleus of ^{60}Ca with the Royer and Rousseau radius, but without isospin dependence. From the figure, we see that for semi-central reactions, the flow decreases further compared to that with actual R_{Royer} having isospin dependence (compare solid squares and cross pentagons). Thus, we see that radius and its isospin dependence play a significant role in the isospin effects of transverse flow.

10.3.6 Isospin Radius: Influence on Nuclear Fragmentation

Next, we investigate the role of radius and its isospin dependence on the fragmentation pattern. Figures 10.7a–d display the mass of the heaviest fragment (A^{max}), and multiplicities of free nucleons (FNs), light-charged particles ($2 \leq A \leq 4$) (LCPs), and intermediate mass fragments ($5 \leq A \leq A/3$) (IMFs). Symbols have the same meaning as in Fig. 10.6. From Fig. 10.7a, we see that the size of A^{max} (yield of free nucleons) is higher (lower) for the reactions of $^{60}Zn+^{60}Zn$ compared to $^{60}Ca+^{60}Ca$ with the liquid-drop radius. This is because of the repulsive Coulomb interaction that will throw more matter out of a participant zone and thus leading to a bigger A^{max} (lesser free nucleons). This effect is dominating at peripheral

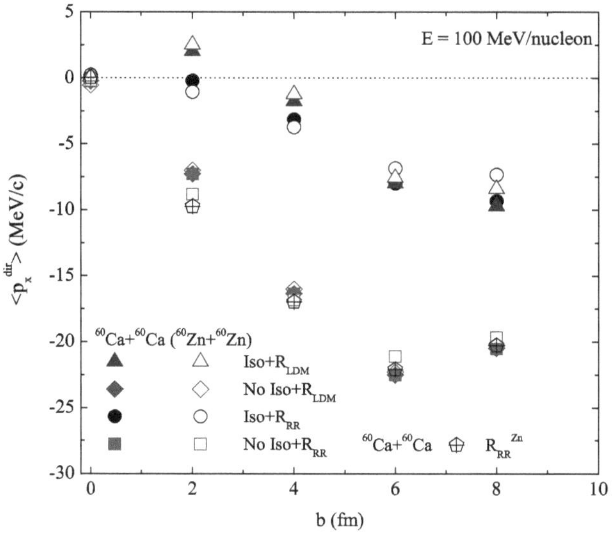

Figure 10.6 The transverse in-plane flow $<p_x^{\text{dir}}>$ as a function of impact parameter for the reactions of ^{60}Ca+^{60}Ca and ^{60}Zn+^{60}Zn at an incident energy of 100 MeV/nucleon. Various symbols are explained in text and in the figure. Reprinted with permission from Sakshi Gautam, Physical Review C, 88, 057603, 2013, Copyright (2019) from the American Physical Society.

collisions. The difference vanishes when we neglect all isospin effects (see diamonds). On the other hand, with the Royer and Rousseau radius, the A^{max} is bigger for the ^{60}Ca+^{60}Ca reactions (circles). It is expected that the total number of collisions should be more in the case of smaller radius, which will lead to a smaller A^{max}. But due to less available phase space, most of the collisions will be Pauli blocked and thus A^{max} would be bigger in the case of a smaller radius. Moreover, in the case of bigger nuclei, the density profile will smoothen out and, therefore, the role of the attractive mean field decreases, which will also lead to a smaller A^{max}. The difference, though small, is because of the counter isospin effect that is operating. When we neglect all isospin effects with the Royer and Rousseau radius initialization (squares), the difference increases, thus pointing toward the significant role of the isospin dependence of radius. The role of the explicit isospin-dependent radius in the

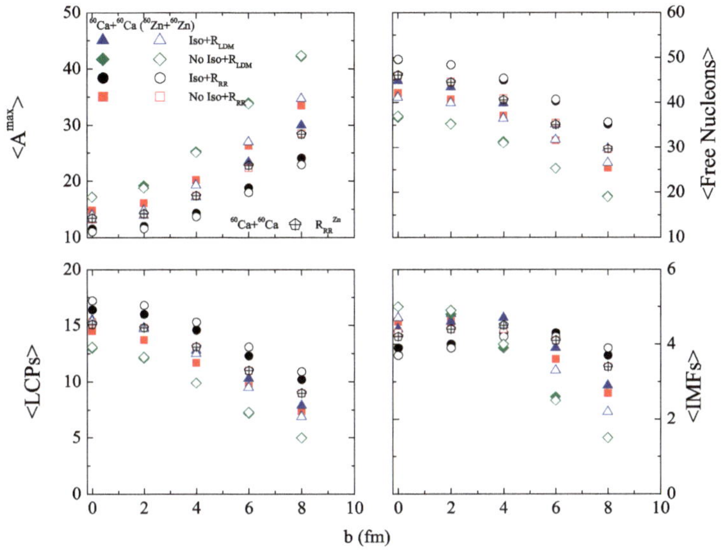

Figure 10.7 The size of the heaviest fragment A^{max}, and yields of free nucleons, light-charged particles (LCPs), and intermediate mass fragments (IMFs) as a function of the impact parameter for the reactions of ^{60}Ca+^{60}Ca and ^{60}Zn+^{60}Zn. Symbols have the same meaning as in Fig. 10.6. Reprinted with permission from Sakshi Gautam, Physical Review C, 88, 057603, 2013, Copyright (2019) from the American Physical Society.

fragmentation of ^{60}Ca + ^{60}Ca reactions is also investigated (crossed pentagons) as done earlier. We notice that A^{max} for the reactions of ^{60}Ca+^{60}Ca decreases compared to that with the actual R_{Royer} for the ^{60}Ca nucleus and becomes same as that for ^{60}Zn+^{60}Zn reactions. Similarly, the multiplicities of light-charged particles and intermediate mass fragments also show the role of radius and behave accordingly. The above-mentioned reactions are highly neutron-rich reactions of ^{60}Ca+^{60}Ca. To see whether the above results persist for experimentally accessible reactions also, we simulate the reactions of ^{48}Ca+^{40}Ca and ^{48}Cr+^{40}Ca at the same reaction conditions. Our findings reveal that the role of the isospin-dependent radius is still there but, as expected, with a lesser magnitude (results not shown).

10.4 Summary

In summary, within the framework of the IQMD model, we demonstrated that the directed transverse flow shows strong dependence on the structural effects via nuclear radii for lighter systems and less sensitivity for heavier systems. Our study indicated that the radii of the colliding nuclei must be treated carefully while extracting the energy of vanishing flow for lighter systems. Our finding revealed that different choices of radii can affect the energy of vanishing flow by 40–50 % for lighter colliding nuclei such as $^{12}C+^{12}C$. The heavier nuclei like $^{197}Au+^{197}Au$, however, remain unaffected by the choice of the radius. Therefore, the disagreement between the findings of simulations and experiments is most probably due to an insufficient reproduction of the density profile in small systems and no new physics has to be evoked to explain the experimental data. Furthermore, we investigated the role of radius and its isospin dependence in reaction dynamics via studying transverse flow and the fragmentation of neutron-rich and neutron-deficient isobaric colliding pairs. Our findings revealed that radius as well as its isospin dependence has a significant effect on isospin physics and, in fact, the role of isospin on flow and fragmentation gets reversed with explicit isospin in radius.

Acknowledgment

The authors are thankful to Prof. Rajeev K. Puri for giving access to various computer programs used for the present study.

References

Abdel-Waged, K. (2004). Influence of medium modified angular distributions and mean field on the quasielastic and inelastic peaks of neutrons induced by 1.2 GeV protons on targets with $27 < A < 232$, *Phys. Rev. C* **70**, pp. 014605.

Aichelin, J. (1991). "Quantum" molecular dynamics: A dynamical microscopic n-body approach to investigate the fragment formation and the nuclear equation of state in heavy-ion collisions, *Phys. Rep.* **202**, pp. 233–360.

Angeli, I. (2004). A consistent set of nuclear rms charge radii: Properties of the radius surface R (N,Z), Atomic Data and Nuclear Data Tables, *At. Data Nucl. Data Tables* **87**, 185, pp. 185–206.

Bansal, R., Gautam, S., and Puri, R. K. (2014). Systematic study of the energy of vanishing flow using IQMD model and comparison with various theoretical models, *J. Phys. G: Nucl. Part. Phys.* **41**, 035103.

Bansal, R., Gautam, S., Puri, R. K., and Aichelin, J. (2013). Role of structural effects on the collective transverse flow and the energy of vanishing flow in nuclear collisions, *Phys. Rev. C* **87**, 061602 (R).

Bass, R. (1977). Nucleus–nucleus potential deduced from experimental fusion cross sections, *Phys. Rev. Lett.* **39**, pp. 265.

Bauer, W. (1988). Nuclear stopping at intermediate beam energies, *Phys. Rev. Lett.* **61**, pp. 2534.

Blocki, J. *et al.* (1977). Proximity forces, *Ann. Phys.* **105**, pp. 427.

Brown, B. A. (2000). Neutron radii in nuclei and the neutron equation of state, *Phys. Rev. Lett.* **85**, pp. 5296.

Christensen, P. R. and Winther, A. (1976). The evidence of the ion–ion potentials from heavy ion elastic scattering, *Phys. Lett. B* **65**, pp. 19–22.

Cugnon, J. (1980). Monte Carlo calculation of high-energy heavy-ion interactions, *Phys. Rev. C* **22**, pp. 1885.

Dai, Z. T. *et al.* (2015). Effect of neutron skin thickness on projectile fragmentation, *Phys. Rev. C* **91**, pp. 034618.

de la Mota, V., Sebille, F., Farine, M., Remaud, B., and Schuck, P. (1992). Analysis of transverse momentum collective motion in heavy-ion collisions below 100 MeV/nucleon, *Phys. Rev. C* **46**, pp. 677–686.

Denisov, V. Y. (2002). Interaction potential between heavy ions, *Phys. Lett. B* **526**, pp. 315–321.

Dhiman, N. K. and Puri, R. K. (2006). The study of fusion of different isotopes/isotones leading to the same compound nucleus, *Act. Phys. Pol. B* **38**, pp. 2133–2141.

Dieperink, A. E. L. *et al.* (2003). Nuclear symmetry energy and the neutron skin in neutron-rich nuclei, *Phys. Rev. C* **68**, pp. 064307.

Dutt, I. and Bansal, R. (2010). A modified proximity approach in the fusion of heavy ions, *Chin. Phys. Lett.* **27**, pp. 112402.

Dutt, I. and Puri, R. K. (2010). Comparison of different proximity potentials for asymmetric colliding nuclei, *Phys. Rev. C* **81**, pp. 064609.

Elton, L. R. B. (1961). *Nuclear Sizes*, Oxford University Press, London.

Elton, L. R. B. (1967). A-dependence of nuclear charge and mass distributions, *Phys. Rev.* **158**, pp. 970.

Fang, D. Q. *et al.* (2011). Effect of neutron skin thickness in peripheral nuclear reactions, *Chin. Phys. Lett.* **28**, 102102.

Feldmeier, H. and Schnack, J. (1997). Fermionic molecular dynamics, *Prog. Part. Nucl. Phys.* **39**, pp. 393.

Gautam, S. and Sood, A. D. (2010). Isospin effects on the mass dependence of the balance energy, *Phys. Rev. C* **82**, 014604.

Gautam, S., Sood, A. D., Puri, R. K., and Aichelin, J. (2011). Isospin effects in the disappearance of flow as a function of colliding geometry, *Phys. Rev. C* **83**, 014603.

Hartnack, C. *et al.* (1998). Modelling the many body dynamics of heavy-ion collisions: Present status and future perspective, *Eur. Phys. J. A* **1**, pp. 151–169.

Hofstadter, R., Fechter, H. R., and McIntyre, J. A. (1953). High-energy electron scattering and nuclear structure determinations, *Phys. Rev.* **92**, pp. 978.

Jancovici, B. G. (1954). Coulomb energy and nuclear radius, *Phys. Rev.* **95**, pp. 389.

Kaur, S. and Puri, R. K. (2014). Role of model ingredients in the production of light particles and entropy, *Phys. Rev. C* **90**, 037602.

Klakow, D., Welke, G., and Bauer, W. (1993). Nuclear flow excitation function, *Phys. Rev. C* **48**, pp. 1982–1987.

Kumar, S., Bansal, R., and. Kumar, S. (2010). Experimental balance energies and isospin-dependent nucleon-nucleon cross-sections, *Phys. Rev. C* **82**, 024610.

Lehmann, E., Faessler, A., Zipprich, J., Puri, R. K., and Huang, S. W. (1996). Study of in-medium effects on the disappearance of the sidewards flow in heavy-ion collisions, *Z. Phys. A* **355**, pp. 55–60.

Li, C. *et al.* (2013). Finite-size effects on fragmentation in heavy-ion collisions, *Phys. Rev. C* **87**, 064615.

Łukasik, J. *et al.* (2005). Directed and elliptic flow in ^{197}Au+^{197}Au at intermediate energies, *Phys. Lett. B* **608**, pp. 223–230.

Magestro, D. J., Bauer, W., and Westfall, G. D. (2000). Isolation of the nuclear compressibility with the balance energy, *Phys. Rev. C* **62**, 041603 (R).

Majestro, D. J. *et al.* (2000). Disappearance of transverse flow in Au+Au collisions, *Phys. Rev. C* **61**, 021602 (R).

Myers, W. D. and Ŝwiąteck, W. J. (2000). Nucleus–nucleus proximity potential and superheavy nuclei, *Phys. Rev. C* **62**, pp. 044610.

Ngô, H. and Ngô, Ch. (1980). Calculation of the real part of the interaction potential between two heavy ions in the sudden approximation, *Nucl. Phys. A* **348**, pp. 140–156.

Ohtsuka, N., Linden, R., Faessler, A., and Malik, F. B. (1987). Real and imaginary parts of the microscopic optical potential between nuclei in the sudden and adiabatic approximation and its application to medium energy ^{12}C+^{12}C scattering, *Nucl. Phys. A* **465**, pp. 550.

Ono, A. and Horiuchi, H. (1995). Flow of nucleons and fragments in ^{40}Ar+^{27}Al collisions studied with antisymmetrized molecular dynamics *Phys. Rev. C* **51**, 299.

Ono, A., Horiuchi, H., and Maruyama, T. (1993). Nucleon flow and fragment flow in heavy ion reactions, *Phys. Rev. C* **48**, 2946.

Ono, A., Horiuchi, H., Maruyama, T., and Ohnishi, A. (1992). Fragment formation studied with antisymmetrized version of molecular dynamics with two-nucleon collisions, *Phys. Rev. Lett.* **68**, pp. 2898.

Pomorska, B. N. and Pomorski, K. (1994). Simple formula for nuclear charge radius, *Z. Phys. A* **348**, pp. 169.

Reisdrof, W. (1994). Heavy-ion reactions close to the Coulomb barrier, *J. Phys. G: Nucl. Part. Phys.* **20**, pp. 1297.

Royer, G. and Rousseau, R. (2009). On the liquid drop model mass formulae and charge radii, *Eur. Phys. J. A* **42**, pp. 541.

Royer, G. (2008). On the coefficients of the liquid drop model mass formulae and nuclear radii, *Nucl. Phys. A* **807**, pp. 105.

Sood, A. D. and Puri, R. K. (2004). Mass dependence of disappearance of transverse in-plane flow, *Phys. Rev. C* **69**, 054632.

Sood, A. D. and Puri, R. K. (2006). Influence of momentum-dependent interactions on balance energy and mass dependence, *Eur. Phys. J. A* **30**, pp. 571–577.

Tsang, M. B. *et al.* (2012). Constraints on the symmetry energy and neutron skins from experiments and theory, *Phys. Rev. C* **86**, pp. 015803.

Sood, A. D., Puri, R. K., and Aichelin, J. (2004). Study of balance energy in central collisions for heavier nuclei, *Phys. Lett. B* **594**, pp. 260–264.

Westfall, G. D. *et al.* (1993). Mass dependence of disappearance of flow in nuclear collisions, *Phys. Rev. Lett.* **71**, pp. 1986–1989.

Winther, A. (1995). Dissipation, polarization and fluctuation in grazing heavy-ion collisions and the boundary to the chaotic regime, *Nucl. Phys. A* **594**, pp. 203.

Yariv, Y. and Fraenkal, Z. (1979). Intranuclear cascade calculation of high-energy heavy-ion interactions, *Phys. Rev. C* **20**, pp. 2227.

Yong, G. C. *et al.* (2011). Initialization effect in heavy-ion collisions at intermediate energies, *Phys. Rev. C* **84**, 034609.

Chapter 11

Symmetry Energy and Its Effect on Various Observables at Intermediate Energies

Sakshi Gautam[a] **and Mandeep Kaur**[b]

[a] *Department of Physics, Panjab University, Chandigarh 160014, India*

[b] *Post Graduate Govt. College for Girls, Chandigarh 160011, India*

sakshigautam@pu.ac.in

11.1 Introduction

A continuous progress in the development of radioactive-ion beam (RIB) facilities around the world has shifted the interest of the community to understand the nature of the equation of state of isospin asymmetric nuclear matter, namely the density dependence of the nuclear symmetry energy [Li (2008)]. The nuclear symmetry energy plays a vital role in the dynamics of neutron-rich reactions as well as in the structure of exotic nuclei [Baran (2005)]. Furthermore, the knowledge of nuclear symmetry energy also helps in understanding the astrophysical phenomena such as neutron stars and supernova explosions [Steiner (2005)]. In 1968, Brueckner *et al.* [Brueckner (1968)] marked the first study regarding the equation of state of isospin asymmetric nuclear matter. Later, a number of studies have been carried out that are based on different many-body

Multifragmentation in Heavy-Ion Reactions: Theory and Experiments
Edited by Rajeev K. Puri, Yu-Gang Ma, and Arun Sharma

ISBN 978-981-4968-69-0 (Hardcover), 978-1-003-38513-4 (eBook)
www.jennystanford.com

theories [Li (2008), Dieperink (2003), Tsang (2011), Sun (2010), Shetty (2007)]. But the predictions about the nature of equation of state of isospin asymmetric nuclear matter differ both at low and at high densities [Li (2008), Dieperink (2003)].

At the same time, heavy-ion reactions at intermediate energy have set an excellent platform to investigate the equation of state of the isospin asymmetric nuclear matter [Li (2008)]. The equation of state for the asymmetric nuclear matter is approximated as [Li (2008)]:

$$E_{sym}(\rho,\alpha) = E_{sym}(\rho,0) + E_{sym}(\rho)\alpha^2, \tag{11.1}$$

where $\alpha = (\rho_n - \rho_p)/(\rho_n + \rho_p)$ is called the isospin asymmetry parameter; $E_{sym}(\rho,0)$ and $E_{sym}(\rho)$ signify the equation of state for symmetric nuclear matter and symmetry energy, respectively.

Various theoretical studies predicted the value of the nuclear symmetry energy at normal nuclear matter density around to be 30 MeV [Meyers (1996), Chen (2007), Chen (2005)]. However, the understanding of nuclear symmetry energy gets feeble as one moves away from the normal nuclear matter and β-stability line [Li (2008), Xu (2010)]. Therefore, the equation below provides the parametrization for the density dependence of nuclear symmetry energy:

$$E_{sym}(\rho) = E_{sym}(\rho_0)\left(\frac{\rho}{\rho_0}\right)^{\gamma}, \tag{11.2}$$

Here γ signifies the strength (or stiffness) of the nuclear symmetry energy at densities below and above the saturation density. The strength of the nuclear symmetry energy is different for different values of γ.

As we know the nuclear symmetry energy is not a directly observable quantity. Various observables have been proposed in the past studies (at sub- and supra-saturation densities) to constrain the nature and form of the nuclear symmetry energy [Li (2008), Li (1997), Huang (2010), Tsang (2011)]. The behavior of nuclear symmetry energy at the sub-saturation density region has been studied using different probes (or observables) such as isospin fractionation [Li (2008), Ono (2003)], isospin diffusion [Li (2008), Tsang (2011)], isobaric ratio of various species [Huang (2010), Ma (2011)], isotopic scaling [Shetty (2007), Colonna (2006)], *N/Z* ratio of various fragments [Tsang (2009), Zhang (2005)], neutron-skin thickness [Chen (2010)], the giant monopole resonance [Li (2010)], etc. The form and strength of nuclear symmetry energy is fairly well constrained below the sub-saturation density region. For example, in Ref [Chen (2005)], the National Superconducting Cyclotron Laboratory (NSCL) group at Michigan State University (MSU) compared the isospin diffusion data for the reaction of $^{124}_{50}$Sn + $^{112}_{50}$Sn at an incident energy of 50 MeV/nucleon with isospin-dependent Boltzmann–Uehling–Uhlenbeck (IBUU04) calculations [Li (2004)] and obtained the symmetry energy of the form E_{sym} = 31.6 $(\rho/\rho_0)^{1.05}$. Li *et al.* [Li (2005)] also compared the NSCL data and found γ to be 0.69. In a similar study done by the NSCL/MSU group [Shetty (2007)], isotopic yield distributions for the reactions of ($^{40}_{18}$Ar, $^{40}_{20}$Ca) + $^{58}_{26}$Fe, $^{58}_{28}$Ni + ($^{58}_{26}$Fe, and $^{58}_{28}$Ni) were compared with Antisymmetrized Molecular Dynamics (AMD) calculations [Ono (2003)] and the results obtained the nuclear symmetry energy of the form 31.6 $(\rho/\rho_0)^{0.69}$. Thus, from the above studies, we can conclude that the form of symmetry energy is well constrained at the sub-saturation density region.

In contrast, a lot of efforts have been carried out to constrain the behavior of the nuclear symmetry energy at supra-saturation densities by using different observables such as collective [Li (2008), Li (2000), Yong (2006)] and elliptic flows [Russotto (2011), Russotto (2014)], *n/p* ratios of free nucleons [Zhang (2014), Yong (2007)], K^+/K^0 [Q. Li (2005), Q. Li (2004)], Σ^-/Σ^+ [Q. Li (2004)], π^-/π^+ ratios [Xiao (2009), Feng (2010)] as well as η meson production [Yong (2013)], etc. It was first observed in Ref. [Li (1996)] that transverse flow is quite sensitive to the isospin asymmetry of the reaction system. Later, these results were also verified experimentally by the NSCL group [Pak (1997), Pak (1997b)]. In addition, Li [Li (2002)] studied the collective flow for the reaction of $^{132}_{50}$Sn + $^{124}_{50}$Sn at an incident energy ranging from 200 to 1000 MeV/nucleon. It was observed that collective flow is quite sensitive to the density dependence of nuclear symmetry energy and pointed out a striking difference of about a factor of 2 (toward density dependence of symmetry energy) at beam energies $E_{beam} \geq 200$ MeV/nucleon. This noticeable difference in collective flow motivated the researchers for further studies (theoretical as well as experimental) to constrain the symmetry energy at the supra-saturation density region. In view of this, a number of studies have been done in the past to pin down the nature and form of nuclear symmetry energy [Li (2008), Yong (2006), Li (2000), Cozma (2011), Feng (2012)].

In addition to collective flow, *n/p* ratios of squeezed-out nucleons have also been proposed as an effective observable to probe the density dependence of nuclear symmetry energy as these nucleons are emitted from the high-density participant zone and are not hindered by the presence of spectators [Li (2006), Yong (2007)]. As an example, Kumar *et al.* [Kumar (2012)] studied the single and double *n/p* ratios from different kinds of fragments for the reactions of $^{112}_{50}$Sn + $^{112}_{50}$Sn, $^{124}_{50}$Sn + $^{124}_{50}$Sn, and $^{132}_{50}$Sn + $^{132}_{50}$Sn at an incident energy varying from 50 to 600 MeV/nucleon. The single *n/p* ratios from free nucleons (FNs) and light-charged particles (LCPs) are found to be sensitive to the symmetry energy, incident energy as well as isospin asymmetry of the reaction system. In another study, Li *et al.* [Li (2006)] also proposed single and double *n/p* ratios from free nucleons to be a sensitive tool to pin down the density dependence of nuclear symmetry energy.

From the above survey, it is quite obvious that various studies (using collective flow as well as fragmentation) have been carried out to probe and understand the density dependence of nuclear symmetry energy [Xie (2015), Kumar (2012), Yong (2007), Zhang (2014), Kohley (2010), Kohley (2012)]. In one such study, the transverse flow of intermediate mass fragments has been investigated for isotopic and isobaric colliding pairs [Kohley (2010), Kohley (2012)] and an observable "R_{flow}" has been proposed, which represents the ratio of the relative transverse flow (of intermediate mass fragments) of isotopic and isobaric colliding pairs. This relative ratio shows up to 50% sensitivity toward nuclear symmetry energy [Kohley (2010), Kohley (2012)]. On the other hand, the sensitivity of bare transverse flow of fragments (i.e., Z_f = 1,2) is found to be up to 30% in the Fermi energy region [Kohley (2011)]. Therefore, here we will present our result of a couple of attempts where we tried to study the density dependence of nuclear symmetry energy using collective flow and multifragmentation. In particular, we will study the sensitivity of transverse flow toward nuclear symmetry energy as well as its detailed mechanism responsible. Second, we will also investigate the yields of different kinds of fragments as well as free particles and put forward a nice observable sensitive to nuclear symmetry energy.

11.2 Results and Discussion

For the first part of the analysis, we have simulated several thousands of events for the neutron-rich colliding nuclei $^{60}Ca+^{60}Ca$ at incident energies of 100, 400, and 800 MeV/nucleon for semi-central collisions at an impact parameter of $\hat{b}$ = 0.2–0.4. We used a soft equation of state along with an isospin- and energy-dependent cross section reduced by 20%, i.e., $\sigma = 0.8\ \sigma_{NN}$. The various forms of symmetry energy used in the present chapter are $E_{\text{sym}} \propto F_1(u)$, $E_{\text{sym}} \propto F_2(u)$, and $E_{\text{sym}} \propto F_3(u)$, where $u = \frac{\rho}{\rho_0}$, $F_1(u) \propto u$, $F_2(u) \propto u^{0.4}$, $F_3(u) \propto u^2$, and F_4 represent calculations without symmetry energy.

11.2.1 Time Evolution of Transverse Flow

The transverse flow is calculated using $< p_x/A >$ and "*directed transverse in-plane flow* $< p_x^{\text{dir}} >$," which averages the transverse momenta of nucleons according to the rapidity of particles [Li (2008)]. In Fig. 11.1, we display $< p_x/A >$ as a function of $Y_{\text{c.m.}}/Y_{\text{beam}}$ at final time (left panels) and the time evolution of $<p_x^{\text{dir}}>$ (right panels) calculated at 100 (top panel), 400 (middle), and 800 MeV/nucleon (bottom) for different density dependencies of the symmetry energy. The solid, dash-dotted, and dotted lines represent the symmetry energy proportional to $\frac{\rho}{\rho_0}$, $\left(\frac{\rho}{\rho_0}\right)^{0.4}$, and $\left(\frac{\rho}{\rho_0}\right)^2$, respectively, whereas dashed lines represent calculations without symmetry energy. Comparing the left and right panels in Fig. 11.1, we find that both the methods show similar behavior for symmetry energy. For example, at an incident energy of 100 MeV/nucleon for $E_{\text{sym}} \propto \left(\frac{\rho}{\rho_0}\right)^{0.4}$, $<p_x^{\text{dir}}> = 0$. Similarly, the slope of $<p_x/A>$ at midrapidity is zero. We also find that the transverse momentum is sensitive to symmetry energy and its various density dependence $F_1(u)$, $F_2(u)$, and $F_3(u)$ in the low-energy region (100 MeV/nucleon). On the other hand, the transverse flow (calculated using both the methods) is insensitive to different strengths of symmetry energy and its different density dependence above Fermi energy. This is because the repulsive nucleon–nucleon scattering dominates the mean field at high incident energies. Therefore, mean field does not alter the results significantly.

11.2.2 Time Evolution of Rapidity Distribution of Nucleons

To understand the sensitivity of transverse momentum to the symmetry energy as well as its density dependence in the Fermi energy region, we calculate the transverse flow as well as rapidity distribution of the nucleons having $\frac{\rho}{\rho_0} < 1$ (denoted as BIN 1) and nucleons having $\frac{\rho}{\rho_0} \geq 1$ (BIN 2) separately at each time step. As mentioned earlier, both methods show similar behavior; therefore, we shall discuss the flow in terms of $< p_x^{\mathrm{dir}} >$ only. In Fig. 11.2, we display the rapidity distributions at 100 MeV/nucleon of all the particles (dotted lines), and particles corresponding to BIN 1 (solid) and BIN 2 (dashed) at 0, 10, 20, 30, 40, and 60 fm/c. We have calculated rapidity distributions for different forms of symmetry energy. We find that they are insensitive to the symmetry energy [Zhang (2011), Lehaut (2014)]. Therefore, we only display the results for linear density dependence of symmetry energy [$F_1(u)$]. From the figure, we find that during the initial Stages, there are two Gaussians at projectile and target rapidity. The peaks of the Gaussians will be prominent at higher energies. The interest for our discussion is in BIN 1 and BIN 2. During the early stage of the reaction (0 fm/c), a higher number of particles lie in BIN 1; i.e., a higher number of particles have $\frac{\rho}{\rho_0} < 1$. As the colliding nuclei begin to overlap, the density increases in the overlap zone. Now, the number of particles increases in BIN 2 (at 10 fm/c). From 10 fm/c to 20 fm/c, the number of particles continues to increase in BIN 2 at midrapidity; i.e., the particles from large rapidity continue shifting to BIN 2 in the midrapidity region. This is expected since in the Fermi energy region, dynamics is governed by the attractive mean field. The dominance of the attractive mean field will prompt the deflection of the particles into negative angles, i.e., toward the participant zone. After 20–30 fm/c, the expansion phase of the reaction begins and the number of

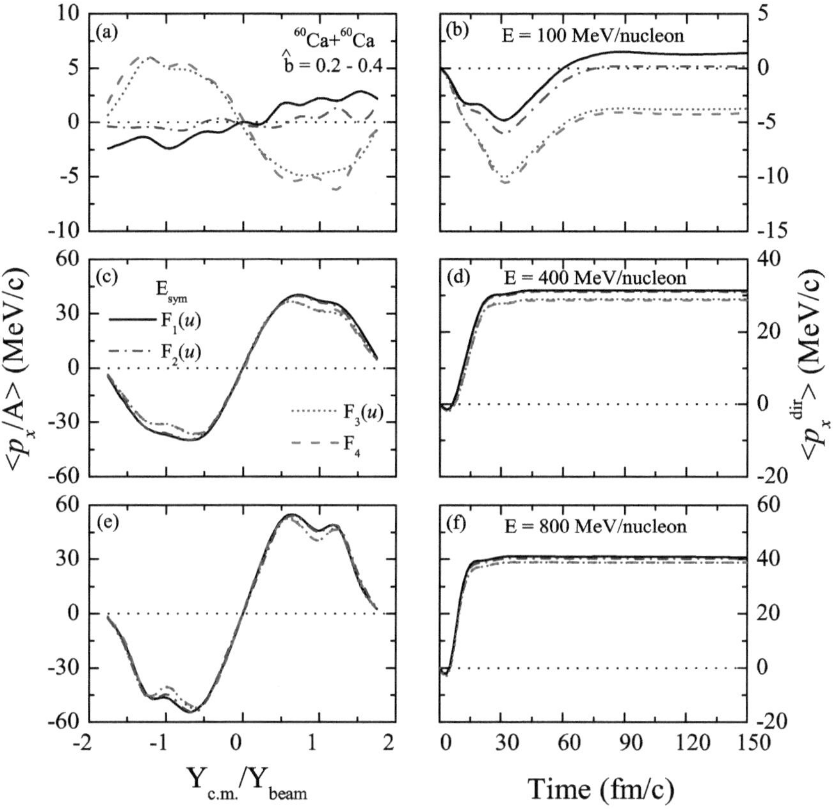

Figure 11.1 Left panel: $< p_x/A >$ as a function of $Y_{\text{c.m.}}/Y_{\text{beam}}$ at different incident energies of 100, 400, and 800 MeV/nucleon using different forms of symmetry energy. Right panel: the time evolution of $< p_x^{\text{dir}} >$ at 100, 400, and 800 MeV/nucleon for different forms of symmetry energy at $\hat{b}$ = 0.2–0.4. The solid, dash-dotted, and dotted lines represent the calculations with symmetry energy of the form $F_1(u)$, $F_2(u)$, and $F_3(u)$, respectively. The dashed lines represent the calculations without symmetry energy (F_4). Reprinted with permission from Sakshi Gautam, Aman D. Sood, Rajeev K. Puri, and J. Aichelin, Physical Review C, 83, 034606, 2011, Copyright (2019) from the American Physical Society.

particles increases in BIN 1, and by 60 fm/c, most of the particles lie in BIN 1 (see dotted and solid lines in the right bottom panel).

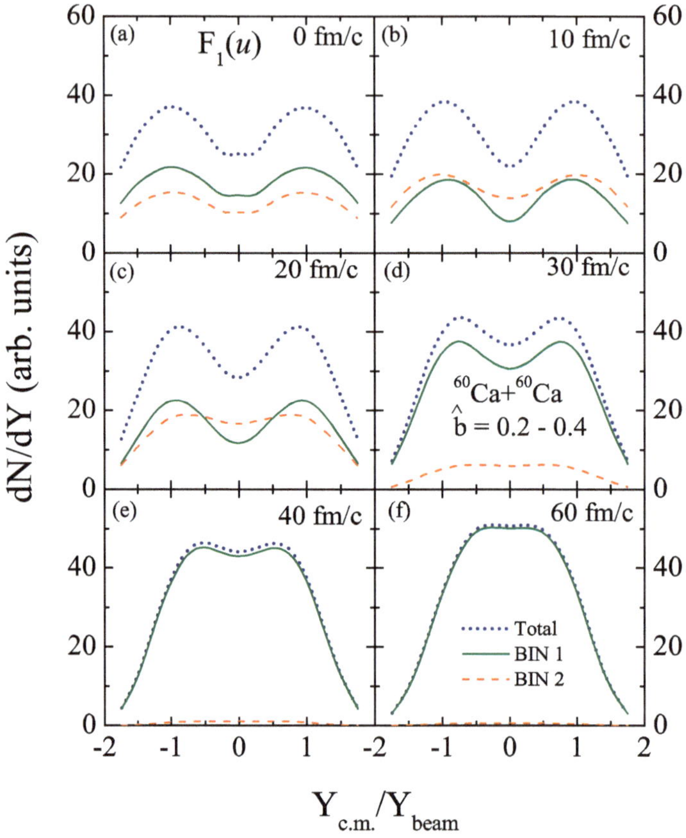

Figure 11.2 The time evolution of the rapidity distribution with symmetry energy of the form $F_1(u)$ for various bins at $\hat{b}$ = 0.2–0.4. Lines are explained in the text. Reprinted with permission from Sakshi Gautam, Aman D. Sood, Rajeev K. Puri, and J. Aichelin, Physical Review C, 83, 034606, 2011, Copyright (2019) from the American Physical Society.

11.2.3 Directed Transverse Momentum of Nucleons Feeling Various Densities

In Fig. 11.3, we display the time evolution of the $<p_x^{\mathrm{dir}}>$ for different forms of symmetry energies at 100 MeV/nucleon for particles lying in the different bins. Lines have the same meaning as used in

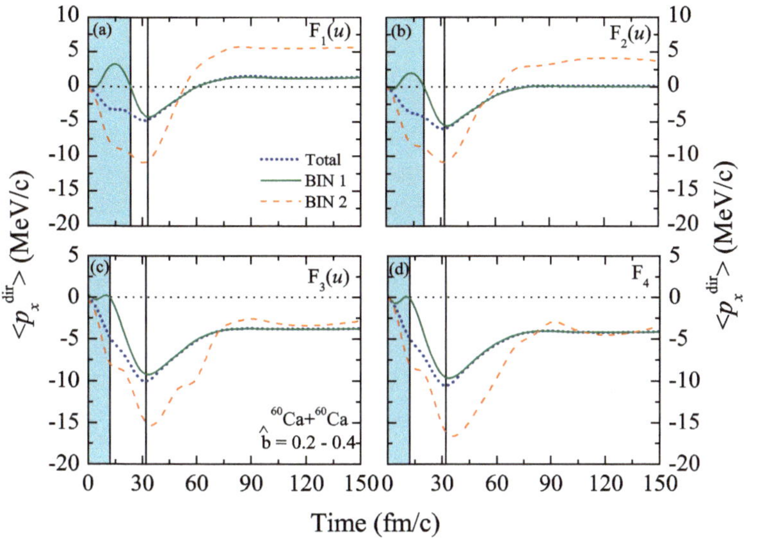

Figure 11.3 The time evolution of $< p_x^{\mathrm{dir}} >$ for different forms of the symmetry energy using different density bins at $\hat{b}$ = 0.2–0.4. Lines have the same meaning as used in Fig. 11.2. Reprinted with permission from Sakshi Gautam, Aman D. Sood, Rajeev K. Puri and J. Aichelin, Physical Review C, 83, 034606, 2011, Copyright (2019) from the American Physical Society.

Fig. 11.2. Panels (a), (b), and (c) are for $E_{\mathrm{sym}} \propto \frac{\rho}{\rho_0}, \left(\frac{\rho}{\rho_0}\right)^{0.4}$, and $\left(\frac{\rho}{\rho_0}\right)^{2}$, respectively. Panel (d) is for calculations without symmetry energy. The total $<p_x^{\mathrm{dir}}>$ is negative during the initial stage and decreases till 30 fm/c, which indicates the dominance of attractive interactions. In panels (a) and (b), it becomes positive, whereas in panels (c) and (d), it remains negative during the course of the reaction. If we look at $<p_x^{\mathrm{dir}}>$ of particles lying in BIN 1 for $F_1(u)$ (Fig. 11.3a) and $F_2(u)$ (Fig. 11.3b) in the time interval 0 to about 20–25 fm/c, we see that it remains positive. It increases up to 15 fm/c and reaches a peak value. This is because in the spectator region (where high rapidity particles lie), the repulsive symmetry energy will accelerate the particles away from the overlap zone in the transverse direction. After 15 fm/c, $<p_x^{\mathrm{dir}}>$ (of particles in BIN 1) begins to decrease. This is because these particles will now be attracted toward the central dense zone. As shown in Fig. 11.2, from 10 to 20 fm/c, the number of particles in BIN 2 increases in the midrapidity region. In the case of $F_1(u)$ and

$F_2(u)$, particles that enter the central dense zone (BIN 2) have already a high positive value of $<p_x^{\rm dir}>$ (i.e., going away from the dense zone). So, the attractive mean field has to decelerate the particles first, make them stop, and then accelerate the particles back toward the overlap zone. At about 20–25 fm/c, particles from BIN 1 have zero $<p_x^{\rm dir}>$ (shaded area in Figs. 11.3a and 11.3b). Up to 30 fm/c, particles feel the attractive mean field potential after which the high-density phase is over; i.e., in the case of $F_1(u)$ and $F_2(u)$ between 0 and 30 fm/c, particles from BIN 1 are accelerated toward the overlap zone only for a short time interval of about 5 fm/c, whereas for the case of $F_3(u)$ (Fig. 11.3c) and F_4 (Fig. 11.3d) between 0 and 30 fm/c, particles from BIN 1 are accelerated toward the overlap zone for a longer time interval of about 20 fm/c between 10 and 30 fm/c. Moreover, the $<p_x^{\rm dir}>$ of particles lying in BIN 1 [for $F_3(u)$ and F_4] follows a similar trend. This is because for $\rho/\rho_0 < 1$, the strength of symmetry energy $F_3(u)$ will be small and so there will be less effect of symmetry energy on the particles, which is evident from Fig. 11.3c where one sees that the $<p_x^{\rm dir}>$ remains about zero during the initial stages between 0 and about 10 fm/c. Therefore, our study pointed out that the transverse flow of nucleons can act as a good probe of nuclear symmetry energy and its density dependence in the Fermi energy region.

11.2.4 Yields of Various Fragments

As mentioned in the introduction, the study of fragmentation dynamics in isotopic and isobaric colliding pairs will also be important as it will shed light on the relative role of symmetry energy and Coulomb interactions in governing the fragmentation pattern [Kohley (2014)].

In view of this, we simulated several thousand events of the reactions of $^{64}_{28}$Ni + $^{64}_{28}$Ni, $^{64}_{30}$Zn + $^{64}_{30}$Zn, $^{70}_{30}$Zn + $^{70}_{30}$Zn, $^{124}_{50}$Sn + $^{124}_{50}$Sn, $^{124}_{54}$Xe + $^{124}_{54}$Xe, and $^{136}_{54}$Xe + $^{136}_{54}$Xe at incident energies between 50 and 400 MeV/nucleon for central ($\widehat{b}$ = 0.2–0.4) colliding geometry. Again, a soft equation of state along with momentum-dependent interactions is used. To investigate the behavior of nuclear symmetry energy, we have used two density dependences of symmetry energy, i.e.,

$F_5(u) \propto u^{0.5}$ (soft symmetry energy) and $F_6(u) \propto u^{1.5}$ (stiff symmetry energy).

In Fig. 11.4, we display the yields of FNs (top panels), LCPs (middle panels), and IMFs (bottom panels) as a function of the N/Z of the colliding system [$(N/Z)_{sys}$] at incident energies of 50 (left panels), 100 (middle panels), and 400 (right panels) MeV/nucleon. Here half-filled circles (magenta) and half-filled triangles (olive) are the calculations with soft [$F_5(u)$] and stiff [$F_6(u)$] symmetry energy, respectively. We notice that the yields of FNs and LCPs increase with energy due to violent collisions at higher energies, which, on the other hand, will reduce the yields of IMFs, as observed. As evident from the figure, the yields of FNs and LCPs are quite sensitive to the density dependence of symmetry energy, whereas the yields of IMFs do not exhibit this sensitivity. Moreover, the sensitivity of free nucleons production to the symmetry energy remains almost same throughout the energy range. Similar results for IMFs have also been reported earlier [Liu (2001)] where their production does not show sensitivity to the density dependence of symmetry energy. We find that the soft symmetry energy leads to higher production of LCPs compared to stiffer one and this sensitivity toward symmetry energy increases at higher beam energies. Thus, production of LCPs can be a useful tool to probe the supra-saturation behavior of symmetry energy.

It is worth mentioning that earlier studies using QMD-type models also reported higher yields of light clusters with soft symmetry potential [Li (2005), Zhang (2005)], though reverse behavior is reported using BUU-type models [Chen (2003)]. The yield of free nucleons follows an opposite trend with stiffer symmetry energy resulting in more emission. We also observe that the yield of LCPs (and FNs) is almost similar for the reactions of $^{64}_{30}Zn + ^{64}_{30}Zn$ and $^{64}_{28}Ni + ^{64}_{28}Ni$, i.e., going from $(N/Z)_{sys} = 1.13$ to 1.29. This indicates that light cluster production is insensitive to the N/Z of the colliding system in isobaric pairs. However, the yields of LCPs increase for the reaction of $^{70}_{30}Zn + ^{70}_{30}Zn$ [$(N/Z)_{sys} = 1.33$], indicating that it is governed by the total colliding mass (or neutron content) of the system in isotopic pairs. It is worth mentioning that the *n/p* ratio of LCPs is reported to be sensitive to the N/Z of the colliding system for isotopic colliding pairs [Kumar (2012)].

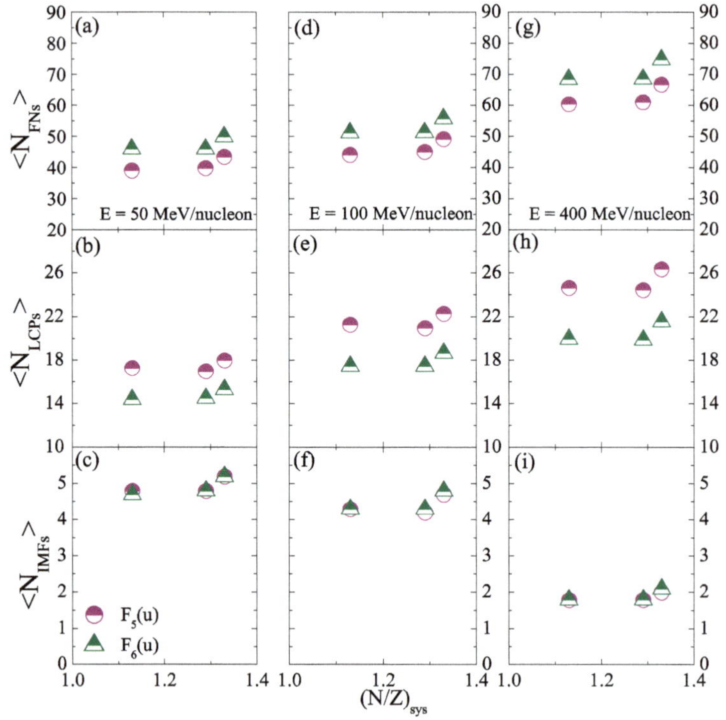

Figure 11.4 The yields of FNs (upper panels), LCPs (middle panels), and IMFs (bottom panels) as a function of the N/Z of colliding system [$(N/Z)_{sys}$] at incident energies of 50 (left panels), 100 (middle panels), and 400 (right panels) MeV/nucleon. Here half-filled circles (magenta) and half-filled triangles (olive) are the calculations with soft [$F_5(u)$] and stiff [$F_6(u)$] symmetry energy, respectively. Reprinted from Mandeep Kaur, Sakshi Gautam, and Rajeev K. Puri, Fragmentation in isotopic and isobaric systems as probe of density dependence of nuclear symmetry energy, Nuclear Physics A, 955, 133–144, Copyright (2019) with permission from Elsevier.

11.2.5 Rapidity Distribution of Fragments

To further explore the different behavior of LCPs and IMFs toward the density dependence of symmetry energy, we have shown the rapidity distribution (dN/dy_{red}) of LCPs (upper panel) and IMFs (bottom panel) at an incident energy of 50 MeV/nucleon for soft symmetry

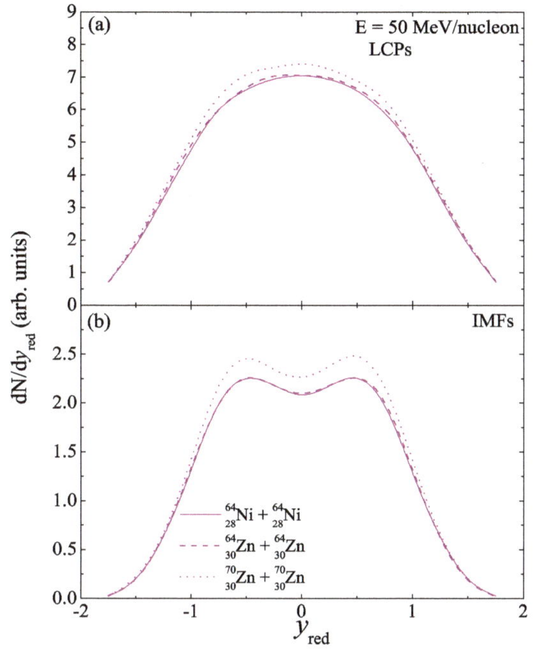

Figure 11.5 The rapidity distribution of LCPs (upper panel) and IMFs (bottom panel) for the reactions of $^{64}_{28}$Ni + $^{64}_{28}$Ni, $^{64}_{30}$Zn + $^{64}_{30}$Zn, and $^{70}_{30}$Zn + $^{70}_{30}$Zn at an incident energy of 50 MeV/nucleon for soft symmetry energy [$F_5(u)$]. Various lines are explained in the text. Reprinted from Mandeep Kaur, Sakshi Gautam, and Rajeev K. Puri, Fragmentation in isotopic and isobaric systems as probe of density dependence of nuclear symmetry energy, Nuclear Physics A, 955, 133–144, Copyright (2019) with permission from Elsevier.

energy in Fig. 11.5. The solid, dashed, and dotted lines represent the calculation for the reactions of $^{64}_{28}$Ni + $^{64}_{28}$Ni, $^{64}_{30}$Zn + $^{64}_{30}$Zn, and $^{70}_{30}$Zn + $^{70}_{30}$Zn with soft symmetry energy. From the figure, we observe that the rapidity distribution of LCPs is peaked around midrapidity [i.e., y_{red} ($=y_{c.m.}/y_{beam}$) = 0] showing their emission from the participant zone, where densities differ from their normal values. Hence, the density-dependent symmetry potential will come into play and govern the dynamics. To the contrary, the IMFs are peaked around target and projectile rapidities, which point their origin from the spectator region where density will be closer to the normal nuclear matter density and hence soft and stiff symmetry energies behave in a similar fashion. Therefore, intermediate mass fragments production is insensitive to symmetry energy. It is worth mentioning here that these results are consistent with various earlier (experiments or theoretical) studies [Thèriault (2006), Sabotka (1997)] where the nuclear symmetry energy has led to different dynamics in the midrapidity (or participant) region and target/projectile-rapidity (or spectator) region. For example, in Ref. [Thèriault (2006)], the neutron-to-proton ratio at quasi-projectile and midrapidity emission was studied for the reaction of $^{64}_{30}$Zn + $^{64}_{30}$Zn at 45 MeV/nucleon. The study revealed higher neutron-to-proton ratio in midrapidity emission compared to that for quasi-projectile because of the preferential transfer of neutrons toward midrapidity due to the density-dependent behavior of symmetry energy [Zhang (2005), Baran (1998)]. Similar findings are also reported in other measurements [Sabotka (1997)] where light clusters emitted at midrapidity have higher neutron content compared to quasi-projectile/quasi-target.

11.2.6 Phase Space of Fragments

To further strengthen our point, in Figs. 11.6 and 11.7, we have displayed the snapshot of a single event (in the reaction plane) for the reaction of $^{64}_{28}$Ni + $^{64}_{28}$Ni at 50 MeV/nucleon for the production of LCPs and IMFs, respectively. We have checked the participant and spectator matter contributions to LCPs and IMFs. Here, we have used the nucleonic concept to differentiate between participant and spectator matter following Refs. [Gautam (2012), Bansal (2015)]. We have checked the participant and spectator content in LCPs and IMFs at freeze-out time (i.e., 300 fm/c), and the corresponding nucleons are back tracked till the start of the reaction (i.e., 0.1 fm/c). Here blue (red) circles correspond to participant (spectator) nucleons belonging to LCPs and IMFs, whereas the rest ones are represented by open circles. From the figures, we observe that well-separated target and projectile at the initial stage form a compressed zone as the reaction proceeds and a significant production of LCPs (Fig. 11.6) and IMFs (Fig. 11.7) is observed at the final stage. We clearly notice a dominance of participant matter in LCPs, whereas IMFs constitute (a majority of) spectator matter, as concluded from the earlier figure. These results clearly demonstrate that different origins of LCPs and IMFs lead to their different behavior toward the density dependence of symmetry energy. It should be noted that experimental findings at Istituto Nazionale di Fisica Nucleare-Laboratori Nazionali del Sud (INFN-LNS) superconducting cyclotron of Catania, Italy, also reported [Filippo (2012)] a correlation between the emission time scale of clusters with their isotopic composition. Their findings pointed out that light clusters that are dynamically emitted during the early stage of the reaction (i.e., participant zone) have larger neutron-to-proton ratios and the isotopic composition of fragments is governed by the density dependence of nuclear symmetry energy [Filippo (2012)].

Recently, Puri and coworkers have carried out a detailed study on the fragmentation of proton-induced asymmetric reactions where even the IMF yield is observed to be influenced by the density dependence of symmetry energy [Sharma (2016)]. It is worth mentioning that though the yield of IMFs (in symmetric and nearly symmetric reactions) does not show sensitivity to the density dependence of symmetry energy, the transverse flow of IMFs is reported to be a good probe of symmetry energy [Kohley (2012), Kohley (2010)].

11.2.7 Relative Yields R_N

As mentioned in the introduction, the ratio of relative transverse flow in isotopic and isobaric colliding pairs is reported to be sensitive to the density dependence of symmetry energy. Therefore, as the next step, we also calculate the ratio of the relative yields of FNs and LCPs in the reactions of ${}^{64}_{28}$Ni + ${}^{64}_{28}$Ni, ${}^{64}_{30}$Zn + ${}^{64}_{30}$Zn, and ${}^{70}_{30}$Zn + ${}^{70}_{30}$Zn. As the IMF yield is not sensitive to the density dependence of symmetry energy, we calculate the ratio of the relative yields of FNs and LCPs only. As the ratio of relative flow is defined in Ref. [Kohley (2010)], we also similarly define the ratio of relative yields $R_{<N>}$ as:

$$R_{<N>} = \frac{<N>_{({}^{64}_{30}Zn+{}^{64}_{30}Zn)} - <N>_{({}^{70}_{30}Zn+{}^{70}_{30}Zn)}}{<N>_{({}^{64}_{28}Ni+{}^{64}_{28}Ni)} - <N>_{({}^{70}_{30}Zn+{}^{70}_{30}Zn)}}, \tag{11.3}$$

where $<N>$ can be the yield of either FNs or LCPs. The results of $R_{<FNs>}$ (upper panel) and $R_{<LCPs>}$ (bottom panel) are displayed

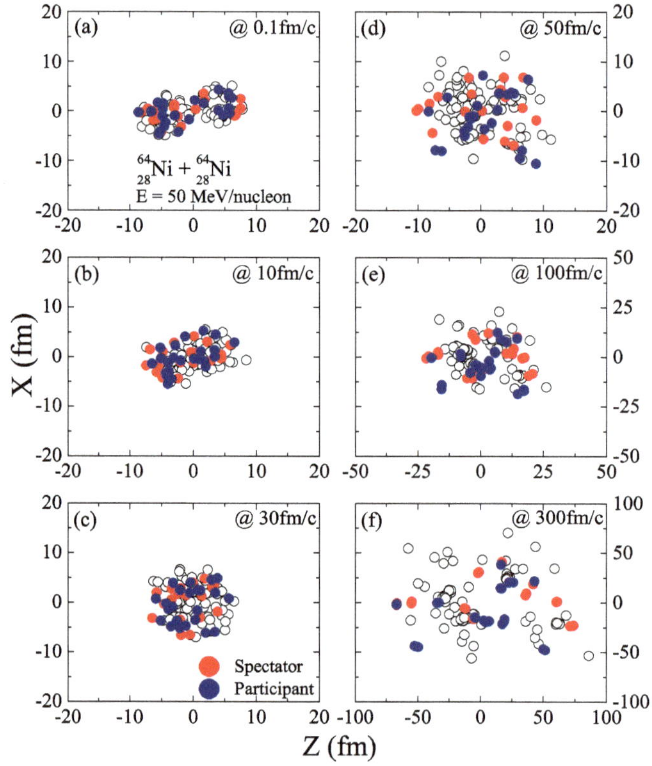

Figure 11.6 The snapshot of a single event for the reaction of $^{64}_{28}$Ni + $^{64}_{28}$Ni at an incident energy of 50 MeV/nucleon for the production of light-charged particles (LCPs). Here blue (red) circles represent participant (spectator) nucleons belonging to LCPs and rest of the nucleons are represented by open circles. Reprinted from Mandeep Kaur, Sakshi Gautam, and Rajeev K. Puri, Fragmentation in isotopic and isobaric systems as probe of density dependence of nuclear symmetry energy, Nuclear Physics A, 955, 133–144, Copyright (2019) with permission from Elsevier.

in Fig. 11.8 as a function of the incident energy. Here, half-filled circles (magenta) and half-filled triangles (olive) are the calculations with soft and stiff symmetry energy, respectively. From the figure, we observe that both R $<_{FNs}>$ and R $<_{LCPs}>$ are sensitive to the density dependence of symmetry energy, though the sensitivity decreases

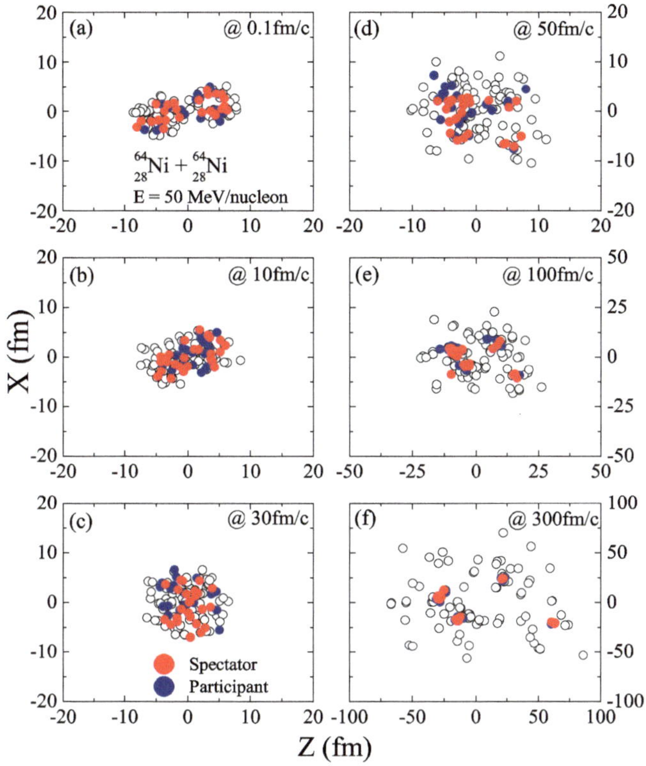

Figure 11.7 Same as Fig. 11.6, but for the production of intermediate mass fragments. Reprinted from Mandeep Kaur, Sakshi Gautam, and Rajeev K. Puri, Fragmentation in isotopic and isobaric systems as probe of density dependence of nuclear symmetry energy, Nuclear Physics A, 955, 133–144, Copyright (2019) with permission from Elsevier.

at very high energy where nucleon–nucleon scattering (rather than mean field) will govern the dynamics. These results are in accordance with the ones reported in Ref. [Zhang (2009)] where the pion ratio in the reactions of $^{48}_{20}$Ca + $^{48}_{20}$Ca, $^{124}_{50}$Sn + $^{124}_{50}$Sn, and $^{197}_{79}$Au + $^{197}_{79}$Au at beam energies from 250 to 600 MeV/nucleon is studied using the IBUU model. The study reported a strong sensitivity of the pion ratio toward symmetry energy at lower beam energies and which gets weakened at higher energies. We also find that the sensitivities

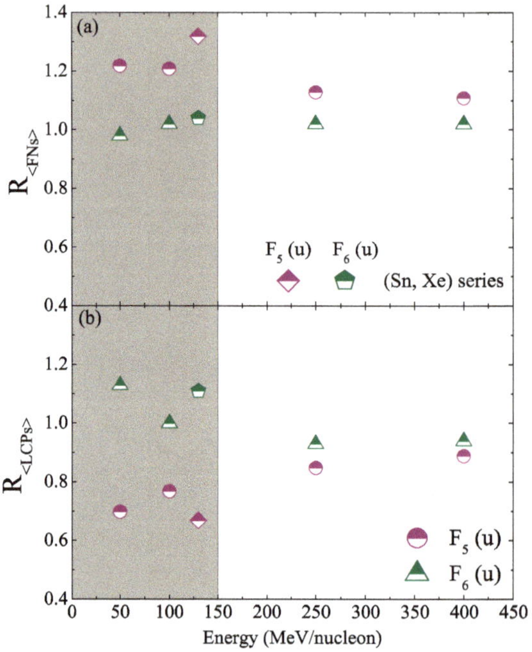

Figure 11.8 R $<_{FNs}>$ (top panel) and R $<_{LCPs}>$ (bottom panel) as a function of the incident energy with soft $[F_5(u)]$ and stiff $[F_6(u)]$ forms of symmetry energy. Various symbols are explained in the text. Reprinted from Mandeep Kaur, Sakshi Gautam, and Rajeev K. Puri, Fragmentation in isotopic and isobaric systems as probe of density dependence of nuclear symmetry energy, Nuclear Physics A, 955, 133–144, Copyright (2019) with permission from Elsevier.

of $R <_{FNs}>$ and bare free nucleons yield are almost same as both have $\sim$ 10–20% sensitivity toward different density dependence of symmetry energy. However, $R <_{LCPs}>$ show up to 60% sensitivity compared to $\sim$ 15–20% sensitivity shown by bare yield of LCPs. Here, it is worth mentioning that enhanced sensitivity is also reported for the ratio of relative flow rather than individual flow in various colliding systems [Kohley (2010)]. Therefore, the ratio of the relative yield of LCPs can act as a better candidate to constrain the density-dependent behavior of symmetry energy in Fermi energy region (shown by the shaded area). Though the ratio of relative yields acts as a promising probe of nuclear symmetry energy, it should be noted that coalescence algorithm (as used in the present study to identify clusters) with transport model calculations normally overestimates (underestimates) the yields of nucleons (clusters) [Wang (2014)]. But this effect gets weakened here as we are calculating the ratio of relative yields and they show almost similar behavior.

Further, to see the effect of the total colliding mass on the sensitivity of $R_{<N>}$ to symmetry energy, we have also performed the calculations for heavier colliding systems of $^{124}_{50}$Sn + $^{124}_{50}$Sn, $^{124}_{54}$Xe + $^{124}_{54}$Xe, and $^{136}_{54}$Xe + $^{136}_{54}$Xe at 100 MeV/nucleon and the results are shown by half-filled diamonds (magenta) and half-filled pentagons (olive) for soft and stiff symmetry energy, respectively. Note that results for heavier colliding systems are displaced horizontally for the clarity of the figure. The ratio of the relative yields of FNs and LCPs in heavier isotopic and isobaric colliding pairs shows higher sensitivity ($\sim 65\%$) compared to medium colliding pairs and, thus, poses as a better candidate to probe and constrain symmetry energy. These results, along with enhanced sensitivity at lower energies, indicate that observables originating from long-time process (or larger time–space volume of collisions) pose better candidates to probe the behavior of nuclear symmetry energy. These findings are in accordance with the studies [Hudan (2012)] that favored long-lived processes to study the density-dependent potential in heavy-ion collisions.

11.2.8 Neutron-to-Proton (*n/p*) Ratio of Free Nucleons

As stated in the introduction, the *n/p* ratio is also found to be another good candidate to probe the density-dependent behavior of nuclear symmetry energy since the production of neutrons (as well as protons) from neutron-rich nuclei has a direct relation with the symmetry energy term (in the semi-empirical mass formula) [Kumar (2012), Yong (2007), Zhang (2014)]. So, as the next step, we will redefine the relative ratio using the *n/p* ratio of free nucleons. Therefore, the equation can be rewritten as:

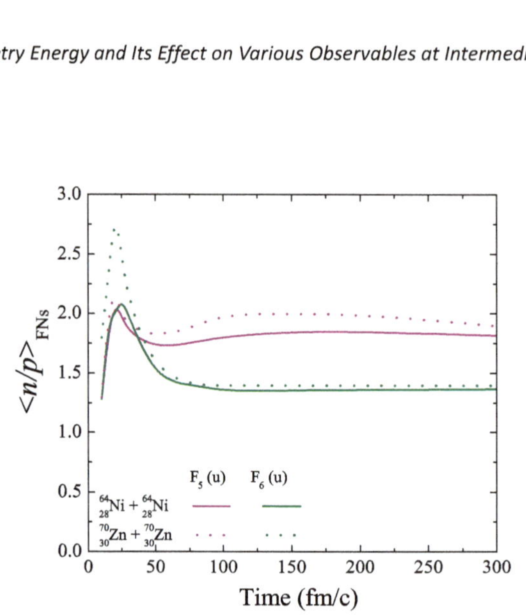

Figure 11.9 The time evolution of the n/p ratio of free nucleons emitted in the reactions of $^{64}_{28}$Ni + $^{64}_{28}$Ni and $^{70}_{30}$Zn + $^{70}_{30}$Zn at incident energies of 50 and 100 MeV/nucleon, respectively. Various lines are explained in the text.

$$R_X = \frac{X_{(^{70}_{30}Zn+^{70}_{30}Zn)}/X_{(^{64}_{30}Zn+^{64}_{30}Zn)}}{X_{(^{64}_{28}Ni+^{64}_{28}Ni)}/X_{(^{64}_{30}Zn+^{64}_{30}Zn)}}, \tag{11.4}$$

where X is the ratio of free neutrons to protons (i.e., $X = n/p$).

In Fig. 11.9, we display the time evolution of the n/p ratio of free nucleons for the reactions of $^{64}_{28}$Ni + $^{64}_{28}$Ni and $^{70}_{30}$Zn + $^{70}_{30}$Zn at incident energies of 50 and 100 MeV/nucleon, respectively. The solid and dotted lines represent the calculation for the reactions of $^{64}_{28}$Ni + $^{64}_{28}$Ni and $^{70}_{30}$Zn + $^{70}_{30}$Zn with soft (magenta) and stiff (olive) symmetry energies, respectively. From the figure, we observe that during the early stage of the reaction, stiff symmetry energy leads to more n/p ratio than the soft one. This is because of the fact that during the compressional stage, the density achieved is greater than the normal nuclear matter density (ρ_0) and, therefore, the stiff symmetry energy will be more repulsive (attractive) for neutrons (protons) than the soft one, thus yielding a higher n/p ratio with the former one. However, as the reaction proceeds, i.e., during the expansion

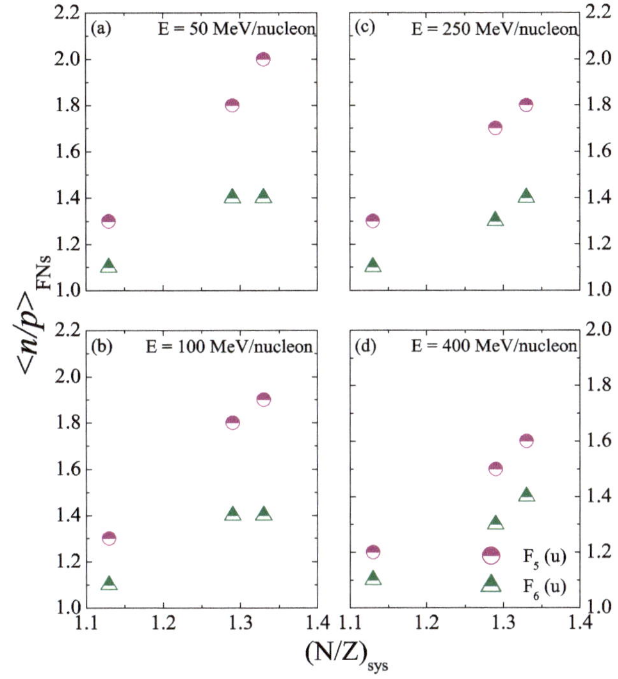

Figure 11.10 The n/p ratio of free nucleons as a function of N/Z of colliding pairs [$(N/Z)_{sys}$] with soft [$F_5(u)$] and stiff [$F_6(u)$] forms of symmetry energy. Various symbols have the same meaning as in Fig. 11.6.

phase, lower densities are achieved and thus soft symmetry potential will be more repulsive (for neutrons), resulting in a higher n/p ratio. We also see that the sensitivity of n/p ratio toward the stiffness of symmetry energy is higher for a neutron-rich system (i.e., ${}^{70}_{30}Zn$ + ${}^{70}_{30}Zn$) as expected.

Next, to see the systematic behavior of the n/p ratio of free nucleons with beam energy, we performed the calculations at 250 and 400 MeV/nucleon also and the results are displayed in Fig. 11.10. The n/p ratios of free nucleons (at the freeze-out stage) for the reactions of ${}^{64}_{30}Zn$ + ${}^{64}_{30}Zn$ (N/Z=1.13), ${}^{64}_{28}Ni$ + ${}^{64}_{28}Ni$ (N/Z=1.29), and ${}^{70}_{30}Zn$ + ${}^{70}_{30}Zn$ (N/Z=1.33) are displayed as a function of the N/Z of the colliding pair at incident energies of 50 (top left panel), 100 (bottom left panel), 250 (top right panel), and 400 (bottom right

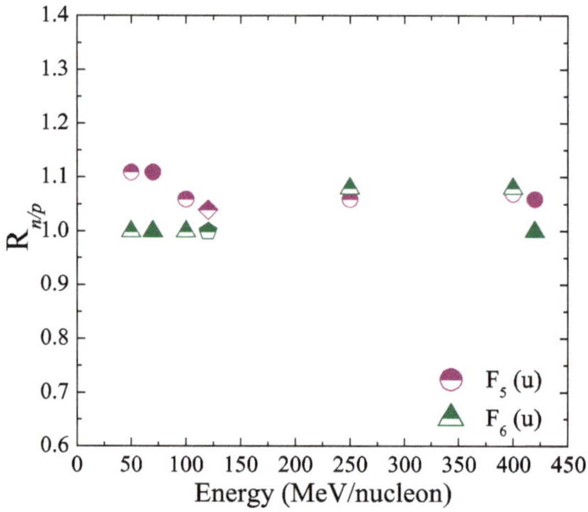

Figure 11.11 $R_{n/p}$ as a function of incident energy using soft [$F_5(u)$] and stiff [$F_6(u)$] forms of symmetry energy. Various symbols are explained in the text.

panel) MeV/nucleon. All symbols have the same meaning as in Fig. 11.4. From the figure, we observe that the *n/p* ratio decreases with incident energy for soft symmetry energy. This happens because of a stronger (repulsive) character of the symmetry energy for neutrons (protons) at lower incident energies, which will enhance the *n/p* ratio. On the other hand, at high beam energies, Coulomb interactions dominate over symmetry energy, which will be repulsive for protons and hence lowering the *n/p* ratio. Therefore, the relative competition between Coulomb and symmetry potentials governs the dynamics [Wu (2015)]. We also find that soft symmetry energy leads to a higher *n/p* ratio of free nucleons at the freeze-out stage compared to stiff symmetry energy for all colliding pairs as discussed earlier and also reported in Ref. [Kumar (2011)]. One also observes that the sensitivity of the *n/p* ratio toward the stiffness of symmetry energy increases as the N/Z of the colliding system increases.

To see the sensitivity of the relative *n/p* ratio of free nucleons toward symmetry energy in isotopic and isobaric colliding pairs, in Fig. 11.11, we display $R_{n/p}$ [given by Eq. (11.4)] as a function of incident energy for the central collisions of $^{64}_{28}$Ni + $^{64}_{28}$Ni, $^{64}_{30}$Zn + $^{64}_{30}$Zn, and $^{70}_{30}$Zn + $^{70}_{30}$Zn. Here, half-filled circles (magenta) and half-filled triangles (olive) are the calculations with soft and stiff symmetry energy, respectively. From the figure, we notice that the sensitivity of $R_{n/p}$ to symmetry energy is reduced compared to that of bare *n/p* ratio. Also, $R_{n/p}$ is almost independent of the beam energy. To further investigate the role of system mass and colliding geometry on $R_{n/p}$, we perform similar calculations by simulating central reactions of the heavier colliding pairs of $^{124}_{50}$Sn + $^{124}_{50}$Sn, $^{124}_{54}$Xe + $^{124}_{54}$Xe, and $^{136}_{54}$Xe + $^{136}_{54}$Xe at an incident of 100 MeV/nucleon. The results are displayed by half-filled diamonds (magenta) and half-filled pentagons (olive) for soft and stiff symmetry energies, respectively. We again notice very less sensitivity of $R_{n/p}$ to symmetry energy. Similarly, the above conclusion remains valid for peripheral collisions also where we simulate the peripheral collisions of $^{64}_{28}$Ni + $^{64}_{28}$Ni, $^{64}_{30}$Zn + $^{64}_{30}$Zn, and $^{70}_{30}$Zn + $^{70}_{30}$Zn at 100 and 400 MeV/nucleon. The results are displayed by solid circles (magenta) and solid triangles (olive) for soft and stiff symmetry energies, respectively. Note that the results of the heavier colliding pairs and peripheral collision have been displaced slightly in the horizontal direction to maintain the clarity of the figure. The above analysis reveals that the relative *n/p* ratio of free nucleons in isotopic and isobaric colliding pairs does not show any significant sensitivity to different density dependence of symmetry energy (independent of incident energy, mass, and colliding geometry) and hence the bare *n/p* ratio serves as a better candidate to probe and constrain the density dependence of symmetry energy.

11.3 Summary

We have studied the sensitivity of directed transverse flow to symmetry energy and its density dependence in the Fermi energy region as well as at higher incident energies for semi-central and peripheral collisions. Our study has revealed that transverse flow shows sensitivity to symmetry energy in the Fermi energy region and is insensitive to its strength and density dependence at higher energies because of the dominance of nucleon–nucleon scattering. We also presented a detailed mechanism showing the behavior of directed transverse flow to symmetry energy and found that the acceleration time of particles toward the central overlap zone depends on the strength of the symmetry energy as well as on its density dependence, which in turn governs the saturation value of the flow.

We have also studied the fragmentation in the central collisions of neutron-rich isotopic and isobaric colliding pairs. Our study revealed that the IMF production is insensitive toward the density dependence of symmetry energy; however, the production of light clusters and free particles is influenced by the density dependence of symmetry energy. In addition, the ratio of the relative yield of LCPs in isotopic and isobaric colliding pairs is a better candidate to probe and constrain the nuclear symmetry energy than the bare yield of lighter fragments. Further, a systematic study on the n/p ratio of free nucleons was performed and relative neutron-to-proton ratio was also calculated for isotopic and isobaric colliding pairs. Our investigations revealed that the bare n/p ratio in individual systems acts as a better probe for different density dependence of nuclear symmetry energy than the relative n/p ratio.

Acknowledgment

The authors are thankful to Prof. Rajeev K. Puri for giving access to various computer programs used for the present study.

References

Bansal, P., Gautam, S., and Puri, R. K. (2015). On the peak mass production of different fragments in intermediate-energy heavy-ion collisions, *Eur. Phys. J. A* **51**, pp. 139.

Baran, V. *et al.* (1998). Spinodal decomposition of low-density asymmetric nuclear matter, *Nucl. Phys. A* **632**, pp. 287.

Baran, V. *et al.* (2005). Reaction dynamics with exotic nuclei, *Phys. Rep.* **410**, pp. 335–466.

Brueckner, K. A. *et al.* (1968). Nuclear symmetry energy, *Phys. Rev.* **168**, pp. 1184.

Chen, L. W, Ko, C. M., and Li, B. A. (2005). Determination of the stiffness of the nuclear symmetry energy from isospin diffusion, *Phys. Rev. Lett.* **94**, pp. 032701.

Chen, L. W. *et al.* (2010). Density slope of the nuclear symmetry energy from the neutron skin thickness of heavy nuclei, *Phys. Rev. C* **82**, pp. 024321.

Chen, L. W., Ko, C. M., and Li, B. A. (2003). Light cluster production in intermediate energy heavy-ion collisions induced by neutron-rich nuclei, *Nucl. Phys. A* **729**, pp. 809.

Chen, L. W., Ko, C. M., and Li, B. A. (2005). Nuclear matter symmetry energy and the neutron skin thickness of heavy nuclei, *Phys. Rev. C* **72**, pp. 064309.

Chen, L. W., Ko, C. M., and Li, B. A. (2007). Isospin-dependent properties of asymmetric nuclear matter in relativistic mean field models, *Phys. Rev. C* **76**, pp. 054316.

Colonna, M. and Tsang, M. B. (2006). Isotopic compositions and scalings, *Eur. Phys. J. A.* **30**, pp. 165.

Cozma, M. D. (2011). Neutron–proton elliptic flow difference as a probe for the high density dependence of the symmetry energy, *Phys. Lett. B* **700**, pp. 139.

Dieperink, A. E. L. *et al.* (2003). Nuclear symmetry energy and the neutron skin in neutron-rich nuclei, *Phys. Rev. C* **68**, pp. 064307.

Feng, Z. Q and Jin, G. M. (2010). Probing high-density behavior of symmetry energy from pion emission in heavy-ion collisions, *Phys. Lett. B* **683**, pp. 140.

Feng, Z. Q. (2012). Effective mass splitting of neutron and proton and isospin emission in heavy-ion collisions, *Nucl. Phys. A* **878**, pp. 3.

Filippo, E. De *et al.* (2012). Correlations between emission timescale of fragments and isospin dynamics in ^{124}Sn + ^{64}Ni and ^{112}Sn + ^{58}Ni reactions at 35 AMeV *Phys. Rev. C* **86**, pp. 014610.

Gautam, S. and Puri, R. K. (2012). Participant-spectator matter and thermalization of neutron-rich systems at the energy of vanishing flow, *Phys. Rev. C* **85**, pp. 067601.

Huang, M. *et al.* (2010). Isobaric yield ratios and the symmetry energy in heavy-ion reactions near the Fermi energy, *Phys. Rev. C* **81**, pp. 044620.

Hudan, S. *et al.* (2012). Tracking saddle-to-scission dynamics using N/Z in projectile breakup reactions, *Phys. Rev. C* **86**, pp. 021603.

Kohley, Z. and Yennello, S. J. (2014). Heavy-ion collisions: Direct and indirect probes of the density and temperature dependence of E_{sym}, *Eur. Phys. J. A* **50**, pp. 31.

Kohley, Z. *et al.* (2010). Investigation of transverse collective flow of intermediate mass fragments, *Phys. Rev. C* **82**, pp. 064601.

Kohley, Z. *et al.* (2011). Transverse collective flow and midrapidity emission of isotopically identified light charged particles, *Phys. Rev. C* **83**, pp. 044601.

Kohley, Z. *et al.* (2012). Sensitivity of intermediate mass fragment flows to the symmetry energy, *Phys. Rev. C* **85**, pp. 064605.

Kumar, S. *et al.* (2011). Probing the density dependence of the symmetry energy via multifragmentation at subsaturation densities, *Phys. Rev. C* **84**, pp. 044620.

Kumar, S. *et al.* (2012). Sensitivity of neutron to proton ratio toward the high density behavior of the symmetry energy in heavy-ion collisions, *Phys. Rev. C* **85**, pp. 024620.

Lehaut, G. *et al.* (2010). Study of nuclear stopping in central collisions at intermediate energies, *Phys. Rev. Lett.* **104**, pp. 232701.

Li, B. A. (2000). Neutron–proton differential flow as a probe of isospin-dependence of the nuclear equation of state, *Phys. Rev. Lett.* **85**, pp. 4221.

Li, B. A. (2002). Probing the high density behavior of the nuclear symmetry energy with high energy heavy-ion collisions, *Phys. Rev. Lett.* **88**, pp. 192701.

Li, B. A. and Chen, L. W. (2005). Nucleon-nucleon cross sections in neutron-rich matter and isospin transport in heavy-ion reactions at intermediate energies, *Phys. Rev. C* **72**, pp. 064611.

Li, B. A. *et al.* (1996). Isospin dependence of collective flow in heavy-ion collisions at intermediate energies, *Phys. Rev. Lett.* **76**, pp. 4492.

Li, B. A. *et al.* (2004). Momentum dependence of the symmetry potential and nuclear reactions induced by neutron-rich nuclei at RIA, *Phys. Rev. C* **69**, pp. 011603 (R).

Li, B. A. *et al.* (2006). Double neutron/proton ratio of nucleon emissions in isotopic reaction systems as a robust probe of nuclear symmetry energy, *Phys. Lett. B* **634**, pp. 378.

Li, B. A., Chen, L. W., and Ko, C. M. (2008). Recent progress and new challenges in isospin physics with heavy-ion reactions, *Phys. Rep.* **464**, pp. 113–281.

Li, B. A., Ko, C. M., and Ren Z. (1997). Equation of state of asymmetric nuclear matter and collisions of neutron-rich nuclei, *Phys. Rev. Lett.* **78**, pp. 1644.

Li, Q. *et al.* (2005). Σ^-/Σ^+ ratio as a candidate for probing the density dependence of the symmetry potential at high nuclear densities, *Phys. Rev. C* **71**, pp. 054907.

Li, Q. *et al.* (2005). Probing the density dependence of the symmetry potential in intermediate-energy heavy ion collisions, *J. Phys. G: Nucl. Part. Phys.* **31**, pp. 1359.

Li, Q. *et al.* (2005). Probing the density dependence of the symmetry potential at low and high densities, *Phys. Rev. C.* **72**, pp. 034613.

Li, T. *et al.* (2010). Isoscalar giant resonances in the Sn nuclei and implications for the asymmetry term in the nuclear-matter incompressibility, *Phys. Rev. C* **81**, pp. 034309.

Liu, J. Y. *et al.* (2001). Isospin effect on the process of multifragmentation and dissipation at intermediate energy heavy ion collisions, *Phys. Rev. C* **63**, pp. 054612.

Ma, C. W. *et al.* (2011). Isobaric yield ratios in heavy-ion reactions, and symmetry energy of neutron-rich nuclei at intermediate energies, *Phys. Rev. C* **83**, pp. 064620.

Meyers, W. D. and Swiatecki, W. J. (1996). Nuclear masses and deformations. *Nucl. Phys. A*, **81**, pp. 1.

Ono, A. *et al.* (2003). Isospin fractionation and isoscaling in dynamical simulations of nuclear collisions, *Phys. Rev. C* **68**, pp. 051601 (R).

Pak, R. *et al.* (1997). Isospin dependence of collective transverse flow in nuclear collisions, *Phys. Rev. Lett.* **78**, pp. 1022.

Pak, R. *et al.* (1997). Isospin dependence of the balance energy, *Phys. Rev. Lett.* **78**, pp. 1026.

Russotto, P. *et al.* (2011). Symmetry energy from elliptic flow in ^{197}Au + ^{197}Au, *Phys. Lett. B* **697**, pp. 471.

Russotto, P. *et al.* (2014). Flow probe of symmetry energy in relativistic heavy-ion reactions, *Eur. Phys. J. A.* **50**, pp. 38.

Sabotka, L. G. *et al.* (1997). Clustered and neutron-rich low density neck region produced in heavy-ion collisions, *Phys. Rev. C* **55**, pp. 2109.

Sharma, A., (2016). Thesis, Panjab University.

Shetty, D. V, Yennello, S. J., and Souliotis, G. A. (2007). Density dependence of the symmetry energy and the nuclear equation of state: A dynamical and statistical model perspective, *Phys. Rev. C* **76**, pp. 024606.

Shetty, D. V. *et al.* (2007). Density dependence of the symmetry energy and the equation of state of isospin asymmetric nuclear matter, *Phys. Rev. C* **75**, pp. 034602.

Steiner, A. W. *et al.* (2005). Isospin asymmetry in nuclei and neutron stars, *Phys. Rep.* **411**, pp. 325–375.

Sun, Z. Y. *et al.* (2010). Isospin diffusion and equilibration for Sn + Sn collisions at E / A = 35 MeV, *Phys. Rev. C* **82**, PP. 051603.

Thèriault, D. *et al.* (2006). Neutron-to-proton ratios of quasiprojectile and midrapidity emission in the ^{64}Zn + ^{64}Zn reaction at 45 MeV/nucleon, *Phys. Rev. C* **74**, pp. 051602.

Tsang, M. B. *et al.* (2009). Constraints on the density dependence of the symmetry energy, *Phys. Rev. Lett.* **102**, pp. 122701.

Tsang, M. B. *et al.* (2011). Constraints on the density dependence of the symmetry energy from heavy-ion collisions, *Prog. Part. Nucl. Phys.* **66**, pp. 400.

Wang, R. S. *et al.* (2014). Time-dependent isospin composition of particles emitted in fission events following ^{40}Ar + ^{197}Au at 35 MeV/u, *Phys. Rev. C* **89**, pp. 064613.

Wu, Q. *et al.* (2015). Competition between Coulomb and symmetry potential in semi-peripheral heavy ion collisions, *Phys. Rev. C* **91**, pp. 014617.

Xiao, Z. *et al.* (2009). Circumstantial evidence for a soft nuclear symmetry energy at suprasaturation densities, *Phys. Rev. Lett.* **102**, pp. 062502.

Xie, W. J. *et al.* (2015). Probing the momentum-dependent symmetry potential via nuclear collective flows, *Phys. Rev. C* **91**, pp. 054609.

Xu, C., Chen, L. W., and Li, B. A. (2010). Symmetry energy, its density slope, and neutron-proton effective mass splitting at normal density extracted from global nucleon optical potentials, *Phys. Rev. C* **82**, pp. 054607.

Yong, G. C. and Li, B. A. (2013). Effect of nuclear symmetry energy on η meson production and its rare decay to the dark U-boson in heavy-ion reactions, *Phys. Lett. B* **723**, pp. 388.

Yong, G. C., Li, B. A., and Chen, L. W. (2006). Double Neutron–proton differential transverse flow as a probe for the high density behavior of the nuclear symmetry energy, *Phys. Rev. C* **74**, pp. 064617.

Yong, G. C., Li, B. A., and Chen, L. W. (2007). The neutron/proton ratio of squeezed-out nucleons and the high density behavior of the nuclear symmetry energy, *Phys. Lett. B* **650**, pp. 344.

Zhang, G. Q. *et al.* (2011). Unified description of nuclear stopping in central heavy-ion collisions from 10 AMeV to 1.2 AGeV, *Phys. Rev. C* **84**, pp. 034612.

Zhang, M. *et al.* (2009). Systematic study of the π^-/π^+ ratio in heavy-ion collisions with the same neutron/proton ratio but different masses, *Phys. Rev. C* **80**, pp. 034616.

Zhang, Y. and Li, Z. (2005). Probing the density dependence of the symmetry potential with peripheral heavy-ion collisions, *Phys. Rev. C* **71**, pp. 024604.

Zhang, Y. *et al.* (2012). Influence of in-medium NN cross sections, symmetry potential, and impact parameter on isospin observables, *Phys. Rev. C* **85**, pp. 024602.

Chapter 12

Can We Constraint Density Dependence of Symmetry Energy Using Halo Nuclei Reactions?

Sucheta,[a] **Rohit Kumar,**[b] **and Rajeev K. Puri**[a]

[a] *Department of Physics, Panjab University, Chandigarh 160014, India*

[b] *Facility for Rare Isotope Beams (FRIB), Michigan State University, East Lansing, Michigan 48824, USA*

drrkpuri@gmail.com

12.1 Introduction

In the recent times, the major provocation in heavy-ion physics is to understand the isospin effects, i.e., symmetry energy, nucleon–nucleon cross section, etc. in the asymmetric nuclear matter. It takes part in a number of phenomena at low, intermediate, and high incident energies, such as fusion–fission, cluster radioactivity, multifragmentation, nuclear stopping, flow, and pion–kaon production. At intermediate energies, a large number of observables have been presented in the literature, which are found to be sensitive probes toward the density dependence of symmetry energy; e.g., isotopic diffusion [Li (2008), Tsang (2004), Sun (2010)], isobaric ratio [Li (2008)], single and double neutron–proton ratios [Li (2008), Kumar (2011), Li (1997), Famiano (2006), Zhang (2008)], isospin

Multifragmentation in Heavy-Ion Reactions: Theory and Experiments
Edited by Rajeev K. Puri, Yu-Gang Ma, and Arun Sharma

ISBN 978-981-4968-69-0 (Hardcover), 978-1-003-38513-4 (eBook)
www.jennystanford.com

diffusion [Tsang (2009)], isospin fractionation [Li (2008), Li (1997)], π^+/π^- [Li (2008), Li (2005)], Kaons K^+/K^0 [Ferini (2005)], and $\sum^+/\sum^-$ [Toro (2007)]. These observables have their relative importance depending on the region of density that one wants to explore.

From the preceding studies, the neutron-to-proton ratios and the isospin fractionation are identified as one of the most promising observables to explore the isospin physics. To mention a few studies, Famiano *et al.* studied the reactions of Sn + Sn at an incident energy of 50 MeV/nucleon and measured the double neutron–proton ratios. They compared their results with the BUU97 model [Li (1997)] and ceased their results with γ = 0.5 [Famiano (2006)] (see Eq. 3 for the definition of γ). A significant progress for the isospin and n/p double ratio was made by Tsang *et al.* They performed the study, and the results are ceased for the values of γ between 0.4 and 1.05 [Tsang (2009)]. Using the Improved Quantum Molecular Dynamics (ImQMD) model, Zhang *et al.* [Zhang (2008)] studied the n/p yields for the reactions of ^{124}Sn+^{124}Sn and successfully explained the experimental data with γ = 0.5. Using the Isospin-dependent Quantum Molecular Dynamics (IQMD) model, Kumar *et al.* studied the effect of the density dependence of symmetry energy on single and double neutron–proton ratios for the reactions of ^{112}Sn + ^{112}Sn, ^{124}Sn + ^{124}Sn, and ^{132}Sn + ^{132}Sn at 50 MeV/nucleon of energy [Kumar (2011)]. It is worth mentioning that one also struggles to isolate isospin effects in symmetry energy and cross section as that most of the observables are found to be sensitive to both.

In addition to the above study, there is an ongoing effort within the community of nuclear physicists in exploring the properties of the poorly known exotic nuclei, which are away from the line of stability, such as halo nuclei. The topic of halo nuclei has generated interest and excitement ever since its discovery in the mid-1980s. The main credit behind their discovery goes to Tanihata *et al.* for his work at the Lawrence Berkeley Laboratory's Bevalac in 1985 [Tanihata (1985)]. This topic of halo nuclei is an approach that arises due to the weak binding of the last one or two nucleons, which are spatially decoupled from a tightly bound inert "core" holding all other nucleons, e.g., ^{6}He, ^{11}Li, ^{11}Be, ^{8}B, ^{14}B, ^{14}Be, ^{19}B, ^{19}C, ^{29}Ne, ^{31}Ne, ^{35}Mg, and ^{37}Mg [Tanihata (1988), Tanihata (1992),

Ozawa (2001), Takechi (2012), Takechi (2014), Kobayashi (2014)]. In the last few decades, some investigations related to fragmentation at intermediate energies have been carried out using exotic nuclei as projectile/target. For instance, using the IQMD model, the ratio of light-charged clusters was studied to understand the proton skin effects for the neutron-deficient system of ^{58}Ni + ^{58}Ni. They found that the yield ratios have exponential dependence on the neutron–proton skin thickness [Yan (2020)]. In another study, Liu *et al.* studied the reactions of halo nuclei and found that the effect of the halo-structured nuclei on fragmentation to be more prominent at lower incident energies that gradually disappears at higher incident energies [Liu (2005)]. In the present study, we performed a detailed investigation for various halo-induced reactions to explore if these reactions can be vital for understanding the isospin effects in heavy-ion reactions.

12.2 The Model

In the present study, we have used the isospin-dependent quantum molecular dynamics (IQMD) model. This model is an n-body theory that successfully explains the heavy-ion collisions from low to relativistic energies. This model treats different charge states of nucleons, deltas, and pions especially as inherited from the Vlasov–Uehling–Uhlenbeck (VUU) model. Here, in this model, the propagation of hadrons is governed by the classical Hamilton's equations of motion:

$$\dot{\mathbf{r}}_i = \frac{\partial \langle H \rangle}{\partial \mathbf{p}_i}; \qquad \dot{\mathbf{p}}_i = -\frac{\partial \langle H \rangle}{\partial \mathbf{r}_i}, \tag{12.1}$$

In this model, the baryon–baryon potential V_{ij} can be expressed as:

$$V_{ij} = t_1\,\delta(\mathbf{r}_i - \mathbf{r}_j) + t_2\,\delta(\mathbf{r}_i - \mathbf{r}_j)\rho^{\gamma' - 1}((\mathbf{r}_i + \mathbf{r}_j)/2) + t_3\,\frac{e^{-|\mathbf{r}_i - \mathbf{r}_j|/\mu}}{|\mathbf{r}_i - \mathbf{r}_j|/\mu}$$
$$+\frac{Z_i Z_j e^2}{|\mathbf{r}_i - \mathbf{r}_j|} + t_6\frac{1}{\rho_o}T_{3i}T_{3j}\delta(\mathbf{r}_i - \mathbf{r}_j). \tag{12.2}$$

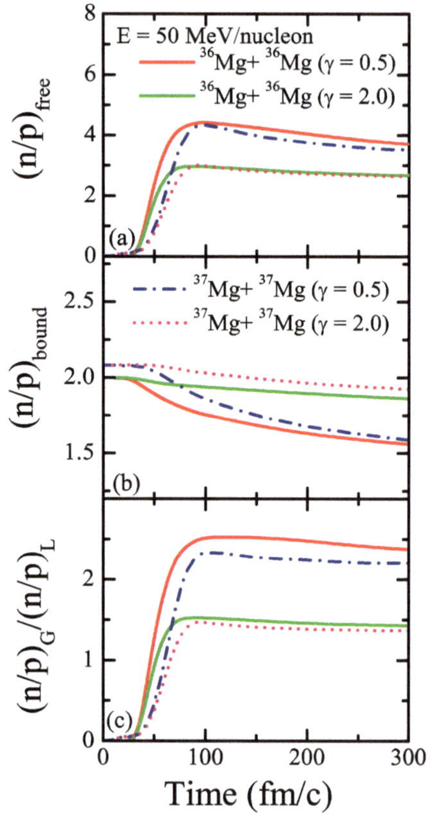

Figure 12.1 Time evolution of $(n/p)_{free}$ (top panels), $(n/p)_{bound}$ (middle panels), and $(n/p)_G/(n/p)_L$ (bottom panels) for the central collisions of ^{36}Mg + ^{36}Mg (solid lines) and ^{37}Mg + ^{37}Mg (dashed line) at 50 MeV/nucleon of energy for the soft and stiff symmetry potentials as explained in the text.

In the above equation, Z_i and Z_j represent the charges of ith and j^{th} baryons, and T_{3i}, T_{3j} are their respective T_3 components, 1/2 for protons and $-1/2$ for neutrons. The parameters t_1, t_2, t_3, ..., t_6 are fitted to the real part of the nucleonic optical potential. The standard Skyrme type parameterization is employed for the density dependence of nucleonic optical potential. The equation for the

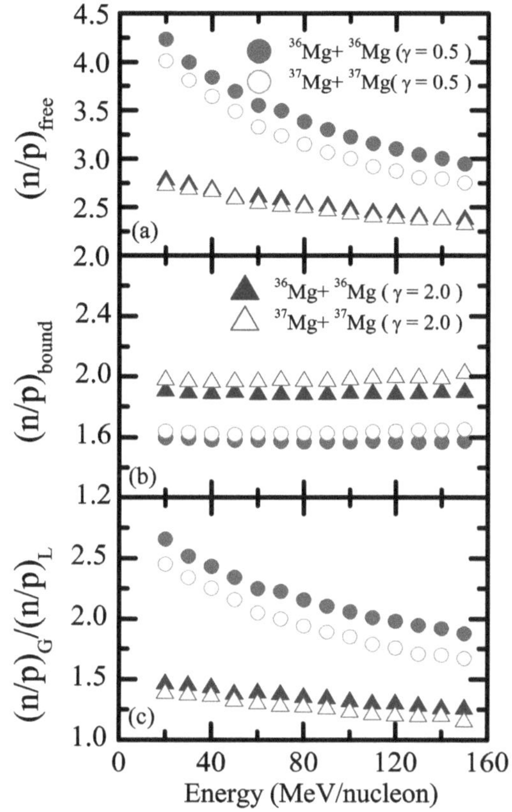

Figure 12.2 The $(n/p)_{free}$, $(n/p)_{bound}$, and $(n/p)_G/(n/p)_L$ as a function of incident beam energy for the central reactions of ^{36}Mg + ^{36}Mg and ^{37}Mg + ^{37}Mg for soft and stiff equation of state. Various symbols are explained in the text.

density dependence of symmetry energy can be written as:

$$E_{sym}(\rho) = E_{sym}(\rho_0)(\frac{\rho}{\rho_0})^{\gamma}, \quad (12.3)$$

where $E_{sym}(\rho_0)$ represents the symmetry energy at normal nuclear density. The strength of the symmetry energy is different for different values of γ, i.e., $\gamma = 0.5$ and $\gamma = 2.0$ corresponds to the soft and stiff density dependence of symmetry energy, respectively. The IQMD model is used to generate the phase space of the nucleons,

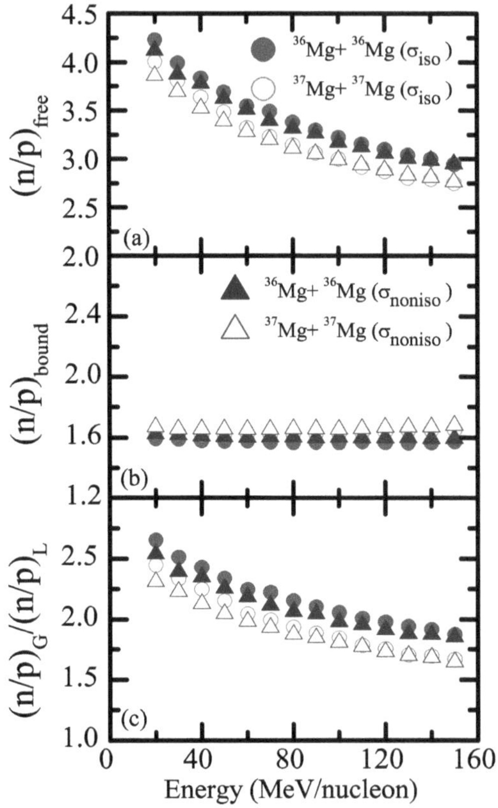

Figure 12.3 Same as Fig. 12.2, but for σ_{iso} and σ_{noniso} nucleon–nucleon cross sections.

which is further subjected to spatial constraints for constructing the fragments.

12.3 Results and Discussion

To perform the present study, thousands of events are generated for the central reactions of stable mass nuclei, i.e., ^{8}Li + ^{8}Li, ^{18}C + ^{18}C, ^{36}Mg + ^{36}Mg, and nearly the same halo mass nuclei, i.e., ^{8}B + ^{8}B, ^{19}C + ^{19}C, ^{37}Mg + ^{37}Mg at different incident energies ranging between 20 and 150 MeV/nucleon. Here, we used the soft equation of state with compressibility K = 200 MeV. These reactions are followed

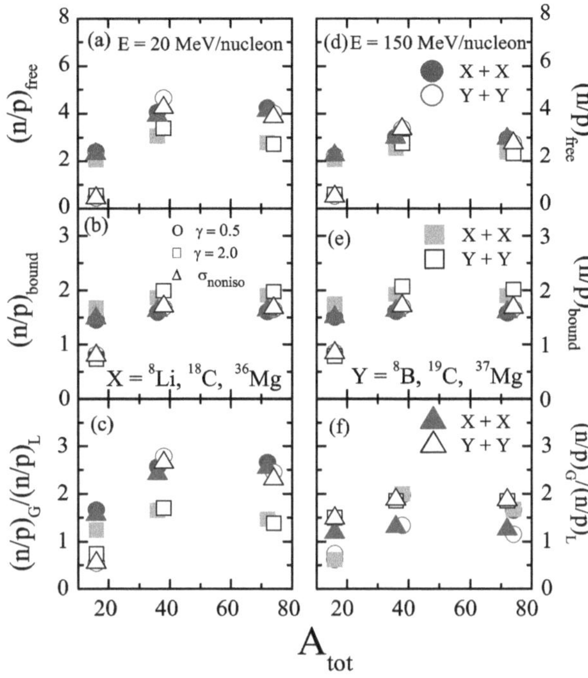

Figure 12.4 The $(n/p)_{free}$ (top panels), $(n/p)_{bound}$ (middle panels), and $(n/p)_G/(n/p)_L$ (bottom panels) as a function of system mass for different reactions involving stable and halo nuclei at incident energies of 20 (left panels) and 150 MeV/nucleon (right panels). Various symbols are explained in the text.

until 300 fm/c, where the fragment structures get saturated and the interactions become negligible.

First, let us understand the ground-state properties of the nucleus in the IQMD model. In this model, the ground-state density of the nucleus is defined as ρ_0 = 0.17 fm^{-3} and using the Fermi-gas model, the Fermi momentum of the nuclei is evaluated. The calculated value of the Fermi momentum for the ^{36}Mg stable nucleus is $\sim$ 0.268 GeV/c, while for ^{37}Mg halo nuclei $\sim$ 0.165 GeV/c. This small momentum value for the halo nucleus is implied by the less ground-state density, which is followed by the extended radius of the considered halo structure as compared to that of the stable nuclei.

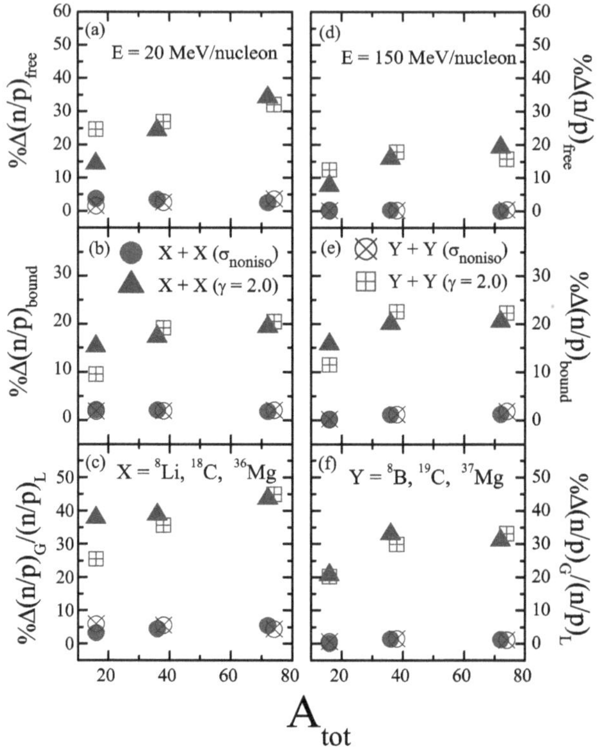

Figure 12.5 The left (right) panels represent the mass dependence of the relative percentage difference of $(n/p)_{free}$ (top panels), $(n/p)_{bound}$ (middle panels), and $(n/p)_G/(n/p)_L$ (bottom panels) at 20 (150) MeV/nucleon of incident energy. Various symbols are explained in the text.

Since the halo nuclei is itself a complex question, we have considered it to have an extended radius to get the upper edge of the role played by the actual halo structure as done in many previous studies [Kumari (2013), M. Sharma (2016), Sucheta (2018)].

In heavy-ion collisions, the nuclear matter is found to exhibit phase transition between liquid and gas, in which the gas phase is neutron rich in comparison to the liquid phase, which is represented by the bound nucleons [Xu (2006)]. In this work, we studied the systematic behavior of the halo-structured nuclei toward the yields of the neutron-to-proton ratios of free nucleons $(n/p)_{free}$, bound nucleons $(n/p)_{bound}$, and isospin Fractionation, i.e., $(n/p)_G/(n/p)_L$. In Fig. 12.1, we display the time evolution for the yield ratios of $(n/p)_{free}$ (top), $(n/p)_{bound}$ (middle), and $(n/p)_G/(n/p)_L$ (bottom panels) for the reactions of ^{36}Mg + ^{36}Mg and ^{37}Mg + ^{37}Mg at an incident energy of 50 MeV/nucleon. To get the effect of the symmetry potential toward halo and stable nuclei reactions, we use γ = 0.5 and γ = 2.0. Here, the red and green lines represent the calculations for the ^{36}Mg + ^{36}Mg reactions, while the blue and pink lines for the ^{37}Mg + ^{37}Mg reactions for γ = 0.5 and γ = 2.0, respectively. The effect of the symmetry potentials and the halo structure of nuclei is clearly seen in the yield ratios after t $\sim$ 40 fm/c. Here, we notice that the $(n/p)_{free}$ (top panel) has slightly different values for ^{36}Mg + ^{36}Mg and ^{37}Mg + ^{37}Mg reactions for both the strengths of the symmetry potentials. We also see that in the case of the stable nuclei reactions, the $(n/p)_{free}$ saturates much earlier compared to that of halo-induced reactions. The reason behind this is the loose structure of the halo nuclei. This loose structure causes the halo nuclei to have lower values of the Fermi momentum. Therefore, the nucleon flow is lesser, which leads to the delay in the saturation of $(n/p)_{free}$ for halo-induced reactions. A similar behavior is seen for $(n/p)_{bound}$ and $(n/p)_G/(n/p)_L$. At the same time, $(n/p)_{free}$ values are greater for stable nuclei reactions compared to halo-nuclei- induced reactions. This behavior is due to the reason that the halo nuclei are weakly bound and easily shatter into more number of free particles. Now, if we see the results for the soft form of symmetry energy for stable and halo nuclei reactions, the values for $(n/p)_{free}$ are enhanced for both the reactions as compared to that of the stiff form of symmetry potentials. As the symmetry potential directly affects the emissions of nucleons, in the soft symmetry potential case, the proton number is enhanced by 19.5% and the neutron number by 16.3%, i.e., the proton yield increases much faster compared to the neutron number. At the time of fragment formation, the density is lower in halo reactions compared to stable reactions, causing the symmetry energy to be weaker than the stable nuclei reactions. This leads to a lesser attractive nature of symmetry energy for protons in the former compared to the later. For neutrons, the symmetry energy is less repulsive in halo nuclei reactions compared to stable ones. This explains the sharper rise for proton yield than to neutron yield for stable to halo nuclei reactions. This behavior of symmetry potential has also been observed in many previous studies.

Next, let us discuss the change in results with the stiffness of the symmetry energy. For the stiff form, the difference between the strength of symmetry energy for halo- and stable-induced reactions further increases. In this case, the symmetry energy has greater strength for stable nuclei reactions case compared to the halo ones. We see a change in protons by 7% between the two reactions, whereas the neutron number changes by 21%. This leads to lesser $(n/p)_{free}$ values and opposite behavior is seen for $(n/p)_{bound}$ values. The trends for $(n/p)_G/(n/p)_L$ follow from these. When we compare the soft symmetry potential for the halo and stable nuclei reactions, the difference for $(n/p)_{free}$ came out to be 5.51% while 1.08% for the stiff symmetry potential.

Next, we observe that the ratio of $(n/p)_{free}$ is greater for both the reactions in the case of soft symmetry potential as compared to the stiffer case. For the stiff form of symmetry energy, a lesser (greater) number of neutrons (protons) are produced compared to the soft form of symmetry energy. The larger values of $(n/p)_{free}$, i.e., gas phase, lead to smaller values of $(n/p)_{bound}$, i.e., liquid phase. The isospin fractionation is derived by the ratio of the (n/p) of the gas phase to the (n/p) of the liquid phase. Here, the isospin fractionation, i.e., $(n/p)_G/(n/p)_L$, is mainly affected by the gas phase. Also, the relative differences between the soft and stiff forms of symmetry energy for halo and stable nuclei reactions at the time of saturation are 24.64% and 28.02% for $(n/p)_{free}$ and 21.12% and 19.33% for $(n/p)_{bound}$ and 37.75% and 39.69% for $(n/p)_G/(n/p)_L$, respectively. The sensitivities toward the different symmetry energy strengths change by a very small amount if one shifts from stable- to halo-induced reactions.

Next, as the sensitivities toward the density dependence of symmetry energy can change with energy, we have extended our study to higher incident energies. In Fig. 12.2, we have shown the results of $(n/p)_{free}$, $(n/p)_{bound}$, and $(n/p)_G/(n/p)_L$ at different incident energies between 20 and 150 MeV/nucleon for both stable and halo nuclei reactions. Here, filled (open) circles and filled (open) triangles represent the calculations for soft ($\gamma = 0.5$) and for stiff ($\gamma = 2.0$) forms of symmetry potential for ^{36}Mg + ^{36}Mg (^{37}Mg + ^{37}Mg) reactions. We see a decrease in $(n/p)_{free}$ with incident energy keeping the total difference between different forms of symmetry energy almost constant. Similar results are obtained for the halo nuclei reactions as that of stable nuclei for both the symmetry energy strengths. But we see $(n/p)_{free}$ has relatively lesser values for both symmetry energy strengths for halo nuclei reactions compared to stable nuclei reactions. The lesser density achieved in halo nuclei reactions is due to weakly bound structure, which causes this difference in results. It is worth mentioning that the behavior of $(n/p)_{free}$ for different forms of symmetry energy at higher incident energy (i.e., at 400 MeV/nucleon) is opposite to what we see presently [Kong (2015)].

In Fig. 12.3, we have studied the role of the isospin dependence of nucleon–nucleon cross sections for both the reactions, including halo and stable nuclei. Here, we took two cross sections, namely isospin-dependent nucleon–nucleon cross section (σ_{iso}), (i.e., a cross section that considers the neutron–proton cross section different from the neutron–neutron and/or proton–proton cross sections) and isospin-independent cross section (σ_{noniso}) (i.e., a cross section having the neutron–proton cross sections the same as the neutron–neutron and/or proton–proton cross sections). Here, the soft form of symmetry energy is fixed for these reactions. It is worth mentioning that the density-dependent nucleon–nucleon cross sections have also been incorporated in many studies (e.g., see Refs. [Yan (2013), M. Sharma (2016), Kaur (2016), Liu (2005)]). Li *et al.* studied the nuclear stopping to explore the isospin effects of nucleon–nucleon cross sections for the symmetric reactions of ^{58}Ni + ^{58}Ni and ^{120}Sn + ^{120}Sn [Li (2002)]. They found a weak dependence of isospin effects on the nucleon–nucleon cross sections. Very interestingly, we have also observed the weak sensitivity of the isospin effects in nucleon-nucleon cross sections for the halo and stable nuclei reactions. It is clearly seen that for σ_{iso} and σ_{noniso} cross sections, both the reactions follow the same trends. Also, the neutron-to-proton ratio of free nucleons is less for halo-induced reactions to that of stable-induced reactions (top panel). The results using an isospin-dependent cross section indicate higher $(n/p)_{free}$ values for the reactions compared to isospin-independent cross sections. Since while including the isospin effects, the neutron–proton cross sections are found to be larger than the neutron–neutron and proton–proton cross sections. The results are consistent with the recent study of isobaric/isotopic colliding pairs on the cross-over energy, which showed more sensitivity to symmetry potential as compared to the isospin effects in nn cross sections [Bansal (2018)]. If we see the neutron–proton ratio in the liquid phase, the $(n/p)_{bound}$ values are larger for halo nuclei reactions than those of stable nuclei reactions. Finally, at 150 MeV/nucleon of energy, the neutron-to-proton ratios and isospin fractionation depict the same values for both the cross sections for the weakly bound halo nuclei as well as for the stable nuclei. We found that the results do not change if one studies the halo-induced reactions instead of stable ones. To ensure that our results are not derived by the choice of the reacting partners, we have extended the present study via including the other available halo nuclei.

Further, we have considered three different halo nuclei reactions of $^{8}B + {}^{8}B$, $^{19}C + {}^{19}C$, $^{37}Mg + {}^{37}Mg$ and nearly the same mass stable nuclei reactions of $^{8}Li + {}^{8}Li$, $^{18}C + {}^{18}C$, $^{36}Mg + {}^{36}Mg$ at two different incident energies of 20 (left) and 150 MeV/nucleon (right panels). Here, the neutron-to-proton ratios are displayed as a function of the system mass. In particular, all the solid (open) symbols (circles, triangles, and squares) correspond to the results of the three halo (stable) colliding systems. The behavior of the neutron-to-proton ratios toward the three halo nucleus structures compared to stable nuclei reactions is similar as observed earlier, i.e., for stable systems, the neutron-to-proton ratio of free nucleons has larger values and for bound nucleons has smaller values with respect to the weakly bound halo system. As we move from the lower mass systems to higher mass systems, we observe that the neutron-to-proton ratios increase. Similar results are reported in Refs. [Bansal (2018), Liu (2005)]. When we compare the halo nuclei reactions with the corresponding stable nuclei at different incident energies, i.e., at 20 (left) and 150 MeV/nucleon (right), we observe that there is significant difference between the halo- and stable-induced reactions at 20 MeV/nucleon, while this difference vanishes at 150 MeV/nucleon. Also, the halo structure influences the results for lighter nuclei reactions compared to heavier ones. The results are derived from the fact that as the incident energy increases, the initial state correlations break due to nucleon–nucleon collisions. Therefore, the structural effects do not influence the results at higher incident energies. This leads to almost the same results for halo as well as stable nuclei reactions [Liu (2005), Sucheta (2018)].

To better understand the results for halo- and stable-induced reactions, we have calculated the relative percentage differences between the different strengths of the symmetry energy and isospin dependence/independent nucleon–nucleon cross section. In Fig. 12.5, the relative percentage differences for the yields of $(n/p)_{free}$, $(n/p)_{bound}$, and $(n/p)_G/(n/p)_L$ at 20 (left) and 150 MeV/nucleon (right panels) are plotted against the system mass. The relative percentage difference of $(n/p)_{free}$ for the stiffer symmetry potential with respect to the soft symmetry potential for the halo and stable nuclei reactions can be represented as:

$$\%\triangle(n/p)_{free} = \left|\frac{(n/p)_{free}(0.5) - (n/p)_{free}(2.0)}{(n/p)_{free}(0.5)}\right| \times 100\%. \quad (12.4)$$

Similarly, we calculate the relative percentage different for $(n/p)_{bound}$ and $(n/p)_G/(n/p)_L$ between the isospin-dependent and independent cross sections. Various symbols are explained in the figure. From this figure, we have noticed that for all the reactions, the relative percentage difference for the halo colliding systems as well as for the stable cases is almost the same at higher incident energies. The free neutron–proton ratios $(n/p)_{free}$ as well as $(n/p)_G/(n/p)_L$ are more sensitive toward the different strengths of symmetry compared to $(n/p)_{bound}$. We see a relative difference of nearly 4% toward the isospin effects in the cross section. At the incident energy of 20 MeV/nucleon (left panels), the $(n/p)_{free}$ shows larger relative difference for the halo nuclei reactions compared to the stable ones toward the different strengths of the symmetry energy for the lighter nuclei. The enhancement is approximately 10%. On the other hand, for $(n/p)_G/(n/p)_L$, we see a decrement of approximately 15% compared to the stable nuclei reactions. We see that $(n/p)_G/(n/p)_L$ is more sensitive toward the halo structure of the colliding nuclei as well as different density dependence of the symmetry energy.

12.4 Summary

We studied the influences of the halo structure nuclei on the density dependence of symmetry potentials and isospin dependence of nucleon–nucleon cross sections at intermediate energies using the IQMD model. We found that the neutron–proton ratios in free nucleons and isospin fractionation show more sensitivity toward the halo structure in the lighter colliding nuclei. The halo structure of the nucleus leads to at most 15% of the change in the relative differences in results compared to a stable nucleus for different strengths of the symmetry energy.

References

Bansal, P. *et al.* (2018). Isospin effects in nuclear multifragmentation of isotopic, isobaric and isotonic reactions, *Phys. Rev. C* **98**, pp. 024604.

Cao, X. G. *et al.* (2009). Density effect of the neutron halo nucleus induced reactions in intermediate energy heavy ion collisions, *Chin. Phys. C* **33**, pp. 49–51.

Famiano, M. A. *et al.* (2006). Neutron and proton transverse emission ratio measurements and the density dependence of the asymmetry term of the nuclear equation of state, *Phys. Rev. Lett.* **97**, pp. 052701.

Ferini, G. (2006). Isospin effects on subthreshold kaon production at intermediate energies, *Phys. Rev. Lett.* **97**, pp. 202301.

Kaur, M. and Gautam, S. (2016). Influence of the constant and density-dependent scaling of the scattering cross-section on reaction dynamics, *J. Phys. G: Nucl. Part. Phys.* **43**, pp. 025103.

Kaur, S. and Puri, R. K. (2013). Isospin effects on the energy of peak mass production, *Phys. Rev. C* **87**, pp. 014620.

Kobayashi, N. *et al.* (2014). Observation of a p-wave one-neutron halo configuration in ^{37}Mg, *Phys. Rev. Lett.* **112**, pp. 242501.

Kong, H. Y. *et al.* (2015). Re-examination of the neutron-to-proton ratio puzzle in intermediate-energy heavy-ion collisions, *Phys. Rev. C* **91**, pp. 047601 .

Kumar, S. *et al.* (2011). Probing the density dependence of the symmetry energy via multifragmentation at subsaturation densities, *Phys. Rev. C* **84**, pp. 044620.

Kumari, R. (2013). Study of fusion probabilities with halo nuclei using different proximity based potentials, *Nucl. Phys. A* **917**, pp. 85–91.

Li, B. A. (2000). Neutron–proton differential flow as a probe of isospin-dependence of the nuclear equation of state, *Phys. Rev. Lett.* **85**, pp. 4221–4224.

Li, B. A. and Ko, C. M. (1997). Chemical and mechanical instabilities in hot, isospin-asymmetric nuclear matter, *Nucl. Phys. A* **618**, pp. 498–508; Baran, V. *et al.* (1998). Spinodal decomposition of low-density asymmetric nuclear matter, *Nucl. Phys. A* **632**, pp. 287–303.

Li, B. A., Chen, L. W., and Ko, C. M. (2008). Recent progress and new challenges in isospin physics with heavy-ion reactions, *Phys. Rep.* **464**, pp. 113–281.

Li, B. A., Ko, C. M., and Ren, Z. (1997). Equation of state of asymmetric nuclear matter and collisions of neutron-rich nuclei, *Phys. Rev. Lett.* **78**, pp. 1644–1647.

Li, Q. *et al.* (2005). Probing the density dependence of the symmetry potential in intermediate-energy heavy ion collisions, *J. Phys. G: Nucl. Part. Phys.* **31**, pp. 1359–1374.

Li, Q. F. *et al.* (2002). Isospin effects on nuclear stopping in intermediate energy heavy ion collisions, *Chin. Phys. Lett.* **3**, pp. 321.

Liu, J. Y. *et al.* (2005). Special roles of loose neutron-halo nucleus structure on the fragmentation and momentum dissipation in heavy ion collisions, *Phys. Lett. B* **617**, pp. 24–32; ibid., (2006) Average property of isospin effects induced by neutron-halo nuclei, *Chin. Phys. Lett.* **23**, pp. 2710–2713.

Ozawa, A. *et al.* (2001). Measurements of interaction cross sections for light neutron-rich nuclei at relativistic energies and determination of effective matter radii, *Nucl. Phys. A* **691**, pp. 599–617.

Sharma, A. *et al.* (2016). Multifragmentation of nearly symmetric and asymmetric reactions within a dynamical model, *Nucl. Phys. A* **945**, pp. 95–111.

Sharma, M. K. *et al.* (2016). Search for halo structure in ^{37}Mg using the Glauber model and microscopic relativistic mean-field densities, *Phys. Rev. C* **93**, pp. 014322.

Sucheta, Kumar, R., and Puri, R. K. (2018). On the study of fragmentation of loosely bound nuclei using dynamical model., *Proc. DAE Symp. Nucl. Phys.* **63**, pp. 556–557; ibid. (2021). Reaction dynamics for stable and halo nuclei reactions at intermediate energies, in *Advances in Nuclear Physics*. Springer Proceedings in Physics, vol 257, pp 93–103. Springer, Singapore.

Sun, Z. Y. *et al.* (2010). Isospin diffusion and equilibration for Sn + Sn collisions at E/A = 35 MeV, *Phys. Rev. C* **82**, pp. 051603(R).

Takechi, M. *et al.* (2012). Interaction cross sections for Ne isotopes towards the island of inversion and halo structures of ^{29}Ne and ^{31}Ne, *Phys. Lett. B* **707**, pp. 357–361.

Takechi, M. *et al.* (2014). Evidence of halo structure in ^{37}Mg observed via reaction cross sections and intruder orbitals beyond the island of inversion, *Phys. Rev. C* **90**, pp. 061305.

Tanihata, I. *et al.* (1985). Measurements of interaction cross sections and nuclear radii in the light p-shell region, *Phys. Rev. Lett.* **55**, pp. 2676–2679.

Tanihata, I. *et al.* (1988). Measurement of interaction cross sections using isotope beams of Be and B and isospin dependence of the nuclear radii, *Phys. Lett. B* **206**, pp. 592–596.

Tanihata, I. *et al.* (1992). Determination of the density distribution and the correlation of halo neutrons in ^{11}Li, *Phys. Lett. B* **287**, pp. 307–311.

Toro, M. D. *et al.* (2007). Heavy ion collisions at relativistic energies: Testing a nuclear matter at high baryon and isospin density, *Nucl. Phys. A* **782**, pp. 267–274.

Tsang, M. B. *et al.* (2004). Isospin diffusion and the nuclear symmetry energy in heavy ion reactions, *Phys. Rev. Lett.* **92**, pp. 062701.

Tsang, M. B. *et al.* (2009). Constraints on the density dependence of the symmetry energy, *Phys. Rev. Lett.* **102**, pp. 122701.

Tsang, M. B. *et al.* (2012). Constraints on the symmetry energy and neutron skins from experiments and theory, *Phys. Rev. C* **86**, pp. 015803.

Xu, H. S. *et al.* (2006). Isospin fractionation in nuclear multifragmentation, *Phys. Rev. Lett.* **85**, pp. 916–919.

Yan, T. Z. and Li, S. (2020). Yield ratios of light particles as a probe of the proton skin of a nucleus and its centrality dependence, *Phys. Rev. C* **101**, pp. 054601.

Zhang, Y. *et al.* (2008). The influence of cluster emission and the symmetry energy on neutron–proton spectral double ratios, *Phys. Lett. B* **664**, pp. 145–148.

Chapter 13

Role of r-Helicity in Antimagnetic Rotational Bands

Sham S. Malik

Department of Physics, Guru Nanak Dev University, Amritsar 143005, India
shammalik@yahoo.com

13.1 Introduction

In nearly spherical nuclei, regular rotation-like bands indicate the unusual type of collectivity, wherein a few high-j valence particle and hole states become available for correlated alignment. At the band-head, due to the shape of their density distribution, the valence particle (hole) angular momentum vector aligns itself toward the nuclear symmetry axis, whereas the hole (particle) angular momentum aligns itself toward an axis perpendicular to it. The resultant angular momentum lies somewhere between the two. Along the band, the angular momentum increases due to a gradual alignment of the particle and hole angular momenta into the direction of the resultant angular momentum. This coupling appears like a closing of a pair of shears; hence, the term shears mechanism [Frauendorf (1993), Baldsiefen (1994), Frauendorf (1994)] was assigned to this type of excitation. In this mechanism, the magnetic dipole moment vector

Multifragmentation in Heavy-Ion Reactions: Theory and Experiments
Edited by Rajeev K. Puri, Yu-Gang Ma, and Arun Sharma

ISBN 978-981-4968-69-0 (Hardcover), 978-1-003-38513-4 (eBook)
www.jennystanford.com

arises mainly from proton particles (holes) and neutron holes (particles) by rotating around the resultant angular momentum vector, and acts as an order parameter, inducing a violation of rotational symmetry. This forms an analogy to a ferromagnet, where the total magnetic dipole moment (equal to the sum of the atomic magnetic dipole moments) is an order parameter.

Parallel to ferromagnetism, anti-ferromagnetism has also been observed in condensed matter physics. In an anti-ferromagnet, one-half of the atomic dipole moments is aligned on one sub-lattice and the other half is aligned in the opposite direction on the second sub-lattice. Although there is no net magnetic moment in an anti-ferromagnet, the state is ordered, i.e., it breaks isotropy like a ferromagnet. In analogy to the spin arrangement in anti-ferromagnetism, a unique proton–neutron spin coupling giving rise to rotational band structures in nearly spherical nuclei was proposed by Frauendorf [Frauendorf (1994)]. Since then the phenomenon called twin-shears mechanism or more commonly, antimagnetic rotation (AMR), has gained much scientific interest.

The AMR is expected to be observed in the same mass region that is also prone to magnetic rotation [Frauendorf (1994)]. This expectation is found to be true only in one mass region $A \sim 100$–110 so far. A number of magnetic rotation bands observed in this mass region have already been interpreted within the framework of shears mechanism [Frauendorf (1994), Clark (1994), Clark (2001), Simons (2003)], wherein the total angular momentum is represented as a vector sum of the angular momentum of individual valence proton ($\pi g_{\frac{9}{2}}$) holes and neutron ($\nu h_{\frac{11}{2}}$) particles. The AMR bands based on the $\pi g_{\frac{9}{2}}^{-2}$ configuration have also been claimed experimentally in $^{105-108,110}Cd$ [Choudhury (2010), Roy (2013), Simons (2005), Roy (2011a), Choudhury (2013), Datta (2005)] and $^{101,104}Pd$ [Singh (2017), Rather (2014)] nuclei. The observed AMR spectrum in each of these nuclei supports the following features.

(1) The magnetic dipole ($M1$) transitions are completely absent in the band because the transverse magnetic moments ($\mu_{\perp}$) of two sub-systems (i.e., consisting of neutron plus di-protons) are anti-aligned and hence cancel each other's contribution.

(2) The antimagnetic rotor is symmetric with respect to a rotation by 180^0 about the rotating axis; as a result, the energy levels

differ in angular momentum by $2\hbar$ and are connected by weak electric quadrupole ($E2$) transitions, reflecting a nearly spherical structure of a system. Moreover, this phenomenon is characterized by a decrease in the $B(E2)$ values with an increase in spin, hence ensuring the decrease in charge asymmetry with an increase in angular momentum.
(3) Signature is a good quantum number for such type of rotating structure. However, one of the signature partner bands is found to be missing in the observed spectrum.

Theoretical progress for the description of AMR was initiated first of all by Frauendorf [Frauendorf (1994)] and later its semiclassical version was developed by Sugawara and co-workers [Sugawara (2009)]. In these approaches, the high-spin states are generated by the so-called two-shears-like mechanism, i.e., by simultaneous closing of two valance protons (neutrons) toward the neutron (proton) angular momentum vector. Further, the mean field approach, based on the geometrical arrangement of angular momentum composition, was also used for understanding the observed features of these rotational bands [Frauendorf (1994), Chiara (2009), Zhang (2016)].

Recently, Zhao and co-workers [Zhao (2011), Zhao (2012), Meng (2012)] have extended the mean field approach using the covariant density functional formalism for investigating the AMR spectrum. For a complete understanding of the spectrum, they have incorporated terms like scalar ($\bar{\psi}\psi$), pseudo-scalar ($\bar{\psi}\gamma^5\psi$), vector ($\bar{\psi}\gamma^\mu\psi$), pseudo-vector ($\bar{\psi}\gamma^\mu\gamma^5\psi$), and anti-symmetric tensor ($\bar{\psi}\frac{i}{2}(\gamma^\mu\gamma^\nu - \gamma^\nu\gamma^\mu)\psi$) in their formalism. They have noticed for the first time a strong contribution of polarization effects on the measured quadrupole moments (i.e., the $B(E2)$ values). In earlier calculations, these effects were either completely absent or were taken into account only partially by minimizing the rotational energy with respect to a few deformation parameters. However, the tilted-axis cranking plus covariant density functional calculations are not so simple. Therefore, an alternative formalism based on r-helicity state is presented in this chapter. This formalism has been proposed based on the following experimental observations.

In a more recent experiment on ^{101}Pd nucleus, a reported negative-parity AMR band forms the yrast line above

$I = \frac{11}{2}\hbar$ [Singh (2017), Sugawara (2012)]. It is interesting enough to notice in this spectrum that the electric-dipole ($E1$) transitions prevail from the $\nu h_{\frac{11}{2}}$ band to the $\nu d_{\frac{5}{2}}$ band. This aspect corresponding to $\nu h_{\frac{11}{2}}$ and $\nu d_{\frac{5}{2}}$ orbitals (i.e., parity change, ΔΠ=-1 and Δ*l*=Δ*j*=3) is related to the possible octupole correlation in this nucleus. Such octupole correlations are possible only if the weak perturbation H_W violates time-reversal ($\hat{T}$) symmetry along with space-inversion (Π) symmetry [Bohr (1975a)]. Both these violations lead to pseudo-Hermitian Hamiltonian, $H_{\Pi T}$, which generates a real discrete spectrum above $I = \frac{11}{2}\hbar$. Zhao and co-workers [Zhao (2011)] have incorporated its (pseudo-character) contribution in their relativistic Lagrangian.

The violation of time-reversal symmetry is also vivid from the following consideration. The observed energy levels in the $\nu(h_{\frac{11}{2}})$ band differ in angular momentum ΔI=2$\hbar$ and are connected by weak electric quadrupole ($E2$) transitions; hence, a spheroidal structure with the axis of rotation perpendicular to the symmetry axis is supported. The signature partners are generally seen for such type of rotating nuclei, and the states with opposite signature are related by time-reversal symmetry. However, one of the signature partner bands is completely missing in the observed $\nu(h_{\frac{11}{2}})$ spectrum and as a result, the time-reversal symmetry gets violated. An individual (i.e., parity and time reversal) violation is supported by the *r*-helicity (hereafter referred to as helicity) formalism. Therefore, its (helicity) introduction in the particle wave function may resolve the paradoxical situation, i.e., how does the AMR band emerge in a given nucleus? This study may open the possibilities of discovering AMR bands in other mass domains.

In this chapter, the concept of helicity has been introduced for the first time in explaining the origin of the AMR spectrum.

Throughout the discussion, I would like to quote an observed $\nu h_{\frac{11}{2}}$ band in ^{101}Pd whose AMR character has already been established by myself and my co-workers [Singh (2017)] using the framework of the semiclassical rotor model as well as the cranked shell model. A description of the helicity formalism together with appropriate symmetries is given in Section 13.2. The results and discussion are given in Section 13.3. Our conclusions are summarized in Section 13.4.

13.2 The Helicity Formalism

An interaction between particles with spins (i.e., $\pi(g_{\frac{9}{2}})^{-2}$) is usually an operator in the spin space. The present generalization of the helicity formalism projects the spin of proton 1 ($\vec{s}_{\pi_1}$) on the direction of $\vec{r}_1$, the spin of proton 2 ($\vec{s}_{\pi_2}$) on the direction $\vec{r}_2$ and replaces the operators in the spin space by their matrix elements between the states of a given helicity. It is worthwhile to mention here that it (helicity) is also the component of the total angular momentum ($\vec{j}$) of a particle along the direction of $\vec{r}$, because the orbital angular momentum ($\vec{l}=\vec{r}\times\vec{p}$) is always perpendicular to $\vec{r}$ and consequently its projection (m_l) on the axis of quantization, i.e., the $\vec{r}$-axis is zero. Keeping these points in mind, a complete formalism is presented below.

Bohr and Mottelson [Bohr (1975b)] have already represented the fermion wave function in helicity bases and pointed out that it is equivalent to the rotational wave function of a system whose intrinsic shape possesses an axial symmetry. Following their footsteps, the bound state of a spin-$\frac{1}{2}$ particle in helicity formalism is carried out. The usual wave function for a spin $\frac{1}{2}$ particle moving in a spherically symmetric and parity conserving potential is given by

$$\mid jm\rangle = f_{lj}(r)\sum_{\mu,\sigma}\langle l\frac{1}{2}\mu\sigma \mid jm\rangle Y_l^{\mu}(\theta,\phi)\mid\sigma\rangle, \tag{13.1}$$

where $\mid\sigma\rangle$ is the spin eigenfunction with a projection σ along the axis of quantization, and $f_{lj}(r)$ is a normalized radial function. The angles θ and ϕ describe the direction of $\vec{r}$ in a fixed reference frame S, as shown in Fig. 13.1. In the helicity representation, the spin orientation refers to a rotated coordinate system S' whose Z'-axis is in the direction of the unit vector $\hat{e}_r$ (see Fig. 13.1).

The orientation of S' with respect to the fixed frame S is represented by three Euler angles, namely, the polar angles θ, ϕ of $\vec{r}$ and third angle ψ. The Euler angle ψ is fixed by choosing the X'- and Y'-axes of S' so that S' is obtained from S by a rotation through an angle θ about an axis with the direction of $Z\times Z'$. Therefore, the Euler angles of S' are ϕ, θ, and ψ=-ϕ. The spherical harmonic can be replaced by a matrix element of the rotation operator (D-function), which takes

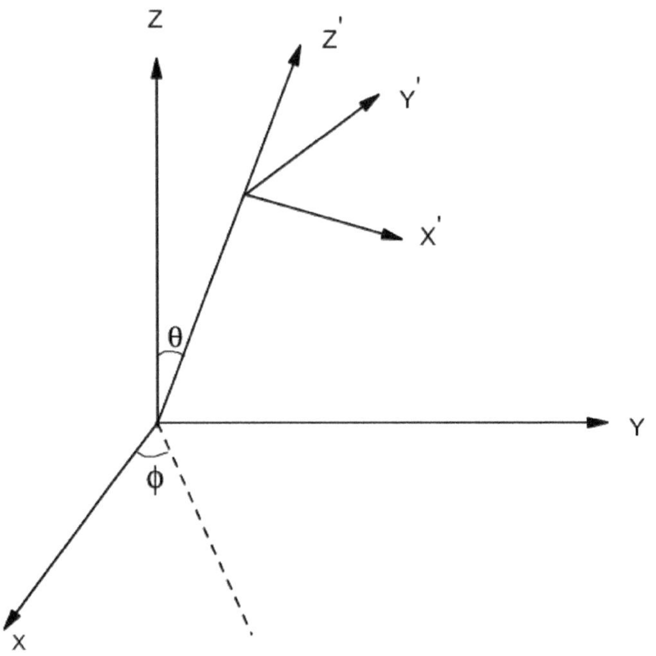

Figure 13.1 The fixed frame of reference X, Y, Z, and the helicity frame X', Y', Z'.

the axis of quantization in the direction of $\vec{r}$, i.e.,

$$Y_l^{\mu}(\theta,\phi) = \sqrt{\frac{2l+1}{4\pi}} D^{l*}_{\mu,0}(\phi,\theta,\psi). \tag{13.2}$$

The Euler angle ψ is related to some origin; when rotated around $\vec{r}$, the spherical harmonic does not depend upon it. Also, $\mid \sigma\rangle$ can be expressed in terms of two helicity functions $\mid h\rangle$, which have a projection $h = \pm\frac{1}{2}$ along $\vec{r}$

$$\mid \sigma\rangle = \sum_h D^{\frac{1}{2}*}_{\sigma,h}(\phi,\theta,\psi) \mid h\rangle. \tag{13.3}$$

Substituting Eqs. (13.2) and (13.3) in Eq. (13.1) and using the Clebsch–Gordan series for the product of two rotation matrix elements together with their orthonormality constraint, the normalized-state $\mid jm\rangle$ in helicity bases becomes

$$\langle h \mid jm\rangle = \sqrt{\frac{2j+1}{16\pi^2}} f_{lj}(r)\big[D^{j*}_{m,-\frac{1}{2}}(\phi,\theta,\psi) \mid h = -\frac{1}{2}\rangle +$$

$$(-1)^{l+j-\frac{1}{2}} D^{j*}_{m,\frac{1}{2}}(\phi,\theta,\psi) \mid h = \frac{1}{2}\rangle\big]. \qquad (13.4)$$

The wave function (13.4) of a particle consists of a linear combination of two states, which differ in helicity. It does not mean that a particle having positive helicity changes itself into that of negative helicity. This implies that reversed helicity does not occur. Under rotation, it is quite possible that the direction of $\vec{r}$ changes, but the helicity of a particle remains unchanged in a system. Thus, the wave function (13.4) clearly represents the rotational state with definite j and h of a system whose intrinsic shape possesses an axial symmetry. Further, the axial symmetry makes it impossible to distinguish orientations differing only in the value of the third Euler angle ψ; this variable is redundant. Instead of treating the Euler angle ψ as a redundant variable, one may constrain ψ to have a definite value, such as ψ=0 or ψ=-ϕ.

13.2.1 Operation of Parity- and Time-Reversal Symmetries on Helicity State

The action of parity operator Π on a presently discussed helicity state is exactly the same as for standard p-helicity [Jacob (1959)] because the vector $\vec{r}$ and the impulsion behave similarly. Therefore

$$\Pi \mid h\rangle = \eta(-)^{s-h} \exp(\imath\pi j_y) \mid -h\rangle, \qquad (13.5)$$

where η is the intrinsic parity of the particle (i.e., η=+1 for nucleon). The operator $\exp(\imath\pi j_y)$ corresponds to a rotation about y-axis through an angle 180^0.

The action of the time-reversal operator $\hat{T}$ on a helicity state is accomplished as

$$\hat{T} \mid h\rangle = (-)^{s+h} \mid -h\rangle, \qquad (13.6)$$

because $\hat{T}$ does not change the vector $\vec{r}$.

An individual violation of parity- and time-reversal symmetry is noticed in the helicity-represented wave function (13.4), but their (parity and time reversal) combined operation remains invariant except the reflection symmetry about a plane. Thus, the helicity representation (13.4) for the wave function of a particle with spin $\frac{1}{2}$ is equivalent to the representation for the rotational wave function of a system whose intrinsic shape possesses an axial symmetry but not the reflection symmetry.

It is quite interesting to notice here that the system represented by the wave function (13.4) possesses an axial symmetric shape in which parity is no longer a good quantum number, i.e., it refers to a pear-shaped system with z-axis as the symmetry axis. For such type of system, a combination of signature- and parity-quantum numbers is conserved rather than the signature alone. Under this combined symmetry (signature plus parity), the rotational states with even parity having plus (minus) signature get separated from odd parity with minus (plus) signature. Further, the states with opposite signature are related by time-reversal symmetry, which also gets violated in the present system represented by the wave function (13.4). As a result, one of the signature partner bands disappears from the rotating system. Thus, the violation of the reflection symmetry ensures a complete missing of one of the signature partner bands and, hence, explains the most prominent feature of the observed spectrum.

13.3 Results and Discussion

At the band-head, the angular momentum vectors $\vec{j}_{\pi 1}$ and $\vec{j}_{\pi 2}$ of two proton holes $(\pi(g_{\frac{9}{2}})^{-2})$ are pointing opposite to each other and are nearly perpendicular to the neutron $(\nu(h_{\frac{11}{2}}))$ angular momentum $\vec{j}_{\nu}$. The reference frame is chosen in such a way that the Euler angles ϕ_1, θ_1, $\psi_1 = -\phi_1$ describe a coordinate system with its z-axis along $\vec{r}_1$ and ϕ_2, θ_2, $\psi_2 = -\phi_2$ another one with its z-axis along $\vec{r}_2$. In order to study the rotational spectra, the axis of quantization is chosen along $\vec{r}_1$ together with a frame of reference for proton 2 given by Euler angles (ϕ_2=0, $\theta_2 = 2\theta$, $\psi_2 = -\phi_2$=0), i.e., the vectors $\vec{r}_1$ and $\vec{r}_2$

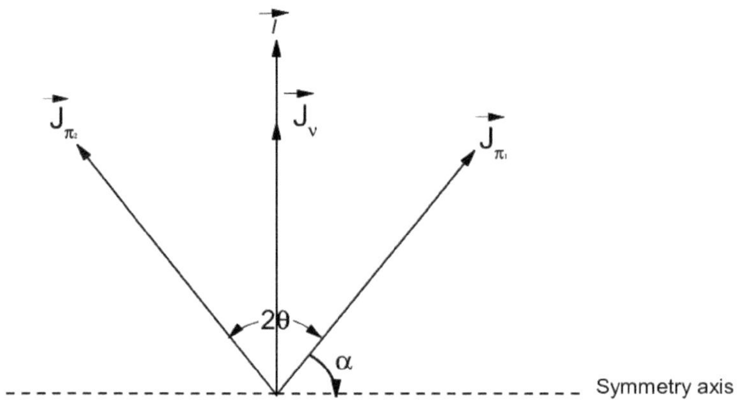

Figure 13.2 Pictorial representation of the total angular momentum generated by di-protons in which the axis of quantization is chosen along particle 1 together with a frame of reference of particle 2 having Euler angles (ϕ_2=0, θ_2=2θ ψ_2=0).

are oriented at an angle 2θ (=180^0) as shown in Fig. 13.2. I would like to mention here that this figure is drawn for the purpose of the next subsection, which is why the angle between the vectors $\vec{j}_{\pi_1}$ and $\vec{j}_{\pi_2}$ is shown to be 2θ. As a result, two protons in the $g_{\frac{9}{2}}$ orbital lying at the position vectors $\vec{r}_i$ (with i=1,2) follow helical orbits (i.e., left and right handed) in a rotating field, whose $+\frac{1}{2}$ spin gives one helicity state, i.e., $\mid h_1\rangle$, while the $-\frac{1}{2}$ gives another helicity state $\mid h_2\rangle$. Both these helicity functions $\mid h\rangle$ have projections $h = +\frac{1}{2}$ along the quantization axis $\vec{r}_1$.

One of the major advantages of using helicity state is that there is no need to separate the total angular momentum $\vec{j}$ into orbital and spin parts, because the resultant orbital angular momentum $\vec{L}$ due to di-protons is always perpendicular to the symmetry axis and generates the centrifugal term $\frac{\hbar^2}{2\Im}\vec{L}^2$. Here, $\Im$ is the inertia tensor. For the observed negative-parity band, the $\Im$ is estimated from the slope of the angular momentum I versus the rotational frequency $\hbar\omega$ plot and it comes out to be 15.5 $\hbar^2 MeV^{-1}$. The neutron acts as a spectator and contributes to the band-head angular momentum only. Thus, a complete rotational spectrum is generated by the centrifugal term. The matrix elements $\langle jmh_1'h_2' \mid \vec{L}^2 \mid jmh_1h_2 \rangle$ of the centrifugal term for a given h_1, h_2 with h=$h_1 - h_2$ are given as follows:

Case I: If h_1'=h_1 and h_2'=h_2, then

$$\langle jmh_1'h_2' \mid \vec{L}^2 \mid jmh_1h_2 \rangle$$
$$= j(j+1) - h^2 + s_{\pi_1}(s_{\pi_1}+1) - {h_1}^2 + s_{\pi 2}(s_{\pi_2}+1) - {h_2}^2. \quad (13.7)$$

Case II: If h_1'=$h_1 \pm 1$ and h_2'=h_2, then

$$\langle jmh_1'h_2' \mid \vec{L}^2 \mid jmh_1h_2 \rangle$$
$$= -\sqrt{j(j+1) - h(h \pm 1)}\sqrt{s_{\pi_1}(s_{\pi_1}+1) - h_1(h_1 \pm 1)}. \quad (13.8)$$

Case III: If h_1'=h_1 and h_2'=$h_2 \pm 1$, then

$$\langle jmh_1'h_2' \mid \vec{L}^2 \mid jmh_1h_2 \rangle$$
$$= -\sqrt{j(j+1) - h(h \mp 1)}\sqrt{s_{\pi_2}(s_{\pi_2}+1) - h_2(h_2 \pm 1)}. \quad (13.9)$$

Case IV: If h_1'=$h_1 \pm 1$ and h_2'=$h_2 \pm 1$, then

$$\langle jmh_1'h_2' \mid \vec{L}^2 \mid jmh_1h_2 \rangle$$
$$= \sqrt{s_{\pi_1}(s_{\pi_1}+1) - h_1(h_1 \pm 1)}\sqrt{s_{\pi_2}(s_{\pi_2}+1) - h_2(h_2 \pm 1)}. \quad (13.10)$$

The calculated rotational energy E versus the angular momentum J is shown in Fig. 13.3 (full line). The observed negative-parity spectrum of ^{101}Pd is also shown as solid circles connected with the solid line in this figure. The experimental energy spectrum is reproduced in an excellent way by the present helicity-based calculations. Therefore, we can conclude that both the helical orbits contribute to the observed rotational excitation energy.

Further, the transition probability $B(E2)$ can be obtained within the semiclassical approximation as

$$B(E2) = \mid \langle j_f m \mid r^2 Y_{20} \mid j_i m \rangle \mid^2 = \frac{5}{4\pi} \mid C^{j_i 2 j_f}_{\frac{1}{2} 0 \frac{1}{2}} \mid^2 \mid \int_0^\infty dr f_{l_f j_f}(r) r^4 f_{l_i j_i} \mid^2 . \tag{13.11}$$

The $B(E2)$ values represented by Eq. (13.11) generally increase with an increase in the angular momentum, whereas the observed spectrum shows a gradual decrease in the reduced transition probability with an increase in spin. This decrease in transition probability can be understood from the following discussion. The quadrupole moment operator in the $B(E2)$ of the rotational spectrum involves the deformation degrees of freedom for a given

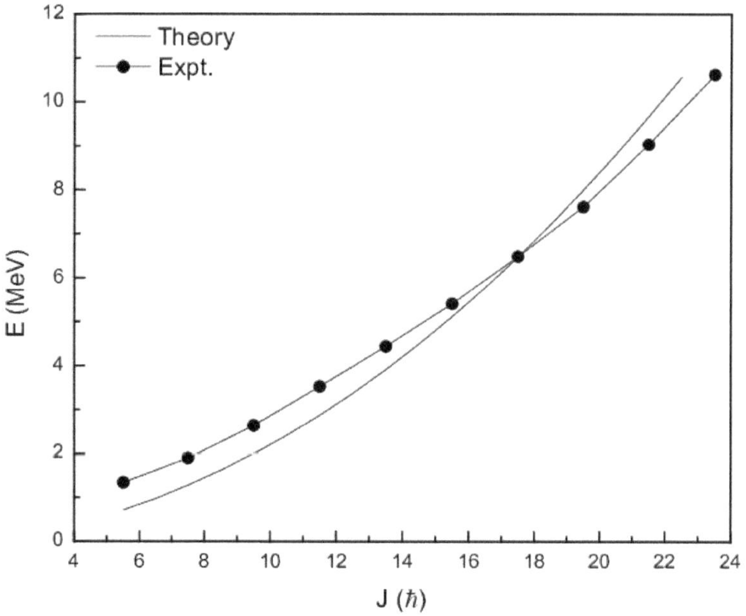

Figure 13.3 The calculated rotational energy versus angular momentum is shown as the full line. Also, the solid circles connected with the solid line show the observed spectrum of the negative-parity AMR band in ^{101}Pd nucleus.

configuration. Also, the helicity operator always commutes with rotation. Under rotation, the fermion having positive helicity state $|+h\rangle$ goes one-way circular motion around the complex plane, while the fermion with $|-h\rangle$ goes the other way. This implies that the phase of each particle quantum helicity state keeps on changing, and hence the degree of polarization also changes (i.e., increases) with a gradual increase in rotation. As a result, an ordered (polarized) state emerges at a critical rotational frequency $\hbar\omega$, which breaks down the isotropy symmetry of the system and an AMR band emerges. This change in polarization also rearranges the charge distribution in a rotating system, which changes the quadrupole moment and hence the $B(E2)$ values accordingly.

Further, in the present model, the total angular momentum is generated along an axis perpendicular to the symmetry axis. For such type of pictures, a gradual increase in rotational frequency generally drives the nucleus toward sphericity and as result a gradual decrease in the $B(E2)$ values may occur with an increase in angular momentum. It is quite evident from Eq. (13.11) that the calculated $B(E2)$ values involve the intrinsic degrees of freedom, which have not been included in the present phenomenological model. Therefore, a complete microscopic development of the helicity-based wave function will pinpoint the exact behavior of the $B(E2)$ values, and this study is in progress.

13.3.1 Relevance with Twin-Shears Mechanism

Finally, it is quite interesting to notice that the present formalism fully supports the twin-shears mechanism, which is popularly known for these bands. In an antimagnetic rotor, two angular momentum vector blades $\vec{j}_{\pi_1}$ and $\vec{j}_{\pi_2}$ of the proton holes are stretched apart, each coupled nearly perpendicular to the angular momentum vector $\vec{j}_\nu$ of a neutron particle, such that the band-head angular momentum is $\vec{j}_\nu$. This type of configuration is shown in Fig. 13.2 with angle 2θ=180^0 at the band-head. The high-spin states in a spectrum are generated due to the gradual closing of the proton blades $\vec{j}_{\pi_1}$ and $\vec{j}_{\pi_2}$ toward the neutron angular momentum $\vec{j}_\nu$ with its resultant angular momentum always pointing toward $\vec{j}_\nu$. It is worthwhile to mention here that in the helicity formalism, the resultant angular momentum is also obtained along an axis perpendicular to the symmetry axis, and hence, it is fully consistent with the twin-shears mechanism.

Further, for an axially symmetric shape, Bohr and Mottelson [Bohr (1975a)] have already identified the K-quantum number with the helicity value. Here, the quantum number K represents the angular momentum of the intrinsic motion and has a fixed value for the rotational band based on a given intrinsic state. For the $\pi(g_{\frac{9}{2}})^{-2} \otimes \nu(h_{\frac{11}{2}})$ configuration, two symmetric shears each having angle θ are formed. It is quite evident from Fig. 13.2 that the quantum number K is equal to $j_\pi \sin\theta$ and its resemblance with helicity can be easily drawn as follows. For a particle of mass m, a complete set of orthogonal states are always characterized by position vector $\vec{r}$ and definite $(2j+1)$ helicity $h = -j, -j+1, ..., +j$ values.

Also, in the semiclassical picture of the twin-shears mechanism [Roy (2011a), Roy (2011b)], the resultant angular momentum is related to the shear angle θ between $\vec{j}_\pi$ and $\vec{j}_\nu$ and is given by an expression

$$J = j_\nu + 2j_\pi \cos\theta + \Im\left(\frac{1.5V_{\pi\nu}\cos\theta}{j_\pi} - \frac{6V_{\pi\pi}\cos 2\theta\cos\theta}{nj_\pi}\right). \quad (13.12)$$

A systematic study of AMR band infers the interaction strengths $V_{\pi\nu}$ and $V_{\pi\pi}$ equal to 1.2 MeV and 0.15 MeV, respectively. One neutron particle and two proton holes fix n=2 with j_ν =5.5 $\hbar$ and j_π =4.5 $\hbar$. The inertia parameter has already been estimated to be 15.5 $\hbar^2 MeV^{-1}$.

Equation (13.12) is a transcendental equation, and its numerical solution fixes θ for each angular momentum and is shown in Fig. 13.4.

Its decreasing trend of θ versus the angular momentum ensures that

$$B(E2) \propto |\, C^{j_i 2 j_f}_{KOK} \,|^2 \propto \sin^4\theta, \quad (13.13)$$

$B(E2)$ values decrease with an increase in angular momentum. On the other hand, the total angular momentum is generated along an axis perpendicular to the symmetry axis and it involves $\cos\theta$. A decrease in θ increases the angular momentum and hence the rotational energy.

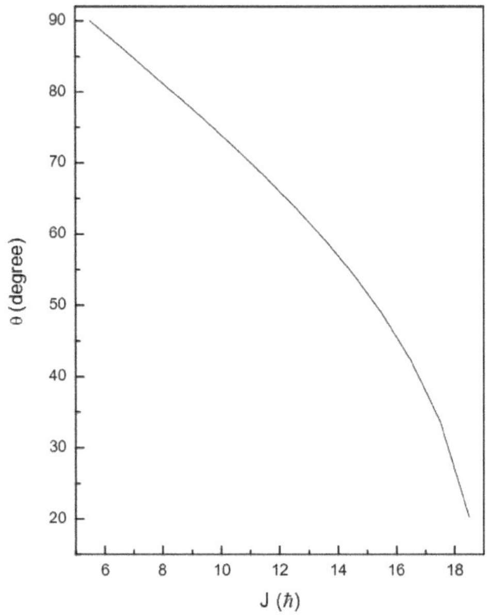

Figure 13.4 Variation of shears angle θ versus the angular momentum.

Thus, the experimental spectrum and the $B(E2)$ values of the recently observed AMR spectrum in ^{101}Pd are fully consistent with the helical represented states. This simple formalism has completely resolved the paradoxical situation of how the AMR emerges in a nucleus. It is worthwhile to mention here that its microscopic version will pinpoint complete features of the AMR spectrum.

13.3.2 Role of Octupole Correlation in AMR Spectrum

As stated above, the octupole correlations likely to play a role in generating the AMR spectrum. This conclusion has been drawn on the basis of the following considerations. Bohr and Mottelson [Bohr (1975a)] have already pointed out that for configurations with particles in unfilled shells, there may occur strong octupole transitions between orbits within a major shell ($1g_{\frac{9}{2}} \longleftrightarrow 2p_{\frac{3}{2}}$ for the shell 28–50; $1h_{\frac{11}{2}} \longleftrightarrow 2d_{\frac{5}{2}}$ for the shell 50–82; $1i_{\frac{13}{2}} \longleftrightarrow 2f_{\frac{7}{2}}$ for the shell 82–126, and $1j_{\frac{15}{2}} \longleftrightarrow 2g_{\frac{9}{2}}$ for the shell above 126). This implies that in normally deformed nuclei, the tendency toward octupole correlation occurs just above the closed shell with particle numbers (i.e., N or Z) $\sim$34 ($1g_{\frac{9}{2}} \longleftrightarrow 2p_{\frac{3}{2}}$ coupling), $\sim$56 ($1h_{\frac{11}{2}} \longleftrightarrow 2d_{\frac{5}{2}}$ coupling), $\sim$88 ($1i_{\frac{13}{2}} \longleftrightarrow 2f_{\frac{3}{2}}$ coupling) and $\sim$134 ($1j_{\frac{15}{2}} \longleftrightarrow 2g_{\frac{9}{2}}$ coupling). The AMR bands have already been confirmed in ^{100}Pd [Zhu (2001)] and ^{101}Pd [Zhang (2016), Singh (2017)]. The octupole transitions have clearly been observed in these nuclei, which (^{100}Pd and ^{101}Pd) fall in the $N \sim$56 region. The cadmium isotopes, which also belong to that region, have been described as the optimal candidates for the observed antimagnetic rotation and their low-lying levels have already been interpreted as quadrupole–octupole oscillations [Garrett (2010)]. Therefore, the role of octupole correlation in the AMR spectrum is discussed with reference to odd-A, i.e., $^{105,107,109}Cd$ isotopes.

In odd-A nuclei, the $E1$ and $E3$ (i.e., the dipole and octupole) transitions between opposite-parity states are considered to be the manifestation of collective and intrinsic degrees of freedom ensuing from the presence of quadrupole–octupole shape deformations [Butler (1996)]. At low angular momenta, the nucleus is characterized by a soft-octupole shape superposed on the top of a stable quadrupole deformation, while with an increasing angular momentum, the resultant quadrupole–octupole shape is stabilized. Then it is considered that at low angular momentum, the system is capable of performing octupole oscillations. At a higher angular momentum, the octupole deformation gets stabilized and the nucleus performs a rotation that is completely governed by the complex quadrupole deformation. Keeping these points in mind, a systematic analysis of each of the odd-A, i.e., $^{105,107,109}Cd$ isotopes is carried out.

The $J^{\Pi} = \frac{11}{2}^{-}$ level appears in all the three odd-A Cd isotopes [Nndc (2020)]. It is noticed that the energy gap between the $\frac{11}{2}^{-}$ state and the ground sate $J^{\Pi} = \frac{5}{2}^{+}$ in $^{105,107}Cd$ is larger than that of ^{109}Cd. It ensures a space for several states of single particle and collective nature to appear in the $^{105,107}Cd$ isotopes. Kisyov and co-workers [Kisyov (2011)] have ensured experimentally that ($^{105,107}Cd$) $\frac{11}{2}^{-}$-level decays via dominant $E1$ branch, whereas

^{109}Cd isotope decays via $M2$ transition. Further, the $E1$-decay mode (i.e., $\frac{31}{2}^{-} \rightarrow \frac{29}{2}^{+}$ with E_{γ}=803 keV) just after the band crossing has also been observed in the ^{109}Cd isotope [Chiara (2009)]. However, a large number of $E1$ transitions could not be observed in these isotopes. A sudden drop of intensity of the populated states at a high angular momentum I could be responsible for such missing $E1$ transitions. Thus, the lowest excited negative-parity band consisting of ΔI=2 transitions in each of the odd-A cadmium isotope is controlled by a soft-octupole shape superposed on the top of a stable quadrupole deformation. Macroscopic–microscopic models [Nazarewicz (1984), Nazarewicz (1987)] have already predicted that the octupole deformation of the ground-state does not persist to high angular momentum I. Smith and co-workers [Smith (1995)] have also pointed out that a sudden drop in the intensity of $E1$ at a high angular momentum is the fingerprint of such a change in shape.

I have tried to understand the absence of octupole contribution at a higher angular momentum along the band using the principal axis cranking (PAC) model. The PAC calculations based on the Nilsson approach are carried out for each odd-A cadmium isotope. Considering the band-head angular momenta, parity, and excitation energy, the configuration $\pi[(g_{\frac{9}{2}})^{-2}]_0 \otimes \nu(h_{\frac{11}{2}})$ has been chosen for each cadmium isotope. In this configuration, $\pi[(g_{\frac{9}{2}})^{-2}]_0$ represents paired quasi-particles and contributes zero angular momentum. In all these cases, the proton and neutron pairing parameters (Δ_p and Δ_n) are chosen as 80% of the odd–even mass difference. The deformation parameters (ϵ_2,ϵ_4,γ) are determined self-consistently by a minimization of the total energy. The values of the calculated Δ_p, Δ_n, and the corresponding deformation parameters ϵ_2, ϵ_4, and γ for each of the three cadmium isotopes are given in Table 13.1.

Figures 13.5a–c show the calculated as well as the experimental data of angular momentum I versus the rotational frequency $\hbar\omega$ for the three Cd isotopes. I We shall first analyze the case of the ^{105}Cd isotope (refer to Fig. 13.5a) in detail and then discuss the results for other isotopes. The observed yrast-band built on the $\nu(h_{\frac{11}{2}})$ band-head configuration is reasonably close to the predicted one by the PAC model. The basic configuration for this band is expected to be the prolate-driving $\nu(h_{\frac{11}{2}})$ orbital. However, the

Table 13.1 Calculated PAC model parameters for the ground-state and band-crossing negative-parity configurations in $^{105,109}Cd$ isotopes. These parameters for the ^{107}Cd isotope for the ground-state negative-parity configuration are also listed.

isotope	configuration	$\Delta_p(MeV)$	$\Delta_n(MeV)$	ϵ_2	ϵ_4	γ
^{105}Cd	$\pi[(g_{\frac{9}{2}})^{-2}]_0 \otimes \nu(h_{\frac{11}{2}})$	1.1578	1.0756	0.123	-0.008	15.2^0
^{105}Cd	$\pi[(g_{\frac{9}{2}})^{-2}]_0 \otimes \nu(h_{\frac{11}{2}} g^2_{\frac{7}{2}})$	1.1578	1.0756	0.162	0.0	4.6^0
^{107}Cd	$\pi[(g_{\frac{9}{2}})^{-2}]_0 \otimes \nu(h_{\frac{11}{2}})$	1.1548	1.0759	0.139	0.01	10.3^0
^{109}Cd	$\pi[(g_{\frac{9}{2}})^{-2}]_0 \otimes \nu(h_{\frac{11}{2}})$	1.1698	1.1136	0.137	0.009	9.6^0
^{109}Cd	$\pi[(g_{\frac{9}{2}})^{-2}]_0 \otimes \nu(h^3_{\frac{11}{2}})$	1.1698	1.1136	0.157	0.01	4.5^0

difference between the observed and calculated values may arise due to octupole oscillations, which this system generally performs at a low angular momentum. Further, the observed spectrum shows band crossing around the rotational frequency $\sim 0.4\ MeV$. The PAC model calculations predict the band crossing resulting from the alignment of the $\nu(g_{7/2})^2$ pair. Thus, after the band crossing, the high-spin states above $\frac{27}{2}\hbar$ are generated due to the alignment of a pair of $\pi(g_{\frac{9}{2}})^{-2}$ proton holes with the configuration $\pi[(g_{\frac{9}{2}})^{-2}] \otimes \nu(h_{\frac{11}{2}} g^2_{7/2})$ along with the parameters listed in Table 13.1. It is quite obvious from the listed parameters that a prolate shape emerges after the band crossing and reproduces the observed angular momentum versus the rotational frequency plot. Thus, a comparison with the observed data reveals that at low angular momenta, the system can perform octupole oscillations. At a higher angular momentum, the octupole deformation gets stabilized and the nucleus performs a rotation completely governed by the complex quadrupole deformation.

Kisyov and co-workers [Kisyov (2011)] have already confirmed that the structure of ^{107}Cd is quite similar to that of ^{105}Cd. Unfortunately, its spectrum after band crossing has not yet been reported. Again, its observed yrast-band built on the $\nu(h_{\frac{11}{2}})$ band-head configuration is reasonably close to the predicted one by the PAC model (Fig. 13.5b). The differences between the observed and the calculated values reveal that at low angular momenta, the system is capable of performing octupole oscillations and these oscillations decrease with an increase in angular momentum and hence ensure the prolate shape just near the band-crossing frequency.

In Fig. 13.5c, the results of ^{109}Cd along with experimental data are shown. The observed back-bending above $\frac{31}{2}\hbar$ is explained with the configuration $\pi[(g_{\frac{9}{2}})^{-2}] \otimes \nu(h^3_{\frac{11}{2}})$ along with the parameters listed in Table 13.1. Again, its comparison with the observed data supports similar conclusions as drawn in the case of ^{105}Cd and ^{107}Cd isotopes.

Further, the experimental and calculated $B(E2)$ values for ^{105}Cd and ^{109}Cd isotopes in the high-spin range are compared in Figs. 13.6a and 13.6b, respectively. Except their decreasing trends, the calculated values are of the same order of magnitude as that of the observed ones. It is worthwhile to mention here that the decrease

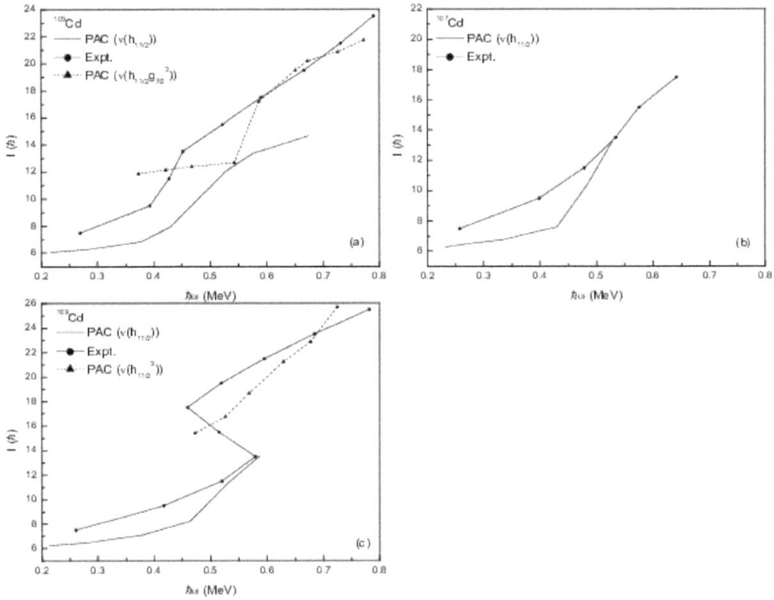

Figure 13.5 The experimental and calculated angular momentum I versus the rotational frequency $\hbar\omega$ for the $^{105,107,109}Cd$ isotopes.

in $B(E2)$ values with an increase in angular momentum is a clear signature of the shears-like mechanism.

These results ensure that the octupole deformation gets stabilized at a higher angular momentum and the nucleus performs an AMR rotation completely governed by the complex quadrupole deformation. Bohr and Mottelson [Bohr (1975a)] have pointed out that the low-frequency quadrupole–octupole oscillations contribute isoscalar and isovector polarizations to a system and these polarizations have been included by Zhao and co-workers [Zhao (2011), Zhao (2012)] in their covariant density functional formalism for explaining the AMR spectrum. These points reveal that the octupole-mode may have some contribution in fixing the antimagnetic character in a system, i.e., why it is one of the rare phenomena that has been observed by using sophisticated detectors.

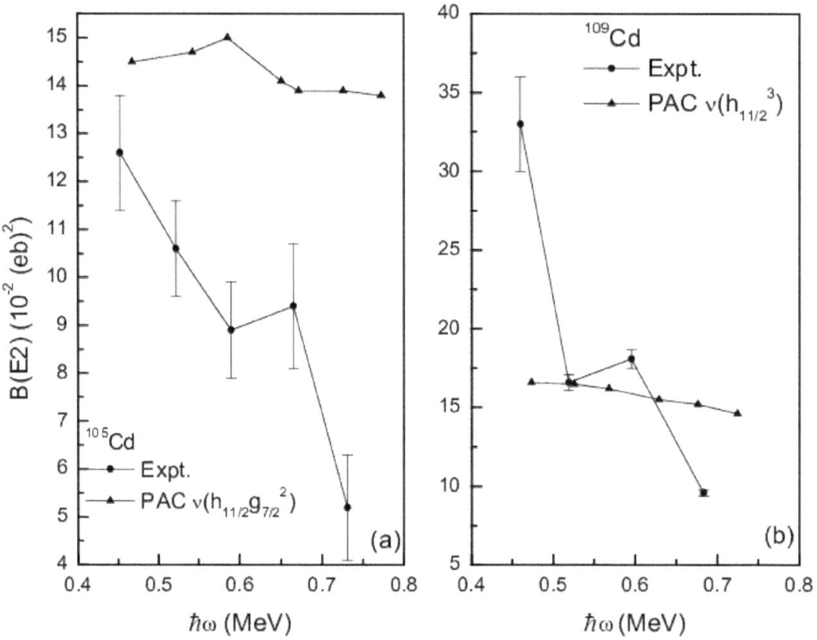

Figure 13.6 The experimental and calculated $B(E2)$ values versus $\hbar\omega$ for the $^{105,109}Cd$ isotopes.

13.3.3 Symmetries Responsible for AMR Spectrum

Zhao and co-workers [Zhao (2011), Zhao (2012)] have noticed for the first time that the combined operation of space reflection (Π), time reversal (T), and reflection in the y-direction remains invariant, i.e., $[H, \Pi_y T] = 0$ and is responsible for explaining the AMR spectrum. Therefore, it is important to understand the role of each of these individual symmetries in the case of a deformed nucleus with an axis of symmetry as the z-axis. Such an axially symmetric shape in which parity alone is no longer a good quantum number generally refers to a pear-shaped structure. Then the combined operation of both parity and reflection in the y-direction (i.e., Π_y) is good symmetry instead of reflection in the y-direction alone. Under this combined symmetry operation, the rotational states with even parity having plus (minus) signature get separated from odd parity with minus (plus) signature. Since $[H, \Pi_y T] = 0$, it implies that both Π_y and time reversal (T) are either conserved or violated separately. The latter choice is likely to be more favorable because the signature partner bands with the opposite parity have not been observed in the AMR spectrum. Therefore, this combined operation of three symmetries does not support two signature partner bands with the same parity. The present formalism (as discussed in section IIA) also ensures a similar combined operation of three symmetry operators and, hence, is fully consistent with that of Zhao and co-workers [Zhao (2011), Zhao (2012)].

It is worthwhile to mention here that the structure of ^{107}Cd isotope is quite similar to that of ^{105}Cd as noticed by Kisyov and co-workers [Kisyov (2011)]. A similar point has also been drawn by Roy and Chattopadhyay [Roy (2011b)] by plotting the aligned angular momenta i of $^{105,107}Cd$ isotopes versus the rotational frequency $\hbar\omega$. It is apparent from their plot (Fig. 2 of Ref. [Roy (2011b)]) that the alignment gain ($i \sim 4$) is similar in both these isotopes. This implies that the neutron-aligned yrast configuration for ^{107}Cd is also similar to that of ^{105}Cd (i.e., $\pi[(g_{\frac{9}{2}})^{-2}]_0 \otimes \nu(h_{\frac{11}{2}} g^2_{7/2})$). Further, the structural similarity in odd-A $^{105-109}Cd$ isotopes can also be understood from the plots shown in Fig. 13.7. The left panel of Fig. 13.7 shows the observed dynamic moment of inertia $\Im^2$ versus the angular momentum I of the negative-parity bands based on the $\nu(h_{\frac{11}{2}}$ band-head configuration. A nearly similar behavior of $\Im^2$ in both the $^{105,107}Cd$ isotopes supports the above conclusions. Also, a comparison among three odd-A isotopes reveals that collectivity persists up to an angular momentum $\frac{27}{2}\hbar$ and then decreases gradually for higher angular momenta where the excited band configurations start to dominate, leading the nucleus toward non-collective shapes. These non-collective shapes are likely to generate antimagnetic rotation in a system. The angular momentum I versus the gamma-ray energy E_γ plot shown in the right panel of Fig. 13.7 also favors the same results. However, the negative-parity yrast sequence of ^{107}Cd nucleus is known only up to $\frac{31}{2}\hbar$ [Jerrestam (1992)], which needs to be extended in order to investigate the possibility of AMR in ^{107}Cd.

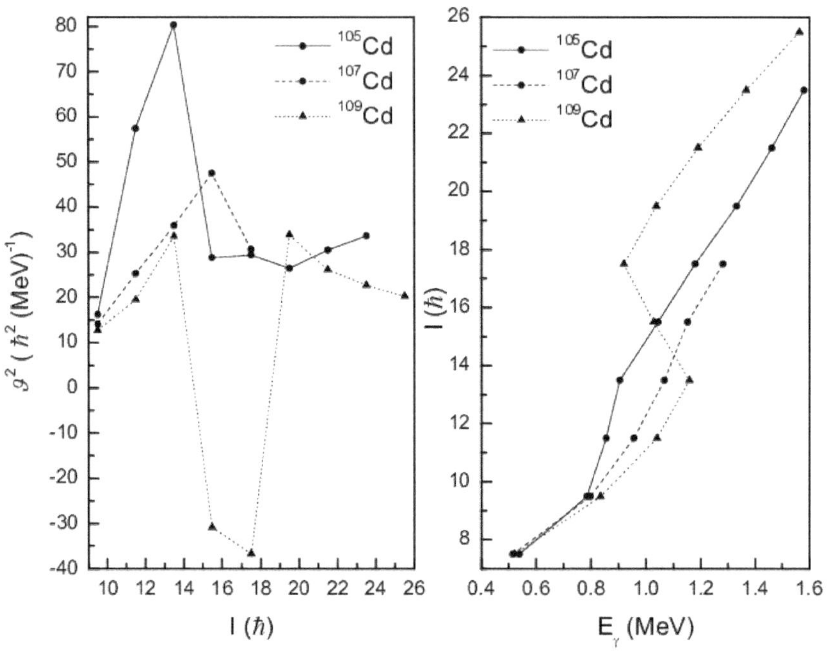

Figure 13.7 Left panel shows the observed dynamic moment of inertia $\Im^2$ versus the angular momentum I, whereas the right panel shows the variation in angular momentum I versus the gamma-ray energy E_γ.

13.4 Summary

To summarize, we have carried out an r-helicity-based analysis of the observed negative-parity AMR band in ^{101}Pd nucleus. The helical modes are found to be responsible for explaining the observed octupole correlations, which force the system to generate antimagnetic spectra. The helicity-based wave function supports a pear-shaped structure of a nucleus having an axis of symmetry. This structure conserves a combined operation of parity and signature quantum numbers instead of signature alone. Simultaneous violation of time reversal symmetry in the helicity-based formalism leads to disappearance of one of the signature partner bands and hence explains one of the prominent observed features in the AMR spectrum. Also, it has been noticed that the degree of polarization breaks down the isotropic symmetry of the nucleus and hence generates an antimagnetic structure. This formalism has been tested successfully for the recently observed negative-parity ΔI=2 antimagnetic spectrum in odd-A ^{101}Pd nucleus. These results establish the importance of helical orbits in the observed AMR spectrum.

Acknowledgment

The author (SSM) is thankful to T.I.F.R. Mumbai, India, for providing a couple of weeks stay for fruitful discussion.

References

Baldsiefen, G. *et al.* (1994). Shears bands in ^{199}Pb and ^{200}Pb, *Nucl. Phys. A* **574**, pp. 521–558.

Bohr, A. and Mottelson, B. R. (1975). *Nuclear Structure*, vol. I (Benjamin, New York, 1975).

Bohr, A. and Mottelson, B. R. (1975). *Nuclear Structure*, vol. II (Benjamin, New York, 1975).

Butler, P. A. and Nazarewicz, W. (1996). Intrinsic reflection asymmetry in atomic nuclei, *Rev. Mod. Phys.* **68**, pp. 349.

Chiara, C. J. *et al.* (2000). Shears mechanism in ^{109}Cd, *Phys. Rev. C* **61**, pp. 034318.

Choudhury, D. *et al.* (2010). Evidence of antimagnetic rotation in odd-A ^{105}Cd, *Phys. Rev. C* **82**, pp. 061308.

Choudhury, D. *et al.* (2013). Multiple antimagnetic rotation bands in odd-A ^{107}Cd, *Phys. Rev. C* **87**, pp. 034304 .

Clark , R. M. and Macchiavelli, A. O. (2001). The shears mechanism, *Nucl. Phys. A* **682**, pp. 415–426.

Clark, R. M. *et al.* (1999). Shears mechanism in the A-110 region, *Phys. Rev. Lett.* **82**, pp. 3220.

Datta, P. *et al.* (2005). Observation of antimagnetic rotation in ^{108}Cd, *Phys. Rev. C* **71**, pp. 041305.

Frauendorf, S. (1993). Tilted cranking, *Nucl. Phys. A* **557**, pp. 259–276 .

Frauendorf, S. (2001). Spontaneous symmetry breaking in rotating nuclei, *Rev. Mod. Phys.* **73**, pp. 463.

Garrett, P. E. and Wood, J. L. (2010). On the robustness of surface vibrational modes: Case studies in the Cd region, *J. Phys. G* **37**, pp. 064028.

Jacob, M. and Wick, G. C. (1959). On the general theory of collisions for particles with spin, *Ann. of Phys.* **7**, pp. 404–428.

Jerrestam, D. *et al.* (1992). Rotational bands in $^{(107)}$Cd, *Nucl. Phys. A* **545**, pp. 835–853.

Kisyov, S. *et al.* (2011). In-beam fast-timing measurements in ^{103,105,107} Cd, *Phys. Rev. C* **84**, pp. 014324.

Meng, J., Peng, J., Zhang, S.Q., and Zhao, P.W. (2013). Relativistic density functional for nuclear structure, *Front. Phys.* **8**, pp. 55–79.

Nazarewicz, W., Olanders, P., Ragnarsson, I., Dudek , J., and Leander, G. A. (1984). High-spin consequences of octupole shape in nuclei around ^{222}Th, *Phys. Rev. Lett.* **52**, pp. 1272.

Nazarewicz, W., Leander, G. A., and Dudek, J. (1987). Octupole shapes and shape changes at high spins in Ra and Th nuclei, *Nucl. Phys. A* **467**, pp. 437–460.

NNDC data base [http://www.nndc.bnl.gov/ensdf].

Rather, N. *et al.* (2014). Antimagnetic rotation in ^{104}Pd, *Phys. Rev. C* **89**, pp. 061303.

Roy, S. and Chattopadhyay, S. (2011). Possibility of antimagnetic rotation in odd- A Cd isotopes, *Phys. Rev. C* **83**, pp. 024305.

Roy, S. *et al.* (2011). Systematics of antimagnetic rotation in even–even Cd isotopes, *Phys. Lett. B* **694**, pp. 322–326.

Roy. S. and Chattopadhyay, S. (2013). Comment on "evidence of antimagnetic rotation in odd-A ^{105}Cd, *Phys. Rev. C* **87**, pp. 059801.

Simons, A. J. *et al.* (2005). Investigation of antimagnetic rotation in light cadmium nuclei: ^{105,106}Cd, *Phys. Rev. C* **72**, pp. 024318.

Simons, A. J., *et al.* (2003). Evidence for a New Type of Shears Mechanism in ^{106}Cd, *Phys. Rev. Lett.* **91**, pp. 162501.

Smith, J. F. *et al.* (1995). Contrasting behavior in octupole structures observed at high spin in ^{220}Ra and ^{222}Th, *Phys. Rev. Lett.* **75**, pp. 1050.

Singh, V. *et al.* (2017). Investigation of antimagnetic rotation in ^{101}Pd, *J. Phys. G*: **44**, pp. 075105 .

Sugawara, M. *et al.* (2015). Lifetime measurement for the possible antimagnetic rotation band in ^{101}Pd, *Phys. Rev. C* **92**, pp. 024309.

Sugawara, M., *et al.* (2009). Possible magnetic and antimagnetic rotations in ^{144}Dy, *Phys. Rev. C* **79**, pp. 064321.

Zhang, Z.-H. (2016). Effects of proton angular momentum alignment on the two-shears-like mechanism in ^{101}Pd, *Phys. Rev. C* **94**, pp. 034305.

Zhao, P. W., Peng, J., Liang, H. Z., Ring P., and Meng, J. (2012). Covariant density functional theory for antimagnetic rotation, *Phys. Rev. C* **85**, pp. 054310.

Zhao, P. W., Peng, J., Liang, H. Z., Ring, P., and Meng, J. (2011). Antimagnetic rotation band in nuclei: A microscopic description, *Phys. Rev. Lett.* **107**, pp. 122501.

Zhu, S. *et al.* (2001). Investigation of antimagnetic rotation in ^{100}Pd, *Phys. Rev. C* **64**, pp. 041302.

Chapter 14

Impact of CFL Locking in Quark Phase on Equations of State of Hybrid Star

Shashi K Dhiman and Suman Thakur

Department of Physics, Himachal Pradesh University, Shimla 171005, India

shashi.dhiman@gmail.com

14.1 Introduction

The investigations of compact stars are an ideal astrophysical laboratories for testing modern theories of dense nuclear matter physics, which provides a connection among various theoretical models of nuclear physics, particle physics, and astrophysics. It has been investigated [Witten (1984), Farhi (1984)] that the material in the core of compact stars may consist of quark matter. Compact stars, with a core made up of quark matter, are entitled as hybrid stars. Compact stars that are absolutely made up of quark matter, but might have a small layer of crust of nuclei, are the so-called quark stars [Haensel (1986), Alcock (1986)]. Quark stars are thought to be confined (bulk) objects containing a plenty of delocalized quarks such as up, down, and strange quarks (the charm, bottom, and top quarks are too massive to appear inside neutron stars) with electrons to maintain charge neutrality. The astrophysical precise

Multifragmentation in Heavy-Ion Reactions: Theory and Experiments
Edited by Rajeev K. Puri, Yu-Gang Ma, and Arun Sharma

ISBN 978-981-4968-69-0 (Hardcover), 978-1-003-38513-4 (eBook)
www.jennystanford.com

experimental observations for compact stars [Antoniadis (2013)] and references therein have provided a reliable and important source of information to constrain the interior composition of plausible equations of state of compact objects.

The composition and properties of a compact star are described by the appropriate equations of state (EOS), which in turn describe its crust and the interior region. The crust part of the compact star is generally described by the hadronic matter. Whereas in the interior region of compact stars, where star matter densities vary in the order of 4–8 times of ρ_0, a phase transition occurs from hadronic to quark matter. It is reasonable to assume that the core of the compact star is dense enough so that it can be treated as a pure quark phase and the boundary between the quark and hadronic matter should be a mixture of hadron and quarks.

Recently, the EOSs of quarks matter have been investigated by various theoretical groups by employing phenomenological non-relativistic [Burgio (2002), Farhi (1984), Zdunik (2000), Chu (2014)] and relativistic quark model [Zacchi (2015), Herbst (2011), Fu (2008), Xin (2014), Ghosh (2015), Huang (2003), Chu (2015), Menezes (2014)]. These models have a relation to some extent with free particle system. In the simplest version of the bag model, for example, the so-called "bag constant" is added to the thermodynamic potential density of the free system to reflect the quark confinement effect. One of the famous models is the MIT bag model. This model has been applied in a vast number of investigations on the properties of quarks star. It is well known, however, that particle masses vary with medium. Such masses are usually called effective masses. In principle, not only masses will change but also the coupling constant will run in a medium. The models with chemical potential and/or temperature-dependent particle masses are known as the quark quasiparticle models (QQPMs) [Schertler (1997), Wen (2009)], which have been explored in great detail over the past two decades. A set of stiff EOS of dense nuclear matter is required to construct a compact star of maximum gravitational mass $\approx 2.0\ M_\odot$. In the present chapter, the maximum masses of pure quark stars are calculated by solving the Tolman–Oppenheimer–Volkoff (TOV) equations.

For the quark phase, we propose to investigate the three-flavor quark matter star (composed of u, d, and s quarks in beta-equilibrium with electrons) within the framework of the self-consistent chemical-potential-dependent QQPM. In the QQPM, the quarks are considered quasiparticles, which gain an effective mass reproduced by the interaction with the other quarks in high dense matter. The medium effect of the coupling constant of the asymptotic freedom (g) is included in the dense quark matter. Therefore, the effective mass of the quarks depends on the strength of the coupling constant of the asymptotic freedom (g) and the chemical potential of the respective quarks. The MIT bag constant B is replaced with the effective bag function, which varies with the chemical potential and effective mass of the quarks. The thermodynamical potential is used to construct a set of plausible EOSs for dense quark matter.

In the present chapter, we construct a set of hybrid EOSs for matter made up of quark matter and hadronic phases and perform mass radius calculation for a hybrid star. The hadronic part of the EOS is evaluated using the field theoretical relativistic mean field (FTRMF) model whose parameters are appropriately calibrated to yield adequately the properties of the finite nuclei and nuclear matter at saturation density [Dhiman (2007)]. The quark matter phase has been treated within the QQPM. For the quark matter phase, we treat quarks as paired quarks in the CFL superconducting phase. But the effects of the quark masses and pairing correlations on the EOS due to the pairing energy associated with color superconductivity and the formation of quark cooper pair have been included [Alford (2001)]. In brief, our motivation in this chapter is to include the quark matter in the EOSs of compact stars to construct the equilibrium sequences. We construct the axially symmetric rotation sequences of the hybrid stars with various rotation frequencies and mass-shedding limit (Kepler frequency) by solving Einstein's field equations of general relativity in 2D. The chapter is organized as follows: in Section 14.2, we describe the models for the hybrid EOSs of the hadronic phase, the quark phase, and the hadron–quark mixed phase and obtain the comparison of hadron matter EOS with the hybrid EOSs. In Section 14.3, we give the results for the stable configurations of static as well as rotating compact stars. Finally, in Section 14.4, we give our conclusions.

14.2 Hybrid Equations of State

14.2.1 Hadronic Phase

We have employed the extended FTRMF model [Dhiman (2007)] to describe the hadronic phase of EOSs of compact stars. In the extended FTRMF model, the effective Lagrangian density consists of self- and mixed interaction terms for σ, ω, and ρ mesons up to the quadratic order in addition to the exchange interaction of baryons with σ, ω, and ρ mesons. The σ, ω, and ρ mesons are responsible for the ground-state properties of the finite and heavy nuclei. The mixed interaction terms involving the ρ-meson field enable us to vary the density dependence of the symmetry coefficient and neutron-skin thickness in heavy nuclei over a wide range without affecting the other properties of the finite nuclei [Furnstahl (2002), Sil (2005)]. In particular, the contribution from the self-interaction of ω-mesons plays an important role in determining the high-density behavior of EOS and the structure properties of compact stars [Dhiman (2007), Muller (1996)]. Whereas with the inclusion of the self-interaction of ρ-meson effect, the ground-state properties of heavy nuclei and compact stars only vary marginally [Muller (1996)]. In the present chapter, we use only BSR3 parameterization to construct the hadronic phase of the hybrid EOS. The Lagrangian density for the extended FTRMF model can be written as

$$\mathcal{L} = \mathcal{L}_{\mathcal{BM}} + \mathcal{L} + \mathcal{L} + \mathcal{L}_{\boldsymbol{\rho}} + \mathcal{L}_{\boldsymbol{\rho}} + \mathcal{L}_{em} + \mathcal{L}_{e\mu} + L_{YY}, \tag{14.1}$$

where the baryonic and mesonic Lagrangian $\mathcal{L}_{\mathcal{BM}}$ can be written as

$$\mathcal{L}_{\mathcal{BM}} = \sum_{B} \overline{\Psi}_B[i\gamma^\mu \partial_\mu - (M_B - g_{\sigma B}\sigma) - (g_{\omega B}\gamma^\mu \omega_\mu + \frac{1}{2} g_{\boldsymbol{\rho} B}\gamma^\mu \tau_B.\boldsymbol{\rho}_\mu)]\Psi_B. \tag{14.2}$$

Here, the sum on B is taken over the baryon octet, which consists of nucleons and Λ, Σ, and Ξ hyperons. A detailed description of each term of the effective Lagrangian density Eq. (14.1) is given in Ref. [Dhiman (2007)]. The equation of motion for baryons, mesons, and photons can be derived from the Lagrangian density defined in Eq.(14.1) by the Euler–Lagrange principle [Dhiman (2007)].

The energy density of the uniform matter within the framework of the extended FTRMF models is given by

$$\mathcal{E} = \sum_{j=B,\ell} \frac{1}{\pi^2} \int_0^{k_j} k^2 \sqrt{k^2 + M_j^{*2}} dk + \sum_B g_{\omega B} \omega \rho_B + \sum_B g_{\rho B} \tau_{3B} \rho + \frac{1}{2} m_\sigma^2 \sigma^2$$
$$+ \frac{\overline{\kappa}}{6} g_{\sigma N}^3 \sigma^3 + \frac{\overline{\lambda}}{24} g_{\sigma N}^4 \sigma^4 - \frac{\zeta}{24} g_{\omega N}^4 \omega^4 - \frac{\xi}{24} g_{\rho N}^4 \rho^4 - \frac{1}{2} m_\omega^2 \omega^2 - \frac{1}{2} m_\rho^2 \rho^2$$
$$- \overline{\alpha_1} g_{\sigma N} g_{\omega N}^2 \sigma \omega^2 - \frac{1}{2} \overline{\alpha_1}' g_{\sigma N}^2 g_{\omega N}^2 \sigma^2 \omega^2 - \overline{\alpha_2} g_{\sigma N} g_{\rho N}^2 \sigma \rho^2 - \frac{1}{2} \overline{\alpha_2}' g_{\sigma N}^2 g_{\rho N}^2 \sigma^2 \rho^2$$
$$- \frac{1}{2} \overline{\alpha_3}' g_{\omega N}^2 g_{\rho N}^2 \omega^2 \rho^2 + \frac{1}{2} m_{\sigma^*}^2 \sigma^{*2} + \sum_B g_{\phi B} \phi \rho_B$$
$$- \frac{1}{2} m_\phi^2 \phi^2. \tag{14.3}$$

The pressure of the uniform matter is given by

$$P = \sum_{j=B,\ell} \frac{1}{3\pi^2} \int_0^{k_j} \frac{k^4 dk}{\sqrt{k^2 + M_j^{*2}}} - \frac{1}{2} m_\sigma^2 \sigma^2 - \frac{\overline{\kappa}}{6} g_{\sigma N}^3 \sigma^3 dk - \frac{\overline{\lambda}}{24} g_{\sigma N}^4 \sigma^4$$
$$+ \frac{\zeta}{24} g_{\omega N}^4 \omega^4 + \frac{\xi}{24} g_{\rho N}^4 \rho^4 + \frac{1}{2} m_\omega^2 \omega^2 + \frac{1}{2} m_\rho^2 \rho^2 + \overline{\alpha_1} g_{\sigma N} g_{\omega N}^2 \sigma \omega^2$$
$$+ \frac{1}{2} \overline{\alpha_1}' g_{\sigma N}^2 g_{\omega N}^2 \sigma^2 \omega^2 + \overline{\alpha_2} g_{\sigma N} g_{\rho N}^2 \sigma \rho^2 + \frac{1}{2} \overline{\alpha_2}' g_{\sigma N}^2 g_{\rho N}^2 \sigma^2 \rho^2$$
$$+ \frac{1}{2} \overline{\alpha_3}' g_{\omega N}^2 g_{\rho N}^2 \omega^2 \rho^2 - \frac{1}{2} m_{\sigma^*}^2 \sigma^{*2} + \frac{1}{2} m_\phi^2 \phi^2. \tag{14.4}$$

14.2.2 Quark Quasiparticle Model (QQPM)

The total thermodynamical potential density of quark matter may written as [Agrawal (2009)]

$$\Omega(\bar{\mu}, \mu_e) = \Omega_{QQPM}(\bar{\mu}) + \Omega(\mu_e) \tag{14.5}$$

where the first term in Eq. (14.5) is obtained from the quarks and electrons and the second term from electrons. The contribution to the first term in Eq. (14.5) is given by [Xia (2014)]

$$\Omega_{QQPM} = \frac{-d_i T}{(2\pi)^3} \sum_i \int_0^\infty \left(\ln\left[1 + e^{\frac{(-\epsilon_i - \mu_i)}{T}}\right] + \ln\left[1 + e^{\frac{-(\epsilon_i + \mu_i)}{T}}\right]\right) d^3\nu_i. \tag{14.6}$$

On further simplification, we get

$$\Omega_{QQPM} = -\frac{1}{\pi^2} \sum_i \int_0^\infty \frac{\nu_i^4}{\sqrt{\nu_i^2 + m_i^{*2}}} [f_{i+} + f_{i-}] d\nu_i, \tag{14.7}$$

where $\epsilon_i = \sqrt{\nu_i^2 + m_i^{*2}}$, m_i^* are effective quark masses, ν_i for fermi momentum, the distribution functions for the quarks and anti-quarks are the Fermi distributions.

$$f_{i\pm} = \frac{1}{(1 + \exp[(\epsilon_i \mp \mu_i)/T])}, \tag{14.8}$$

with μ_i($-\mu_i$) being the chemical potential for quarks (anti-quarks) of type i. These equations apply at nonzero temperatures. For $T = 0$, there are no anti-particles, and the particle distribution functions become the usual step function $\Theta(\mu - \nu)$:

$$\Theta(\mu - \nu) = \begin{cases} 1, if & \nu \leq \mu \\ 0, if & \nu > \mu. \end{cases}$$

For $T = 0$ MeV, the chemical potential μ must be identified with the Fermi energy ϵ_f (the energy of the highest occupied level), d_i is the degeneracy factor (=2(spin)×3(color)=6 for quarks), and B^* is the total effective bag function. The limit of expression for T→ 0 is made for momenta ranging from zero to the Fermi one, since the Fermi–Dirac distribution at $T = 0$ presents a sharp cutoff at the Fermi momentum being useless to perform the integration to $\nu \to \infty$. At finite temperature, however, the broadening of the Fermi–Dirac distribution occurs; hence, the integration has to be extended as well. For $T = 0$ MeV, thermodynamical potential density can be simplified and solved analytically. The thermodynamical potential density becomes

$$\Omega_{QQPM} = -\frac{1}{\pi^2}\sum_i \int_0^{\nu_f} \frac{\nu_i^4}{\sqrt{\nu_i^2 + m_i^{*2}}} d\nu_i. \tag{14.9}$$

Also, it is

$$\Omega_{QQPM} = -\frac{3}{\pi^2}\sum_i \int_0^{\nu_f} (\mu_i - \epsilon_i)\nu_i{}^2 d\nu_i. \tag{14.10}$$

This result is calculated from Eq. (14.6) by using the relation

$$\lim_{T\to 0} \ln\left[1 + \exp\left(-\frac{x}{T}\right)\right] = -x\theta(-x). \tag{14.11}$$

The integrals in Eq. (14.7) and Eq. (14.9) have an analytical solution:

$$\Omega_{QQPM} = -\frac{1}{\pi^2}\sum_i \{\frac{1}{8}\left[\nu_i\sqrt{\nu_i^2 + m_i^{*2}}(2\nu_i^2 - 3m_i^{*2}) + 3m_i^{*4}\ln(\frac{\sqrt{\nu_i^2 + m_i^{*2}} + \nu_i}{m_i^*})\right]\}. \tag{14.12}$$

Medium effects play an important role in describing the properties of quark matter via the concept of effective masses.

The effective mass is taken to be [Schertler (1997), Pisarski (1989)]

$$m_i^* = \frac{m_{i0}}{2} + \sqrt{\frac{m_{i0}^2}{4} + \frac{g^2\mu_i^2}{6\pi^2}}, \tag{14.13}$$

where m_{i0}, μ_i, and g are the current quark mass, quark chemical potential, and strong interaction coupling constant, respectively. In actual practice, a strong coupling constant is running, [Shirkov (1997), Wen (2010), Patra (1996)] which has a phenomenological expression such as

$$g^2(T, \mu_i) = \frac{48}{29}\pi^2\left[\ln\left(\frac{0.8\mu_i^2 + 15.622T^2}{\Lambda^2}\right)\right]^{-1}, \tag{14.14}$$

where Λ is the QCD scale-fixing parameter. In the present calculation, the value is taken to be 120 MeV. And at finite temperature

[Stocker (1984)], number densities for all the three flavors of quarks considered are the same and can be obtained as

$$n_i = \frac{d_q \nu^3}{6\pi^2} + \frac{2\Delta^2 \bar{\mu}}{\pi^2}, \tag{14.15}$$

with i = u, d, and s. At finite temperature, quark number density is

$$n_i = 3 \times 2 \int_0^\infty \frac{d^3 \nu_i}{(2\pi)^3} (f_{i+} - f_{i-}) + \frac{2\Delta^2 \bar{\mu}}{\pi^2}. \tag{14.16}$$

At $T = 0$ MeV,

$$n_i = \frac{3}{\pi^2} \int_0^{\nu_f} \nu^2 d\nu_i + \frac{2\Delta^2 \bar{\mu}}{\pi^2}, \tag{14.17}$$

where $\nu_f = \nu_u = \nu_d = \nu_s = \left[\left(2\mu - \sqrt{\mu^2 + \frac{m_s^2 - m_u^2}{3}} \right)^2 - m_u^2 \right]$ for $\mu = \frac{\mu_u + \mu_d + \mu_s}{3}$ is the common Fermi momentum of the quark system, which depends on the mass of the three quark flavors.
For $m_u = m_d = 0$, the common fermi momentum becomes

$$\nu = 2\mu - \sqrt{\mu^2 + \frac{m_s^2}{3}}. \tag{14.18}$$

The energy density is

$$\begin{aligned} \mathcal{E}_{QM} &= \frac{\partial(\beta\Omega)}{\partial\beta} + \mu_i n_i + B^* \\ &= \sum_i \frac{3}{\pi^2} \int_0^\infty d\nu_i \nu_i^2 \sqrt{\nu_i^2 + m^{*2}} (f_{i+} + f_{i-}) + B^*. \end{aligned} \tag{14.19}$$

At $T = 0$ MeV, the energy density becomes

$$\mathcal{E}_{QM} = \sum_i \frac{3}{\pi^2} \int_0^{\nu_f} d\nu_i \nu_i^2 \sqrt{\nu_i^2 + m^{*2}} + B^*. \tag{14.20}$$

This integral has an analytical solution:

$$\mathcal{E}_{QM} = \frac{1}{\pi^2} \sum_i \frac{3}{8} \left[\nu_i \sqrt{\nu_i^2 + m_i^{*2}} (2\nu_i^2 + m_i^{*2}) - m_i^{*4} \ln\left(\frac{\sqrt{\nu_i^2 + m_i^{*2}} + \nu_i}{m_i^*} \right) \right] + B^*. \tag{14.21}$$

Thus, the pressure and energy density becomes

$$P_{QP} = -\Omega(\bar{\mu}, \mu_e) - B^*$$
$$\mathcal{E}_{QM} = \frac{\partial(\beta\Omega)}{\partial\beta} + \mu_i n_i + B^*,$$

where B* is the effective bag function, which can be divided into two parts: the μ_i-dependent part and the definite integral constant.

$$B^* = \sum_i B_i(\mu_i) + B_0. \tag{14.22}$$

The introduction of B^* is done to show the automatic confinement characteristic in the model, where i = u, d, and s, and B_0 is similar to the conventional MIT bag constant.
The μ_i-dependent part is calculated by

$$B_i = -\int_{m_i^*}^{\mu_i} \frac{\partial\Omega_i}{\partial m_i^*}\frac{\partial m_i^*}{\partial\mu_i} d\mu_i. \tag{14.23}$$

With the above quark mass formulas and thermodynamic treatment, one can get the properties of bulk CFL quark matter.
In the CFL phase at $T = 0$, the three flavors of quarks satisfy the following conditions:
(1) They have equal Fermi momenta, which minimizes the free energy of the system.
(2) They have an equal number of densities n_i, as a consequence of the first condition, which means that $n_i = n_B$ and $\mu_i = \mu$, for i = u, d, s. Same is the case with finite temperature, except there is difference in particle number densities due to temperature dependence. So, there must be some presence of electrons in order to satisfy charge equilibrium condition.

14.2.3 Construction of Hadron–Quark Mixed Phase

In order to study the properties of a rapidly rotating hybrid neutron star, we should first construct the EOS of the star. So, we first do separate modeling of hadronic and quark phases within different theoretical approaches. Then, both phases are joined by a thermodynamic phase transition. We assumed that the hadron–quark phase transition is of first order. In the purely hadronic phase, which consists of baryons (n,p) and leptons (e,μ), the conditions of beta stability and charge neutrality can be expressed as

$$\mu_n - \mu_p = \mu_e = \mu_\mu, \tag{14.24}$$

$$\rho_p = \rho_e + \rho_\mu, \tag{14.25}$$

where μ_i are the chemical potentials and ρ_i are the particle number densities. Similarly, the pure quark matter phase, which contains three-flavor quarks (u, d, s) should satisfy the beta-equilibrium along with electric charge neutrality. The weak beta-equilibrium condition can be expressed as

$$\mu_u = \mu_d = \mu_s \tag{14.26}$$

where μ_i (i = u, d, s) is the chemical potential of the particles in quark matter star. Furthermore, the electric charge neutrality condition is

$$\frac{2}{3}n_u = \frac{1}{3}n_d + \frac{1}{3}n_s. \tag{14.27}$$

We construct the mixed phase of EOS made up of the hadron matter and quark matter by employing the Glendenning construction [Glendenning (2000)] for a hybrid compact star. The equilibrium chemical potential of the mixed phase corresponding to the intersection of the two surfaces representing hadron and quark phases can be calculated from Gibb's condition for mechanical and chemical equilibrium at zero temperature, which reads as

$$P_{HP}(\mu_e, \mu_n) = P_{QP}(\mu_e, \mu_n) = P_{MP}, \tag{14.28}$$

In the mixed phase, the local charge neutrality condition is replaced by the global charge neutrality, which means that both hadron and quark matter are allowed to be charged separately. The condition of the global charge neutrality can be expressed as

$$\chi\rho_c^{QP} + (1-\chi)\rho_c^{HP} = 0, -m_i \tag{14.29}$$

where χ is the volume fraction occupied by quark matter in the mixed phase in terms of charge density ρ_c. The value of the χ increases from zero in the pure hadron phase to $\chi = 1$ in the pure quark phase. The energy density $\mathcal{E}_{MP}$ and the baryon density ρ_{MP} of the mixed phase can be calculated as

$$\mathcal{E}_{MP} = \chi\mathcal{E}_{\mathcal{QP}} + (1-\chi)\mathcal{E}_{HP}, \tag{14.30}$$

$$\rho_{MP} = \chi\rho_{QP} + (1-\chi)\rho_{HP}. \tag{14.31}$$

However, in the present calculation, we consider all the three quarks masses as $m_u = m_d = 4\text{MeV}/c^2$ and $m_s = 95\text{MeV}/c^2$

14.2.4 Rotating Neutron Stars

The structure of a rapidly rotating neutron star is different from the static one, since the rotation can strongly deform the star. We assume that neutron stars are steadily rotating and have axially symmetric structure. Therefore, the space–time metric used to model a rotating star can be expressed as

$$ds^2 = -e^{\gamma+\rho}dt^2 + e^{2\beta}(dr^2 + r^2d\theta^2)e^{\gamma-\rho}r^2\sin^2\theta(d\phi - \omega dt)^2 \tag{14.32}$$

where the potentials γ, ρ, β, and ω are functions of r and θ only. The matter inside the star is approximated by a perfect fluid and the energy-momentum tensor is given by

$$T^{\mu\nu} = (\mathcal{E} + P)u^\mu u^\nu - Pg^{\mu\nu} \tag{14.33}$$

where $\mathcal{E}$, P, and u^μ are the energy density, pressure, and four-velocity, respectively. In order to solve Einstein's field equation for the potentials γ, ρ, β, and ω, we adopt the KEH method and use the public RNS code for calculating the properties of a rotating star.

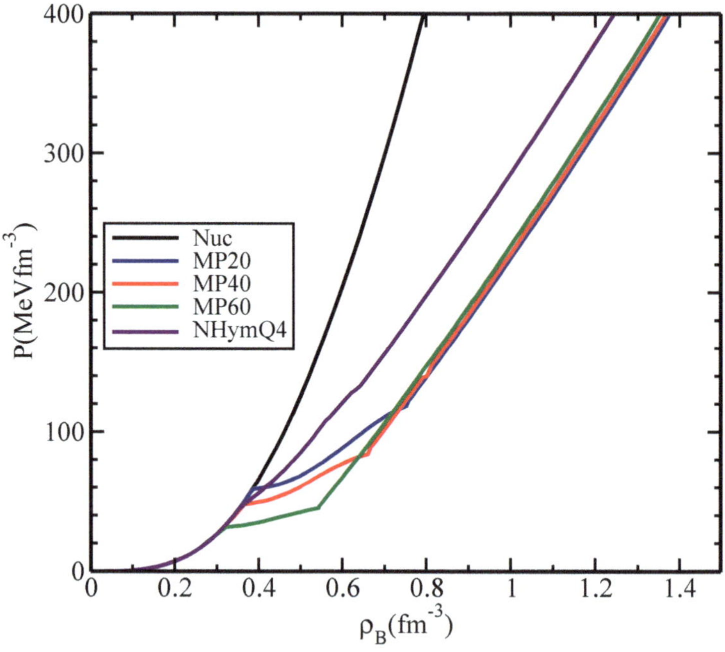

Figure 14.1 The pressure versus energy density for static hybrid star compared with a pure hadron matter star.

14.3 Results and Discussions

14.3.1 Equations of State and Static Sequences of Hybrid Star

In order to study the properties of a hybrid star composed of a CFL quark matter core for different values of the CFL gap parameter, we have considered the three values for CFL gap parameter in the range of 20–60 MeV. Figure 14.1 presents the results of the EOSs for the hybrid star and the pure neutron star in the form $P = P(\mathcal{E})$. In Table 14.1, we have shown the description of the selected EOS models along with composition and other defining parameters. In Fig. 14.2, we display the gravitational mass as a function of the central energy density and of the equatorial radius, using the EOS for the static configurations. In all cases, the maximum masses of a hybrid neutron star are lower than those of a neutron star, because the appearance

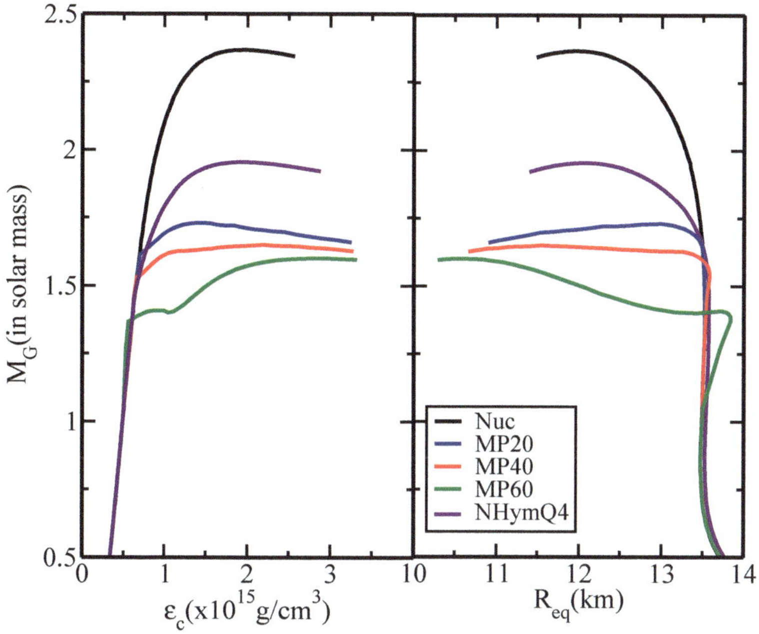

Figure 14.2 The gravitation mass versus central energy density and radius relationship for a static hybrid star compared with a pure hadron matter star.

of quark matter in the core of the star results in a softening of the very hard nucleonic EOS.

14.3.2 Keplerian Limit

In Fig. 14.3, we show the gravitational mass as a function of the central energy density and of the equatorial radius, using the EOSs for rapidly rotating at the Keplerian frequency. Comparing Keplerian and static sequences, rotations increase the maximum mass and equatorial radius substantially.

The rotational frequency is a directly measurable quantity of pulsars, and the Keplerian (mass-shedding) frequency f_K is one of the most studied physical quantities for rotating stars. In Table 14.2, we presented the structural properties of static as well as rotating compact stars at the Keplerain frequency. We presented

Table 14.1 List of used EOS models with particle composition and other parameters used.

S.No	EOS	Particle Production	CFL pairing & Bag constant
1	Nuc	n,p	
2	MP20	n,p,u,d and s quarks	Δ=20MeV,$(B_0)^{1/4}$=170MeV
3	MP40	n,p,u,d and s quarks	Δ=40MeV,$(B_0)^{1/4}$=170MeV
4	MP60	n,p,u,d and s quarks	Δ=60MeV,$(B_0)^{1/4}$=170MeV
5	NHymQ4	n, p, hyperons, u, d and s quarks	Δ=50MeV,$(B_0)^{1/4}$=150MeV

Table 14.2 Several properties of static and rotating neutron star for the selected EOSs: maximum gravitational mass M_{max} in terms of solar mass, corresponding equatorial radius R_{eq}(km), central energy density $\mathcal{E}_c(\times 10^{15} g/cm^{-3})$, and the maximum Keplerian frequency f_K(Hz).

EOS		Nuc	MP20	MP40	MP60	NHymQ4
	$M_{max}/M_\odot$	2.37	1.73	1.65	1.60	1.95
Static	R_{eq}(km)	11.97	12.99	11.55	10.62	12.08
	$\mathcal{E}_c(\times 10^{15} g/cm^{-3})$	1.96	1.39	2.20	2.80	1.95
	$M_{max}/M_\odot$	2.85	2.15	2.05	1.86	2.37
Keplerian	R_{eq}(km)	16.01	18.30	18.57	14.65	16.75
	$\mathcal{E}_c(\times 10^{15} g/cm^{-3})$	1.66	1.17	1.116	2.49	1.61
	f_K(Hz)	1487	1086	1038	1393	1288

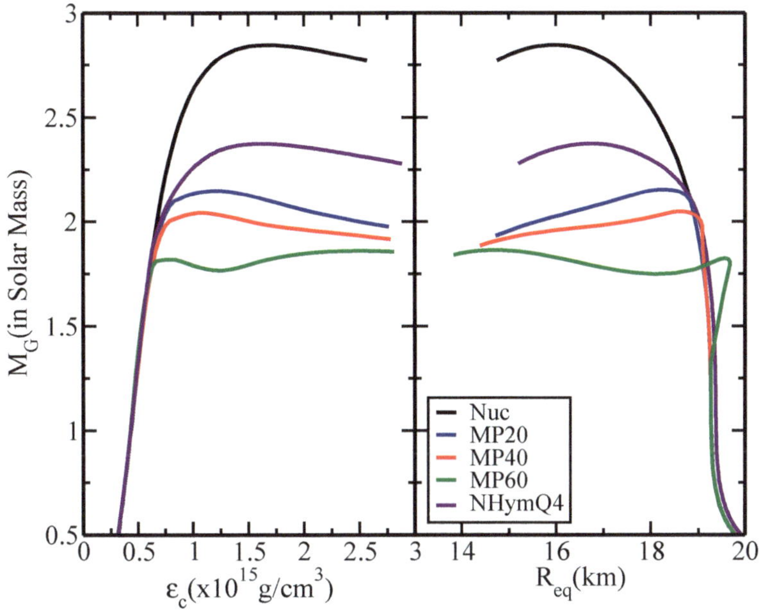

Figure 14.3 The mass–radius relationship for a rotating hybrid star compared with a pure hadron matter star.

the values of the maximum gravitational mass $M_{max}(M_{\odot})$, its corresponding equatorial radius R_{max}(km), central energy density $\mathcal{E}_c(\times 10^{15} g/cm^{-3})$, and the Keplerian frequency f_K(Hz). The value of the Keplerian frequency for the hybrid star at the maximum mass is in the range of 1038–1396 Hz, whereas the Keplerian frequency for the hadronic stars at the maximum mass is about 1487 Hz.

In Fig. 14.4, we presented Keplerian frequency as a function of the gravitational mass for some selected EOSs. We observe that it increases monotonically for both a hadronic star and a hybrid star. The Keplerian frequency of a hybrid star increases more rapidly after the onset of quark matter and is larger than that of a hadron star with the same gravitational mass, because the stellar radius is smaller in the former case due to the presence of a very dense quark matter core. However, due to the lower maximum mass of a hybrid star, the maximum Keplerian frequency of a hybrid star is lower than the hadron star.

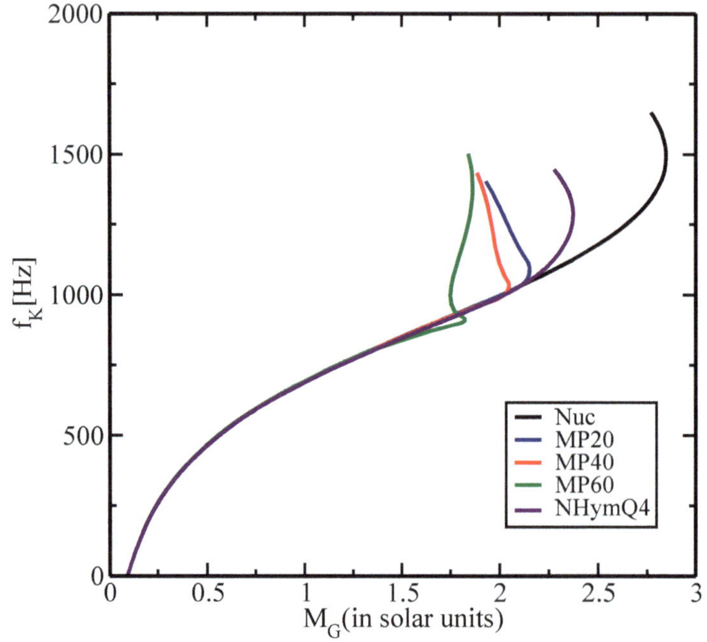

Figure 14.4 Keplerian frequency versus the gravitational mass for a rotating hybrid star EOSs compared with a pure hadron matter.

14.3.3 Back Bending and Stability Analysis in J(f) Plane

The search for the back-bending phenomenon with simultaneous testing of rotating configurations can be most conveniently carried out by plotting the dimensionless angular momentum (cJ/GM_G^2) versus the rotational frequency (f). In Fig. 14.5, we present the cJ/GM_G^2 versus frequency (f) curves with EOS NHymQ4. This EOS is chosen because it is composed of nucleons, hyperons, mixed phase followed by quark phase. These curves represent the evolution of an isolated pulsar of a fixed baryonic mass M_B as it loses its moment of inertia and angular momentum due to the radiation of electromagnetic waves. Along each curve, the central density ρ_c increases monotonically when one moves downward. For stable configurations, J is a monotonic function along its path. Any minimum indicates the onset of instability with respect to axis symmetric perturbations.

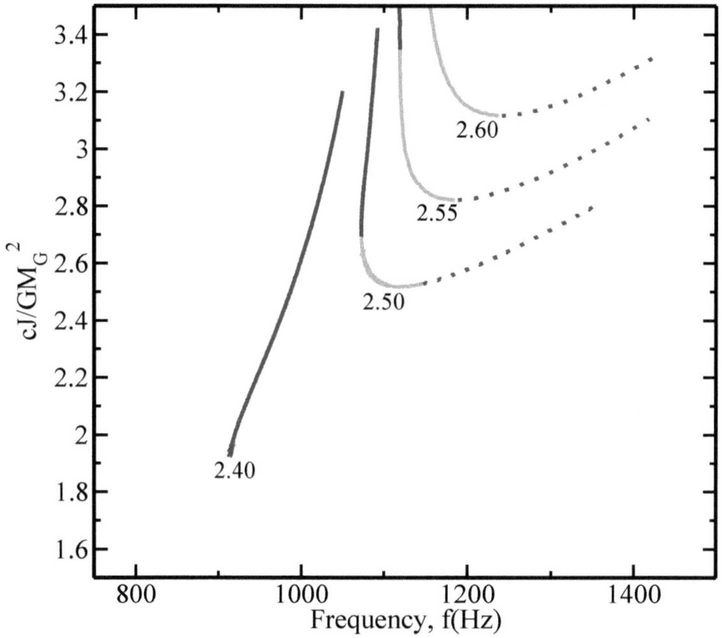

Figure 14.5 Dimensionless angular momentum versus spin frequency for EoS NHymQ4. Each curve corresponds to a fixed M_B, whose value in $M_\odot$ is displayed. Along each curve, the central density increases downward.

14.3.4 Radii of Millisecond Pulsars

In Fig. 14.6, we present the theoretical limits of the radii of various well-known millisecond pulsars as a function of their rotating frequencies. The theoretical limits of the radii and the gravitational mass of these pulsars have been computed by using the NHymQ4 equation of state.

14.4 Summary

We constructed the static as well as Keplerian sequences of hybrid stars for different values of the CFL pairing gap. The hybrid star considered here is assumed to be composed of nuclear matter in the crust, a mixed phase in the intermediate, and CFL quark matter

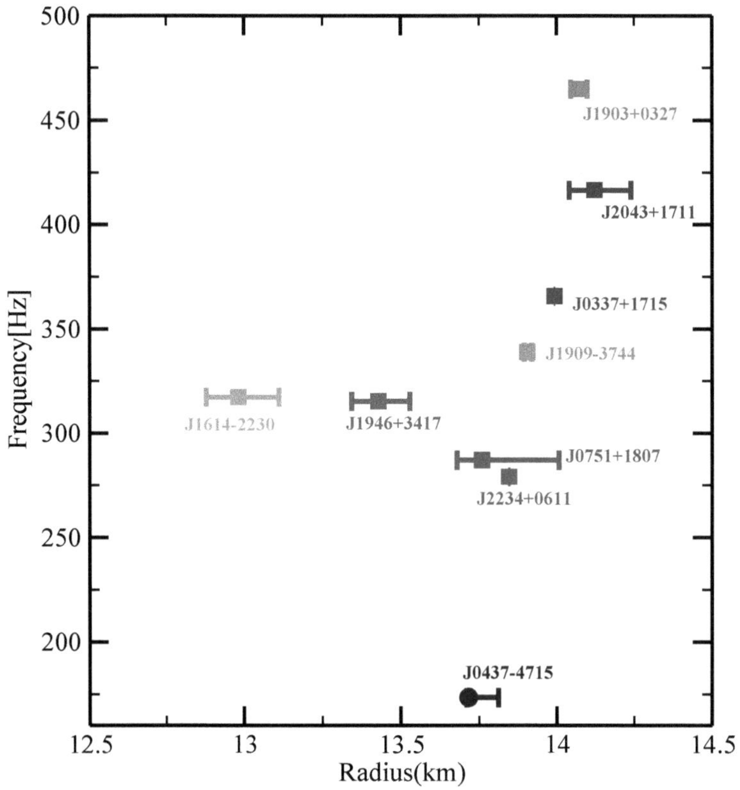

Figure 14.6 The theoretical limits of the radii of various well-known millisecond pulsars are presented as a function of their rotating frequencies. The theoretical limits of the radii and the gravitational mass of these pulsars have been computed by using the NHymQ4 equation of state.

in the core. The hadronic part of EOS is obtained by using the EFTRMF model, whereas the quark matter equation state is obtained by using the QQPM model with different values for the CFL pairing gap. With an increase in the CFL pairing gap, the width of the mixed phase region decreases and there is an early onset of the mixed phase as well as the quark phase. Due to an early onset of the mixed phase region as well as the quark phase region, there is a considerably decrease in the gravitational maximum mass of a compact star. Also, in the case of recently observed pulsars with known gravitational mass, we have extracted the theoretical limits of the radii of various well-known millisecond pulsars as a function of their rotating frequencies by using the NHymQ4 EOS of a hybrid star.

References

Agrawal, B. K. and Dhiman, S. K. (2009). Stable configurations of hybrid stars with color-flavor-locked core, *Phys. Rev. D* **79**, pp. 103006.

Alcock, C., Farhi, E., and Olinto, A. (1986). Strange stars, *Astrophys. J.* **310**, pp. 261–272.

Alford, M. G. (2001). Color-superconducting quark matter, *Annu. Rev. Nucl. Part. Sci.* **51**, pp. 131.

Antoniadis, J. (2013). A massive pulsar in a compact relativistic binary, *Science* **340**, pp. 6131.

Burgio, G. F., Baldo M., Sahu, P. K., and Schulze, H.-J. (2002). Hadron–quark phase transition in dense matter and neutron stars, *Phys. Rev. C* **66**, pp. 025802.

Chu, P. C. and Chen, L. W. (2014). Quark matter symmetry energy and quark stars, *Astrophys. J.* **780**, pp. 1–11.

Chu, P. C., Wang, X., Chen, L. W., and Huang, M. (2015). Quark magnetar in the three-flavor Nambu Jona-Lasinio model with vector interactions and a magnetized gluon potential. *Phys. Rev. D* **91**, pp. 023003.

Dhiman, S. K., Kumar, R., and Agrawal, B. K. (2007). Rotating star, *Phys. Rev. C* **76**, pp. 045801.

Farhi, E. and Jaffe, R. L. (1984). Strange matter, *Phys. Rev. D* **30**, pp. 2379–2390.

Fu, W. J., Zhang, Z., and Liu, Y. X. (2008). 2 + 1 flavor Polyakov Nambu Jona-Lasinio model at finite temperature and nonzero chemical potential, *Phys. Rev. D* **77**, pp. 014006.

Furnstahl, R. (2002). Neutron radii in mean-field models, *Nucl. Phys. A* **706**, pp. 85.

Ghosh, S. K., Raha, S., Ray, R., Saha, K., and Upadhaya, S. (2015). Shear viscosity and phase diagram from Polyakov Nambu Jona-Lasinio model, *Phys. Rev. D* **91**, pp. 054005.

Glendenning, N. K. (2000). *Compact Stars: Nuclear Physics, Particle Physics, and General Relativity*. Springer-Verlag, New York.

Haensel, P., Zdunik, J. L., and Schaeffer, R. (1986). Strange quark stars, *Astron. Astrophys.* **160**, pp. 121–128.

Herbst, T. K., Pawlowski, J. M., and Schaefer, B. J. (2011). The phase structure of the Polyakov-quark-meson model beyond mean field, *Phys. Lett. B* **696**, pp. 58–67.

Huang, M. and Shovkovy, I. (2003). Gapless color superconductivity at zero and at finite temperature, *Nucl. Phys. A* **729**, pp. 835.

Menezes, D. P., Pinto, M. B., Castro, L. B., Costa, P., and Providência, C. M. C. (2014). Repulsive vector interaction in three-flavor magnetized quark and stellar matter *Phys. Rev. C* **89**, pp. 055207.

Müller, H. and Serot, B. D. (1996). Relativistic mean-field theory and high-density nuclear EOS, *Nucl. Phys. A* **606**, pp. 508.

Patra, B. K. and Singh, C. P. (1996). Temperature and baryon-chemical-potential-dependent bag pressure for a deconfining phase transition, *Phys. Rev. D* **54**, pp. 3551.

Pisarski, R. (1989). Renormalized fermion propagator in hot gauge theories, *Nucl. Phys. A* **498**, pp. 423.

Schertler, K., Greiner, C., and Thoma, M. (1997). Medium effects in strange quark matter and strange stars, *Nucl. Phys. A* **616**, pp. 659.

Shirkov, D. V. and Solovtsov, I. L. (1997). Analytic model for the QCD running coupling with universal $\overline{\alpha}_s(0)$ value, *Phys. Rev. Lett.* **79**, pp. 1209.

Sil, T., Centelles, M., Vinas, X., and Piekarewicz, J. (2005). Atomic parity non-conservation, neutron radii, and effective field theories of nuclei, *Phys. Rev. C* **71**, pp. 045502.

Stocker, H. (1984). Deconfinement in the baryon-rich region, *Nucl. Phys. A* **418**, pp. 587c.

Wen, X. J., Feng, Z. Q., Li, N., and Peng, G. X. (2009). Strange quark matter and strangelets in the quasiparticle model, *J. Phys. G: Nucl. Part. Phys.* **36**, pp. 025011.

Wen, X. J., Li, J. Y., Liang, J. Q., and Peng, G. X. (2010). Medium effects on the surface tension of strangelets in the extended quasiparticle model, *Phys. Rev. C* **82**, pp. 025809.

Witten, E. (1984). Cosmic separation of phases, *Phys. Rev. D* **30**, pp. 272–285.

Xia, C. J., Peng, G. X., Chen S. W., Lu, Z. Y., and Xu, J. F. (2014). Thermodynamic consistency, quark mass scaling, and properties of strange matter, *Phys. Rev. D* **89**, pp. 105027.

Xin, X. Y., Qin, S. X., and Liu, Y. X. (2014). Improvement on the Polyakov Nambu Jona-Lasinio model and the QCD phase transitions, *Phys. Rev. D* **89**, pp. 094012.

Zacchi, A., Stiele, R., and Schaffner-Bielich, J. (2015). Compact stars in a SU (3) quark-meson model, *Phys. Rev. D* **92**, pp. 045022.

Zdunik, J., Bulik, T., Kluzniak, W., Haensel, P., and Gondek-Rosinska, D. (2000). On the mass of moderately rotating strange stars in the MIT bag model and LMXBs, *Astron. Astrophys.* **359**, pp. 143.

Chapter 15

Investigation of Light Particle and Intermediate Mass Fragment Production Cross Sections of Excited Compound System $^{44}Ti^*$ Formed in $^{32}S + ^{12}C$ and $^{28}Si + ^{16}O$ Reactions

K. P. Santhosh and P. V. Subha

School of Pure and Applied Physics, Kannur University, Swami Anandatheertha Campus, Payyanur 670327, Kerala, India

drkpsanthosh@gmail.com

The total decay cross section, the intermediate mass fragment (IMF) production cross section, and the cross section for the formation of light particle (LP) for the decay of the compound system $^{44}Ti^*$ formed in the reactions $^{32}S + ^{12}C$ and $^{28}Si + ^{16}O$ have been evaluated by taking the scattering potential as the sum of Coulomb and nuclear proximity potential, with and without incorporating deformation effects. The computed results have been compared with the available experimental data of fragment cross section corresponding to the $Z = 6$ fragment at $E_{CM} = 60$ MeV for the entrance channel $^{32}S + ^{12}C$. Hence, we have extended our studies and have thus computed the total decay cross section, IMF

Multifragmentation in Heavy-Ion Reactions: Theory and Experiments
Edited by Rajeev K. Puri, Yu-Gang Ma, and Arun Sharma

ISBN 978-981-4968-69-0 (Hardcover), 978-1-003-38513-4 (eBook)
www.jennystanford.com

cross section, and LP cross section for the decay of $^{44}Ti^*$ formed through the other entrance channel ^{28}Si + ^{16}O. Hence, we hope that our predictions on the evaluations of the IMF cross sections and the LP cross sections for the decay of $^{44}Ti^*$ formed through these channels can be used for further experimental studies.

15.1 Introduction

In the last few decades, there have been extensive studies [1–11] for understanding the fragment emission mechanisms for low-energy nucleus–nucleus collisions. These studies reveal that for low energy (≤10 MeV/nucleon), light heavy-ion (A_{proj} + A_{target} ≤ 60) collisions, the two dominant mechanisms are fusion followed by asymmetric fission (FF) [12–17] and deep inelastic orbiting [8–11], which contribute to the observed full energy-damped yields of the fragments. Apparently, the decay process is found to depend on temperature and angular-momentum-dependent potential barriers [1]. In the past few years, several experimental and theoretical studies have been done on the decay of light compound nuclear systems formed through heavy-ion reactions [4, 18–21]. Simple statistical theory says that a compound nucleus (CN) is formed after complete equilibration of all the degrees of freedom, which then decays into various exit channels. The decay probability for CN is governed by the available phase space.

The deep inelastic orbiting process is based on the assumption that instead of forming a CN as in the FF process, a long-lived, dinuclear molecular complex [11] is formed, which has a strong memory of the entrance channel. In addition to this, for the light heavy-ion systems, the shapes of the orbiting dinuclear systems are quite similar to the saddle and scission shapes obtained in the course of evolution of the FF process. Moreover, the occurrence of both orbiting and fusion–fission processes has similar timescales and hence the distinction between the signatures of the two processes becomes a real challenge. Planeta *et al.* [22], in a detailed study of fragment emission from the compound system ^{44}Ti, produced via the α-cluster system ^{32}S + ^{12}C at 280 MeV, established that fragments (7 ≤ Z ≤16) were emitted due to symmetric splitting followed by evaporation. On the other hand, the energy-damped yield of binary fragments and quasielastic emission from the system $^{28}Si + ^{16}O$, which produces the same composite ^{44}Ti at two different energies—$E_{C.M.}$ = 39.10 and 50.5 MeV, respectively—was measured by Oliveira *et al.* [23] and led to the finding that the *Q*-value-integrated angular distributions follows ~1/sinθ c.m.-type behavior, indicating a long-lived intermediate state. But the observations such as the mass distribution peaks near to the projectile and target mass, the rather large ratio between the oxygen and the carbon cross sections, and the significantly larger total kinetic energy (TKE) values than the Coulomb repulsion have conjectured the presence of the noncompound orbiting-like mechanisms for the energy-damped yield of the fragments from the system $^{28}Si + ^{16}O$. From the same composite system $^{44}Ti^*$ produced at different excitations via the reaction ^{16}O (76, 96, 112 MeV) + ^{28}Si [24], a large deformation has also been observed in the study of LP emission. The large deformation observed may be associated with the orbiting process. So, a more detailed study of this system is necessary to delineate the fragment emission mechanism. Pandey *et al.* [25] studied the complex fragment emission from the decay of fully energy-relaxed composite $^{44}Ti^*$ formed via the $^{32}S + ^{12}C$ reaction at two excitation energies. Inclusive energy distributions of the fragments ($3 \leq Z \leq 8$) emitted in the reaction $^{32}S + ^{12}C$ have been measured in the angular range ~16°–28°, at two incident energies, 200 and 220 MeV, respectively.

Gupta and collaborators developed the dynamical cluster-decay model (DCM) to study the excited state decay of hot and rotating compound nucleus [26–31]. Using this model, Gupta *et al.* [28, 32] studied the decay of $^{56}Ni^*$ formed in the $^{32}S + ^{24}Mg$ reaction at two incident energies $E_{C.M.}$ = 51.6 and 60.5 MeV. Karthikraj *et al.* [33] applied the reformulated DCM to study the decay of odd-A and non-α-structured $^{59}Cu^*$ formed in the $^{35}Cl + ^{24}Mg$ reaction at E_{lab} = 275 MeV. The decay of light mass alpha and non-alpha conjugate composite systems $^{20,21,22}Ne^*$, $^{28}Si^*$, $^{39}K^*$, $^{40}Ca^*$ formed in light heavy-ion reactions has been studied within the collective clusterization approach of DCM, which revealed the presence of FF and DIO competing reaction mechanisms in different exit channels [34]. One of us K.P.S studied the decay properties of various even–even isotopes of Ba [35] and the decay of the excited compound system $^{48}Cr^*$ [36] formed through $^{24}Mg + ^{24}Mg$, $^{36}Ar + ^{12}C$, and $^{20}Ne + ^{28}Si$ reactions and also the decay of $^{56}Ni^*$ [37] formed through the channels $^{32}S + ^{24}Mg$, $^{36}Ar + ^{20}Ne$, $^{40}Ca + ^{16}O$, and $^{28}Si + ^{28}Si$ at various $E_{C.M.}$ values within the barrier penetration model taking the scattering potential as the sum of the Coulomb and nuclear proximity potential of Blocki *et al.* [38, 39]. We have also studied the decay of various isotopes of $^{26-29}Al^*$ formed through the channels $^{16}O + ^{10}B$, $^{16}O + ^{11}B$, $^{18}O + ^{10}B$, and $^{18}O + ^{11}B$, respectively, for various $E_{C.M.}$ values. In the present manuscript, we have performed an extensive study on the total decay cross section, the production cross section of IMF, and the light particle (LP) formation cross section of $^{44}Ti^*$ formed through the channels $^{32}S + ^{12}C$ and $^{28}Si + ^{16}O$ using Wong's formula [40], the Glas and Mosel formula [41] by incorporating the temperature-dependent diffuseness parameter, and the radius for the nucleus–nucleus potential taken from Royer *et al.* [42], taking the scattering potential as the sum of the deformed Coulomb and deformed nuclear proximity potential. The cross-section calculations for all the entrance channels are done using the ℓ-summed Wong formula for cross section, the approximated Wong formula for relatively large values of E, and the Glas and Mosel formula for cross section with and without incorporating deformation effects. The computed values are compared with the available experimental data [25].

Section 15.2 presents the details of the formalisms used for the evaluation of the scattering potential, the sum of Coulomb and proximity potentials, and the formalisms used for calculating the production cross section. Section 15.3 gives the results and discussion of the study, and Section 15.4 gives the conclusion of the study.

15.2 Theory

15.2.1 The Potential

The cross-section calculations were made by taking the scattering potential as the sum of the Coulomb and proximity potentials of Blocki *et al.* [38, 39]. The interaction potential energy barrier for spherical fragments, which is the sum of Coulomb potential, proximity potential, and centrifugal potential, is given as

$$V = \frac{Z_1 Z_2 e^2}{r} + V_P(z) + \frac{\hbar^2 \ell(\ell+1)}{2\mu r^2}, \text{ for } z > 0 \quad (15.1)$$

where Z_1 and Z_2 are the atomic numbers of the projectile and the target nuclei, ℓ is the angular momentum, r is the distance between the centers of the interacting fragments, and z is the separation between the near surface of the fragments. μ is the reduced mass, which is given as $\mu = mA_1A_2/A$ where A, A_1, and A_2 are the mass numbers of the compound nucleus and the interacting fragments, respectively, and m is the mass of nucleon. The proximity potential V_P is given by Blocki *et al.* [38] as

$$V_P(z) = 4pgb\frac{C_1C_2}{C_1+C_2}j\left(\frac{z}{b}\right) \tag{15.2}$$

Here, $r = z + C_1 + C_2$, where C_1 and C_2 are the Süssman central radii. The relation for the nuclear surface tension coefficient is given as

$$g = 0.9517[1-1.7826(N-Z)^2/A^2] \tag{15.3}$$

Here, N and Z are the neutron and proton numbers of the parent nuclei. The universal proximity potential ϕ is given as [39]

$$j(e) = -4.41e^{\frac{-e}{0.7176}}, \text{ for } \varepsilon > 1.9475 \tag{15.4}$$

$$j(e) = -1.7817 + .9270e + 0.0169e^2 - 0.05148e^3, \text{ for } 0 \le \varepsilon \le 1.9475 \tag{15.5}$$

with $\varepsilon = z/b$ and the relation between the Süssman central radii C_i and the sharp radii R_i is given by $C_i = R_i - \frac{b^2}{R_i}$ and $b \sim 1$ fm is the surface width. The change in the surface width with the inclusion of temperature T is [42]

$$b(T) = 0.99[1+0.009T^2] \tag{15.6}$$

Similarly, we can reformulate the Süssman central radii [42] on the inclusion of temperature T as

$$C_i = R_i(T) - \frac{0.99[1+0.009T^2]^2}{R_i(T)} \tag{15.7}$$

The semi-empirical formula in terms of the mass number A_i for R_i is given as

$$R_i = 1.28A_i^{1/3} - 0.76 + 0.8A_i^{-1/3} \tag{15.8}$$

For an excited compound system,

$$R_i(T) = [1.28A_i^{1/3} - 0.76 + 0.8A_i^{-1/3}][1+0.0007T^2] \tag{15.9}$$

The barrier penetrability P for an excited compound system is given as

$$P = \exp\left\{-\frac{2}{\hbar}\int_a^b \sqrt{2m(V-Q_{eff})}dz\right\} \tag{15.10}$$

with $V(a) = V(b) = Q_{\text{eff}}$. The effective Q value, Q_{eff}, is given as

$$Q_{\text{eff}} = Q + E^{*} \tag{15.11}$$

The excitation energy E^{*} is related to the nuclear temperature T (in MeV) [43] as

$$E^{*} = \frac{1}{9}AT^{2} - T \tag{15.12}$$

For the two deformed and oriented nuclei, the Coulomb interaction, which is taken from Ref. [44] and which includes higher multipole deformation [45–48], is given as

$$V_{\text{C}} = \frac{Z_1 Z_2 e^2}{r} + 3Z_1 Z_2 e^2 \sum_{l,i=1,2} \frac{1}{2l+1} \frac{R_i^{l}(a_i)}{r^{l+1}} Y_l^{(0)}(q_i)\left[b_{li} + \frac{4}{7} b_{li}^{2} Y_l^{(0)}(q_i) d_{l,2}\right] \tag{15.13}$$

$$R_i(a_i) = R_{0i}\left[1 + \sum_l b_{li} Y_l^{(0)}(a_i)\right] \tag{15.14}$$

where $R_{0i}(T) = [1.28A_i^{1/3} - 0.76 + 0.8A_i^{-1/3}][1 + 0.0007T^2]$. Here, θ_i is the angle between the symmetry axis and the axis of collision and α_i is the angle between the radius vector and the symmetry axis of the ith nuclei (see Fig. 1 of Ref. [46]), and here the quadrupole interaction term proportional to $\beta_{21}\beta_{22}$, due to its short-range character, is neglected.

For the proximity potential, the deformation comes only in the mean curvature radius, $V_{\text{P}}(z) = 4pgb\bar{R}\Phi(e)$. The mean curvature radius has been defined as $\bar{R} = \frac{C_1 C_2}{C_1 + C_2}$, for spherical nuclei. The mean curvature radius $\bar{R}$ for two deformed nuclei lying in the same plane can be found by the relation [46]

$$\frac{1}{\bar{R}^{2}} = \frac{1}{R_{11}R_{12}} + \frac{1}{R_{21}R_{22}} + \frac{1}{R_{11}R_{22}} + \frac{1}{R_{21}R_{12}} \tag{15.15}$$

where the four principal radii of curvature R_{i1} and R_{i2}, with $i = 1, 2$, at the two points D and E (see Fig. 1 of Ref. [46]) of the closest approach of the interacting nuclei are given by Baltz and Bayman [49] as

$$R_{i1} = \left| \frac{\left\{R_i^2(a_i) + [R_i'(a_i)]^2\right\}^{3/2}}{R_i''(a_i)R_i(a_i) - 2[R_i'(a_i)]^2 - R_i^2(a_i)} \right| \tag{15.16}$$

$$R_{i2} = \left| \frac{R_i(a_i)\sin a_i \left[R_i^2(a_i) + (R_i'(a_i))^2 \right]^{1/2}}{R_i'(a_i)\cos a_i - R_i(a_i)\sin a_i} \right| \tag{15.17}$$

Here, $R'(\alpha)$ and $R''(\alpha)$ represent the first and second derivatives of $R(\alpha)$ with respect to α, respectively.

15.2.2 Decay Cross Section of Compound Nucleus

Using the probability for the absorption of the ℓth partial wave given by the Hill–Wheeler formula [50], Wong arrived at the expression for the total decay cross section by the quantum mechanical penetration of the potential barrier for an energy E_ℓ given by

$$S = \frac{p}{k^2} \sum_\ell \frac{2\ell + 1}{1 + \exp\left[2p\left(E_\ell - E\right) / \hbar W_\ell \right]} \tag{15.18}$$

where $k = \sqrt{\dfrac{2mE}{\hbar^2}}$. Here, E is the incident energy. The curvature of the inverted parabola is $\hbar\omega_\ell$, and E_ℓ is the interaction barrier for the ℓth partial wave.

$$\hbar W_\ell = \hbar \left[\left. \frac{d^2V(r)}{dr^2} \right|_{R_\ell} \Big/ m \right]^{1/2} \tag{15.19}$$

Here, the radial separation R_ℓ can be obtained from the condition

$$\left. \frac{d^2V(r)}{dr^2} \right|_{R_\ell} = 0 \tag{15.20}$$

In the region of $\ell = 0$, using some parametrizations,

$$E_\ell \cong E_0 + \frac{\hbar^2 \ell(\ell + 1)}{2mR_0^2} \tag{15.21}$$

$$\hbar W_\ell \cong \hbar W_0 \tag{15.22}$$

Wong arrived at an expression for the decay cross section by using Eqs. (15.21) and (15.22) and also by replacing the sum in Eq. (15.18) by integration, which is given as

$$S = \frac{R_0^2 \hbar W_0}{2E} \ln\left\{ 1 + \exp\left[\frac{2p\left(E - E_0\right)}{\hbar W_0} \right] \right\} \tag{15.23}$$

For relatively large values of E, the above equation reduces to

$$\sigma = \pi R_0^2\left[1-\frac{E_0}{E}\right] \tag{15.24}$$

Lefort and his collaborators showed that for the fusion of two complex nuclei [51], not a critical angular momentum but a critical distance of approach may be the relevant quantity. To substantiate the finding of critical approach, it is necessary to check the linear dependence of σ on $1/E$ in the region of higher energies. The critical distance is given by the relation

$$R_c = r_c\left(A_1^{1/3} + A_2^{1/3}\right),\ r_c = 1.0 \pm 0.07 \text{ fm} \tag{15.25}$$

Gutbrod, Winn, and Blann, from their analysis of low energy data [52], obtained the fusion distance as

$$R_B = r_B\left(A_1^{1/3} + A_2^{1/3}\right),\ r_B = 1.4 \text{ fm} \tag{15.26}$$

To understand the difference between the two distances given by Eqs. (15.25) and (15.26), Glas and Mosel [41] set σ as

$$\sigma = \pi ƛ^2 \sum_{\ell=0}^{\infty}(2\ell+1)T_i P_i \tag{15.27}$$

where P_i gives the probability for fusion to take place once the barrier has been passed and $P_i = 1$ for $\ell \leq \ell_c$ and $P_i = 0$ for $> \ell_c$ and the penetration probability through the interaction barrier is T_i.

Approximating the frequencies $\hbar\omega_\ell$ and the position of the interaction barrier by constant values $\hbar\omega$ and R_B, respectively, and then replacing the sum in Eq. (15.25) by integration, σ is obtained as

$$\sigma = \frac{\hbar\omega}{2}R_B^2\frac{1}{E}\ln\left\{\frac{1+\exp[2\pi\{E-V(R_B)\}/\hbar\omega]}{1+\exp[2\pi\{E-V(R_B)-(R_C/R_B)^2[E-V(R_C)]\}/\hbar\omega]}\right\} \tag{15.28}$$

For oriented nuclei, the decay cross section σ is found to depend on the orientation angle θ and hence, here in our calculations, we have evaluated the decay cross sections for different orientations and we got $\sigma(E_{\text{C.M.}})$ on integrating over the angles θ_i.

15.3 Results and Discussion

In our earlier works, the total cross section, the intermediate mass fragment (IMF) production cross section, and the cross section for the formation of light particle (LP) for the decay of $^{48}Cr^*$ [36] formed through the entrance channels $^{24}Mg + {}^{24}Mg$, $^{36}Ar + {}^{12}C$, and $^{20}Ne + {}^{28}Si$ have been evaluated using the barrier penetration model, taking the scattering potential as the sum of the Coulomb and nuclear proximity potential, for various $E_{C.M.}$ values. The computed results have been compared with the available experimental data of the total cross section and LP production cross section corresponding to $E_{C.M.}$ = 44.4 MeV for the entrance channel $^{24}Mg + {}^{24}Mg$ and were found to be in good agreement. Also, the computed total cross sections for the entrance channel $^{36}Ar + {}^{12}C$ with $E_{C.M.}$ = 47 MeV agree well with the corresponding experimental values.

We have also studied the decay of compound system $^{56}Ni^*$ [37] formed through the entrance channels $^{32}S + {}^{24}Mg$, $^{36}Ar + {}^{20}Ne$, $^{40}Ca + {}^{16}O$, and $^{28}Si + {}^{28}Si$ by taking the scattering potential as the sum of the deformed Coulomb and deformed nuclear proximity potentials for various $E_{C.M}$ values. The computed results have been compared with the available experimental data of the total cross section corresponding to $E_{C.M^.}$ = 60.5 and 51.6 MeV for the entrance channel $^{32}S + {}^{24}Mg$, which were found to be in good agreement. The experimental values for the LP production cross section and IMF cross section for the channel $^{32}S + {}^{24}Mg$ were also found to agree with our calculations.

The concept of cold reaction valley is related to the minima in the so-called driving potential, which is defined as the difference between the interaction potential V and the Q value of the reaction. For all the possible combinations of projectile and target in

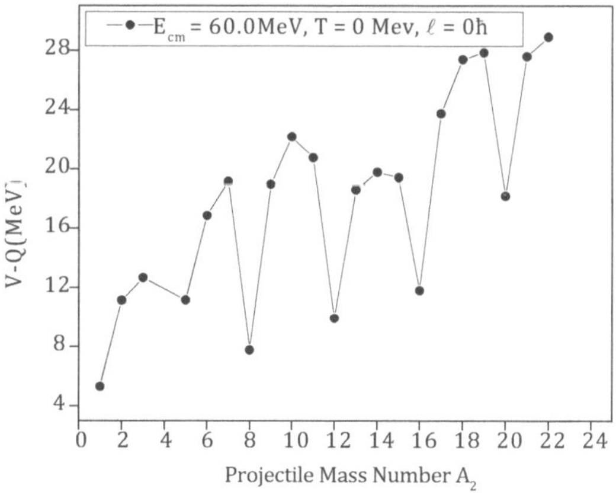

Figure 15.1 Cold reaction valley plot of ^{44}Ti.

touching configuration, the driving potential has been calculated. The minima in the driving potential represent the most probable combination for the formation of CN, which is due to the shell closure of projectile or target or both. The entrance channel ^{32}S + ^{12}C for the formation of ^{44}Ti compound nucleus has been experimentally determined [25]. The cold reaction valley plot of ^{44}Ti is shown in Fig. 15.1, where the projectile mass number A_2 is taken along the *X*-axis and the driving potential $V-Q$ is taken along the *Y*-axis corresponding to T = 0.0 MeV and ℓ = 0 ħ.

Figure 15.2 represents the plot of $V-Q_{\text{eff}}$ versus fragment mass number, which corresponds to the exit channels for the decay of $^{44}Ti^*$. The potential V is found to depend on the distance between the fragment centers "*r*," where $r = z + C_1 + C_2$, where C_1 and C_2 are the Süssman central radii of fragments and z is the distance between the near surface of the fragments. The plot corresponds to the experimental $E_{\text{C.M.}}$ value 60.0 MeV (T = 3.69 MeV) with corresponding ℓ_c values.

The LP ($A \leq 4$) emissions are found to be energetically more favorable at lower energies (lower ℓ values). The collisions with low angular momenta lead to higher compound nucleus temperatures favoring the evaporation of light fragments. The emission of IMFs ($A > 4$) begins only beyond a certain ℓ value (where the cross section has the maximum value), i.e., IMFs in the

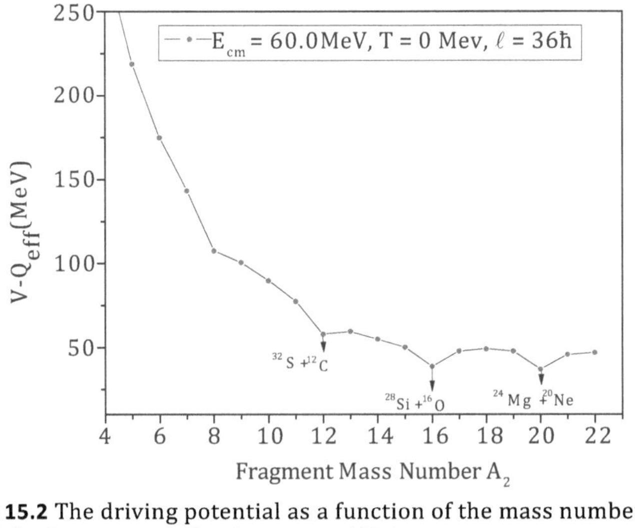

Figure 15.2 The driving potential as a function of the mass number of one of the fragments A_2 for the system $^{44}Ti^*$.

decay of the excited compound nuclei $^{44}Ti^*$ are favored at higher ℓ values. The critical angular momentum value ℓ_c is obtained from the condition that the effective radial potential $V(r,\ell)$, which is the sum of Coulomb, nuclear, and centrifugal potentials, reaches a local maximum equal to the incoming center-of-mass energy $E_{\text{C.M.}}$:

$$V(r,\ell) = E_{\text{C.M.}}, \quad \frac{\partial V(r,\ell)}{\partial r} = 0, \text{ for } \ell = \ell_c, r = R_m \tag{15.29}$$

Here, R_m is the separation distance at which the radial potential reaches the local maximum, corresponding to the critical angular momentum. We have plotted the variation in the interaction barrier with distance between the centers of fragments, and hence the values of E_B and R_B for the calculations of the total decay cross section were determined. The total decay cross sections, with the inclusion of deformations, were calculated using ℓ-summed Wong's formula, Eq. (15.18); approximated Wong's formula for relatively large values of E, Eq. (15.24); and the Glas and Mosel formula for cross section, Eq. (15.28). The total decay cross section, IMF production cross section, and light particle production cross sections were calculated separately by taking into account the temperature effects. The total decay cross section is theoretically given by $S_{\text{Total}} = S_{\text{LP}} + S_{\text{IMF}}$. The behaviors of LPs and IMFs are very different. Also, lower ℓ values contribute to the LP cross section (S_{LP}) and higher ℓ values to fission like, IMF cross section (S_{IMF}).

Table 15.1 shows the calculated value of fragment cross section corresponding to the $Z = 6$ fragment (^{12}C). The experimental value of the FF cross section corresponding to $E_{\text{C.M.}} = 60.0$ MeV is 10 ± 1.4 mb and our calculated value is 61.75 mb for spherical nuclei and 65.29 mb with deformation effects included, which is shown in Table 15.2. The value of the FF cross section computed using DCM [53] is 10.73 mb. The angular momentum ℓ' represents the maximum angular momentum fixed for the vanishing LP cross section, i.e., S_{LP} is negligibly small at $\ell = \ell'$.

The total decay cross sections calculated using Eq. (15.18), Eq. (15.24), and Eq. (15.28) are also given in the table and the computed values of LP cross section are given in the last column. Hence, we have calculated the total decay cross section, IMF cross section, and LP cross section for the fragment ^{16}O for the same $E_{\text{C.M.}}$ value, which is also given in Table 15.1 for the spherical nuclei. The calculations are done for the same channel with

Table 15.1 The computed cross sections S_{Total}, S_{IMF}, S_{LP} for the decay of ^{44}Ti* formed through the ^{32}S + ^{12}C, ^{28}Si + ^{16}O channels without including deformation effects.

Entrance	$E_{C.M.}$	ℓ'	S_{Total}			S_{IMF}		S_{LP}
Channel	(MeV)	(ħ)	Eq. (15.18)	Eq. (15.24)	Eq. (15.28)	Sph.	Exp.	Sph.
3 ^{32}S + ^{12}C	6 60	3 34	1838.4	1838.4	1250.0	61.75	10 ± 1.4	1150
^{28}Si + ^{16}O	6 60	337	1790.1	1790.1	1253.5	34.26	—	1215

Table 15.2 The computed cross sections S_{Total}, S_{IMF}, S_{LP} for the decay of ^{44}Ti* formed through the ^{32}S+^{12}C, ^{28}Si+^{16}O channels by incorporating deformation effects.

Entrance	$E_{C.M.}$	ℓ'	S_{Total}			S_{IMF}		S_{LP}
Channel	(MeV)	(ħ)	Eq. (15.18)	Eq. (15.24)	Eq. (15.28)	Def.	Exp.	Def.
3 ^{32}S + ^{12}C	6 60	332	1645.1	1645.1	1250.0	65.29	10 ± 1.4	1180
^{28}Si + ^{16}O	6 60	330	1109.1	1109.1	1109.1	48.67	—	1060

the inclusion of deformation effects, which are represented in Table 15.2.

15.4 Conclusion

The decay of ^{44}Ti*, formed through various entrance channels ^{32}S + ^{12}C and ^{28}Si + ^{16}O, has been studied using the ℓ-summed Wong's formula, approximated Wong's formula for relatively large values of E, and the Glas and Mosel formula for cross section. The total decay cross section, the intermediate mass fragment (IMF) compared with the available experimental data. Hence, we have studied the total decay cross section, the IMF cross section, and the LP cross section for the decay of ^{44}Ti* formed through other entrance channel ^{28}Si + ^{16}O for the same $E_{C.M.}$ value, which can be used for further experimental studies.

References

1. Sanders, S. J., De Toledo, A. S., and Beck, C. (1999). Binary decay of light nuclear systems, *Phys. Rep.* **311**, pp. 487–551.
2. Bhattacharya, C. *et al.* (2002). Dissipative collisions in $^{16}O + ^{27}Al$ at E_{lab} = 116 MeV, *Phys. Rev. C* **66**, pp. 047601.
3. Bhattacharya, C. *et al.* (2004). Dissipation of angular momentum in light heavy-ion collision, *Phys. Rev. C* **69**, pp. 024607.
4. Bhattacharya, C. *et al.* (1996). Intermediate mass fragments emission in the reaction 96 MeV ^{19}F on ^{12}C, *Phys. Rev. C* **54**, pp. 3099.
5. Beck, C. *et al.* (1998). Study of the fusion-fission process in the $^{35}Cl + ^{24}Mg$ reaction, *Eur. Phys. J. A Hadrons Nuclei* **2**, pp. 281–293.
6. Beck, C. *et al.* (1996). $^{35}Cl + ^{12}C$ asymmetrical fission excitation functions, *Phys. Rev. C* **54**, pp. 227.
7. Farrar, K. A. *et al.* (1996). Fission decay of ^{48}Cr at $E_{CN}^* \approxeq 60$ MeV, *Phys. Rev. C* **54**, pp. 1249.
8. Shapira, D. *et al.* (1979). Orbiting in the $^{12}C + ^{20}Ne$ system, *Phys. Rev. Lett.* **43**, pp. 1781.
9. Shapira, D., Ford Jr, J. L. C., and del Campo, J. G. (1982). Reactions of ^{20}Ne with ^{12}C, *Phys. Rev. C* **26**, pp. 2470.
10. Shapira, D. *et al.* (1982). Deep-inelastic back-angle yields and orbiting in $^{28}Si + ^{12}C$, *Phys. Lett. B* **114**, pp. 111–114.
11. Shivakumar, B., Ayik, S., Harmon, B. A., and Shapira, D. (1987). Equilibrium model for fusion and orbiting, *Phys. Rev. C* **35**, pp. 1730.
12. Moretto, L. G. (1975). Statistical emission of large fragments: A general theoretical approach, *Nucl. Phys. A* 247, pp. 211–230.
13. Sanders, S. J. (1991). Fusion-fission in nuclear systems with $40 \leq A_{CN} \leq 80$, *Phys. Rev. C* **44**, pp. 2676.
14. Matsuse, T., Beck, C., Nouicer, R., and Mahboub, D. (1997). Extended Hauser–Feshbach method for statistical binary decay of light-mass systems, *Phys. Rev. C* **55**, pp. 1380.
15. Dhara, A. K., Bhattacharya, C., Bhattacharya, S., and Krishan, K. (1993). Mechanism of intermediate mass fragment emission at low energy, *Phys. Rev. C* **48**, pp. 1910.
16. de Toledo, A. S., Sanders, S. J., and Beck, C. (1997). Statistical description of strongly damped binary yields from the $^{28}Si + ^{16}O$ reaction, *Phys. Rev. C* **56**, pp. 558.
17. Yu, H., Zuo, T., and Bandrauk, A. D. (1996). Molecules in intense laser fields: Enhanced ionization in a one-dimensional model of H_2, *Phys. Rev. A* **54**, pp. 3290.
18. Sanders, S. J. *et al.* (1987). Asymmetric fission of ^{56}Ni, *Phys. Rev. Lett.* **59**, pp. 2856.

19. Beck, C. *et al.* (1989). Fusion–fission in the ^{35}Cl +^{12}C reaction at E_{lab} = 180 MeV, *Zeitschrift für Physik A Hadrons and Nuclei* **334**, pp. 521–522.
20. Ray, A. *et al.* (1991). Compound nucleus origin of back-angle yields in the ^{31}P+^{16}O and ^{35}Cl + ^{12}C reactions, *Phys. Rev. C* **44**, pp. 514.
21. Betts, R. R. *et al.* (1981). Selectivity in two-particle exclusive heavy-ion reactions, *Phys. Rev. Lett.* **46**, pp. 313.
22. Płaneta, *et al.* (1986). Element distributions after binary fission of ^{44}Ti, *Phys. Rev. C* **34**, pp. 512.
23. Oliveira Jr, *et al.* (1996). Orbiting features in the strongly damped binary decay of the ^{28}Si + ^{16}O system, *Phys. Rev. C* **53**, pp. 2926.
24. Papka, P. *et al.* (2003). Ph.D. thesis, IReS and Université Louis Pasteur Strasbourg, *Acta Phys. Pol. B* **34**, pp. 2343.
25. Pandey, R. *et al.* (2017). Fragment emission mechanism in the ^{32}S + ^{12}C reaction, *Phys. Rev. C* **95**, pp. 064603.
26. Sharma, M. K., Gupta, R. K., and Scheid, W. (2000). α-nucleus structure in fusion–fission and cluster decay modes of ^{56}Ni formed in heavy-ion reactions, *J. Phys. G: Nucl. Part. Phys.* **26**, pp. 45.
27. Gupta, R. K., Balasubramaniam, M., Mazzocchi, C., La Commara, M., and Scheid, W. (2002). Decay of excited ^{116}Ba* formed in the ^{58}Ni+^{58}Ni reaction via the emission of intermediate mass fragments, *Phy. Rev. C* **65**, pp. 024601.
28. Gupta, R. K., Kumar, R., Dhiman, N. K., Balasubramaniam, M., Scheid, W., and Beck, C. (2003). Cluster decay of hot ^{56}Ni* formed in the ^{32}S + ^{24}Mg reaction, *Phys. Rev. C* **68**, pp. 014610.
29. Balasubramaniam, M., Kumar, R., Gupta, R. K., Beck, C., and Scheid, W. (2003). Emission of intermediate mass fragments from hot ^{116}Ba* formed in low-energy ^{58}Ni + ^{58}Ni reaction, *J. Phys. G: Nucl. Part. Phys.* **29**, pp. 2703.
30. Gupta, R. K. (2003). Collective clusterization in excited light nuclear systems, *Acta Phys. Hungarica A: Heavy Ion Phys.* **18**, pp. 347.
31. Gupta, R. K., Balasubramaniam, M., Kumar, R., Singh, D., and Beck, C. (2004). Collective clusterization effects in light heavy ion reactions, *Nucl. Phys. A* **738**, pp. 479–482.
32. Sanders, S. J, *et al.* (1986). Fission-like yields in ^{16}O + ^{40,44}Ca reactions, *Phys. Rev. C* **34**, pp. 1746.
33. Karthikraj, C. and Balasubramaniam, M. (2013). Decay studies of ^{59}Cu* formed in the ^{35}Cl + ^{24}Mg reaction using the dynamical cluster-decay model, *Phys. Rev. C* **87**, pp. 024608.
34. Kaur, M., Singh, B., Patra, S. K., and Gupta, R. K. (2017). Clustering effects and decay analysis of the light-mass N = Z and N ≠ Z composite systems formed in heavy ion collisions, *Phys. Rev. C* **95**, pp. 014611.

35. Santhosh, K. P., Subha, P. V., and Priyanka, B. (2016). Cluster decay of $^{112-122}Ba$ isotopes from ground state and as an excited compound system, *Pramana J. Phys.* **86**, pp. 819–836.
36. Santhosh, K. P., Subha, P. V., and Priyanka, B. (2016). Decay of the excited compound system $^{48}Cr^*$ formed through $^{24}Mg + ^{24}Mg$, $^{36}Ar + ^{12}C$ and $^{20}Ne + ^{28}Si$ reactions, *Eur. Phys. J. A* **52**, pp. 125.
37. Santhosh, K. P. and Subha, P. V. (2017). Decay of the excited compound system $^{56}Ni^*$ formed through various channels using deformed Coulomb and deformed nuclear proximity potentials, *Phys. Rev. C* **95**, pp. 064607.
38. Błocki, J., Randrup, J., Światecki, W. J., and Tsang, C. F. (1977). Proximity forces, *Ann. Phys.* **105**, pp. 427–462.
39. Blocki, J. and Świątecki, W. J. (1981). A generalization of the proximity force theorem, *Ann. Phys.* **132**, pp. 53–65.
40. Wong, C. (1973). Interaction barrier in charged-particle nuclear reactions, *Phys. Rev. Lett.* **31**, pp. 766.
41. Glas, D. and Mosel, U. (1974). Limitation on complete fusion during heavy-ion collisions, *Phys. Rev. C* **10**, pp. 2620.
42. Royer, G. and Mignen, J. (1992). Binary and ternary fission of hot and rotating nuclei, *J. Phys. G: Nucl. Part. Phys.* **18**, pp. 1781.
43. Le Couteur, K. J. and Lang, D. W. (1959). Neutron evaporation and level densities in excited nuclei, *Nucl. Phys.* **13**, pp. 32–52.
44. Santhosh, K. P. and Jose, V. B. (2014). Heavy-ion fusion cross sections of weakly bound ^{9}Be on ^{27}Al, ^{64}Zn and tightly bound ^{16}O on ^{64}Zn target using Coulomb and proximity potential, *Nucl. Phys. A* **922**, pp. 191–199.
45. Gupta, R. K. *et al.* (2005). Optimum orientations of deformed nuclei for cold synthesis of superheavy elements and the role of higher multipole deformations, *J. Phys. G: Nucl. Part. Phys.* **31**, pp. 631.
46. Malhotra, N. and Gupta, R. K. (1985). Proximity potential for deformed, oriented collisions and its application to $^{238}U + ^{238}U$, *Phys. Rev. C* **31**, pp. 1179.
47. Carlson, B. V., Chamon, L. C., and Gasques, L. R. (2004). Accurate approximation for the Coulomb potential between deformed nuclei, *Phys. Rev. C* **70**, pp. 057602.
48. Takigawa, N., Rumin, T., and Ihara, N. (2000). Coulomb interaction between spherical and deformed nuclei, *Phys. Rev. C* **61**, pp. 044607.
49. Baltz, A. J. and Bayman, B. F. (1982). Proximity potential for heavy ion reactions on deformed nuclei, *Phys. Rev. C* **26**, pp. 1969.
50. Hill, D. L. and Wheeler, J. A. (1953). Nuclear constitution and the interpretation of fission phenomena, *Phys. Rev.* **89**, pp. 1102.
51. Galin, J., Guerreau, D., Lefort, M., and Tarrago, X. (1974). Limitation to complete fusion during a collision between two complex nuclei, *Phys. Rev. C* **9**, pp. 1018.

52. Gutbrod, H. H., Winn, W. G., and Blann, M. (1973). Measurement and interpretation of heavy ion fusion excitation functions, *Nucl. Phys. A* **213**, pp. 267–284.
53. Kaur, M., Singh, B. B., and Kaur, S. (2017). Investigation of decay of ^{44}Ti* compound system formed in ^{32}S + ^{12}C reaction, *Proc. DAE Symp. Nucl. Phys.* **62**, pp. 386–387.

Chapter 16

Equilibrium Decay Stage in Proton-Induced Spallation Reactions

Nikolaos G. Nicolis

Department of Physics, The University of Ioannina, Ioannina 45110, Greece

nnicolis@uoi.gr

The study of nucleon-induced spallation reactions is a topic of extensive experimental and theoretical interest due to its importance in basic and applied nuclear science. In such a reaction, an incident energetic hadron induces an intranuclear cascade, which produces a number of promptly emitted nucleons and pions, depending on the bombarding energy. The resulting heavy and highly excited pre-fragments undergo equilibrium decay. In this chapter, we give a brief discussion of models and codes for spallation reactions and the intranuclear cascade. Then, we focus on the evaporation formalisms of Weisskopf and Hauser–Feshbach, employed in statistical model codes for the description of the equilibrium decay stage. In proton-induced spallation reactions on medium to low mass targets, the equilibrium decay is described well in the framework of a generalized Weisskopf evaporation formalism. We illustrate the effectiveness of this

Multifragmentation in Heavy-Ion Reactions: Theory and Experiments
Edited by Rajeev K. Puri, Yu-Gang Ma, and Arun Sharma

ISBN 978-981-4968-69-0 (Hardcover), 978-1-003-38513-4 (eBook)
www.jennystanford.com

method in the description of mass and charge distributions of a typical spallation reaction, described as a two-stage process.

16.1 Importance of Spallation Reactions

Spallation nuclear reactions occur when an energetic hadron (a proton, a neutron, or a pion) interacts with an atomic nucleus. This process may be thought to proceed in two stages. In the first stage, the incident particle interacts with the nucleons of the target in a sequence of collisions. As a result, we have the formation of an intranuclear cascade (INC) of high-energy protons and neutrons within the nucleus. In inelastic nucleon collisions, pion production is possible. This fast process lasts approximately 10^{-22} s. During the INC, energetic hadrons escape from the target. Others deposit their kinetic energy in the nucleus, leaving it in an excited state or proceed with new collisions. In the second stage, the heavy excited nuclear species deexcite. Sequential evaporation with the emission of nucleons, protons, α-particles, and γ-rays occurs with a typical time scale of $10^{-18} - 10^{-16}$ s. The emission of heavier nucleon clusters in their ground or excited states is also possible. If the target is heavy enough, high-energy fission may compete with sequential evaporation. The deexcitation products of target-like and/or fission fragments may be radioactive. In the case of thick target experiments, the secondary high-energy particles produced in the INC phase move roughly in the forward direction and induce secondary reactions. In such a case, a hadronic cascade is observed as a result of an accumulation of all reactions initiated by the primary and secondary particles.

Spallation reactions play an important role in applied and fundamental research. On heavy mass target nuclei, they are instrumental for the operation of neutron sources with applications in the energy production in accelerator-driven systems (ADS) and the accelerator transmutation of nuclear waste (ATW). Spallation neutron sources (SNS) have been developed or they are under construction in Europe, the United States, Russia, Japan, and China. Besides ADS and ATW, neutron sources have numerous applications in the production of exotic isotopes, materials science, and biology [1–3].

Spallation reactions on medium to low mass target nuclei provide useful information for nucleosynthesis studies, cosmic-ray effects on biological and semiconductor systems, and radiation shielding. From the theory point of view, they provide a testing ground for nuclear reaction codes and feed with information simulation codes employed in the design of experiments and the analysis of experimental data [1–4].

16.2 Description of Spallation Reactions

The Monte-Carlo method is best suited for the description of high-energy nuclear reactions, due to their complexity and the desire to obtain results subject to various experimental constraints. N. Metropolis and his collaborators pioneered with Monte-Carlo calculations of internuclear cascades [5]. Their work formed the basis of many follow-up developments. Models developed for the description of spallation reactions comprise

(i) Semi-empirical expressions based on a parametric fit of experimental data. Formulas obtained by Rudstam [6] as well as by Silberberg and Tsao [7, 8] have been used extensively in astrophysics. Recent improvements are incorporated in the semi-empirical parametrizations EPAX3 [9] and SPACS [10].

(ii) Two-stage models based on the hypothesis of Serber [11]. They require a model for the INC coupled with one for the equilibrium decay of the excited pre-fragments.

(iii) Microscopic approaches have also been developed. On one hand, we have mean field models of Boltzmann–Uehling–Uhlenbeck (BUU) type [12, 13]. On the other hand, we have molecular dynamics models, such as the constrained molecular dynamics model (CoMD) [14] and the isospin-dependent quantum molecular dynamics model (IQMD) [15–17].

16.3 Intranuclear Cascade Models

Intranuclear cascade (INC) models may be classified in two categories. The first one includes the model of Bertini [18] and its successor ISABEL [19]. In ISABEL, the target nucleus is considered a continuous medium with a diffuse surface approximated with 16 concentric constant density regions. Collisions between the incident and target nucleons occur with a criterion based on the mean free path. Nucleons involved in collisions follow straight-line trajectories. The Fermi motion of the target nucleons is taken into account. Free nucleon–nucleon cross sections are used. Inelastic nucleon–nucleon collisions occur involving the excitation of the delta resonance and pion production. A Pauli blocking mechanism, taking into account the occupation rate, inhibits collisions leading to occupied states in the target nucleus. The propagation is followed as a function of time and stops when all particle energies fall below a cutoff value.

On the other hand, the INCL model of Cugnon [20] considers the target nucleons as point particles confined in a spherical volume with a sharp surface. The target nucleons move in straight-line trajectories according to a Fermi distribution. They collide as soon as they reach their minimum distance of approach or are reflected on the walls of the nuclear potential. The cascade propagation is followed as a function of time. Inelastic nucleon–nucleon collisions may occur. The cascade stops at a time when thermalization has been reached. The Pauli blocking is taken into account. In INCL (Version 5.3) [21] emission of complex fragments is considered besides the emission of nucleons and pions during the INC. The emission of complex fragments with $Z < 5$ and $A < 9$ is treated as a surface coalescence process, which is realized whenever an emitted nucleon is found in close proximity with others in coordinate and momentum space. The C++ version of the code, named INCL++ [22], offers the possibility to describe collisions between projectiles with $A \leq 18$ and a heavier target.

16.4 Pre-fragment Deexcitation

16.4.1 Statistical Decay

Till the end of the intranuclear cascade stage, pre-equilibrium emissions are possible. Eventually, highly excited and equilibrated pre-fragments are formed. They may be considered compound nuclei, in the Niels Bohr sense. In accordance with the Bohr independence hypothesis [23, 24], it is assumed that the so-formed compound nuclei live long enough so that they have lost memory of the way they were formed. They are found in a state of complete statistical equilibrium. Their excitation energy is shared equally among the constituent nucleons. Decays may occur into any of the available open channels. The larger the number of open channels, the smaller the average flux into each of them. In each decay, the minimum possible energy is removed in a process resembling the evaporation of molecules from a heated liquid [25, 26]. The deexcitation process may involve the emission of particles, nucleon clusters, gamma rays, or fission.

Weisskopf and Ewing [27] developed a theory for the calculation of particle emission and evaporation residue cross sections. Their theory was based on the assumption of complete statistical equilibrium, the Bohr independence hypothesis, and the principle of detailed balance. This forms the basis of the statistical model of the compound nucleus decay [28–30]. As a simplifying assumption, the effect angular momentum in the decay process was ignored. In the following, we describe the basic points of this theory.

Consider a reaction that involves a projectile a interacting with a target A. A compound nucleus C is formed with intrinsic excitation E^* and decays by emitting a particle or fragment b, producing the residual nucleus B. According to the Bohr independence hypothesis, the decay process is independent of the way the compound system was formed. Thus, we have a two-stage process:

$$a + A \rightarrow C^* \rightarrow b + B$$

We denote the entrance channel by $a + A \equiv \alpha$ and the exit channel by $b + B \equiv \beta$. The residual nucleus B may be formed in an excited state. Neglecting spin, the independence hypothesis gives the cross section of channel β via the entrance channel α:

$$\sigma_{\alpha\beta} = \sigma_{CN} \frac{\Gamma_\beta}{\Gamma} = \sigma_{CN} G_\beta$$

where $\sigma_{CN} = \sigma_{\alpha \rightarrow C}$ is the cross section for the formation of the compound nucleus. Here, the probability G_β for the emission of particle or fragment b is expressed as the ratio of the decay width Γ_β into channel β over the sum of all possible decay modes $\Gamma = \sum_\beta \Gamma_\beta$.

We are interested in the absolute value of the decay width Γ_b for the emission of fragment b with kinetic energy E. This can be obtained with the principle of detailed balance. Assume that we have two systems α and β in statistical equilibrium. The two systems have level densities ρ_α and ρ_β, respectively. The system α may undergo a transition to β and vice versa. According to the principle of detailed balance, any depletion of states of the system α due to transitions to β equals their increase in the time-reversed process $\alpha \to \beta$. The decay rate, i.e., the probability per unit time, for transitions from α to β is $W_{a \to b} = 1/t_{ab} = \Gamma_{ab}/\hbar$, where t_{ab} is the lifetime of the process. Similarly, the decay rate of the inverse process is $W_{b \to a} = 1/t_{ba} = \Gamma_{ba}/\hbar$. Equality of the decay populations implies that $r_a W_{a \to b} = r_b W_{b \to a}$ or $r_a \Gamma_{a \to b} = r_b \Gamma_{b \to a}$. Applying the last equality to the process $C^* \to b + B^*$, we get

$$r_{C^*} \Gamma_b = r_b \Gamma_{b \to C^*}$$

Here, ρ_β can be written as the product of the density ρ_b of continuum states of b with the level density ρ_B of the residual nucleus: $r_b = r_b r_B$.

The density of continuum states of b is given by the Fermi gas expression

$$r_b(E) = \frac{g_b m_b E \Omega}{p^2 \hbar^3 u_b}$$

where $g_b = 2s_b + 1$ is the spin degeneracy, $m_b = m_b m_B / (m_b + m_B)$ is the reduced mass, E is the kinetic energy, u_b is the emission velocity of the fragment b, and Ω is the laboratory volume.

The decay width of the inverse process leading to C^* is

$$\Gamma_{b \to C^*} = W_{b \to C^*} \hbar = \frac{u_b s_{\text{inv}}(E)}{\Omega} \hbar$$

where $s_{\text{inv}}(E)$ is the cross section of the inverse process of compound nucleus formation at the excitation energy E^* from channel β.

From the previous equations, we find

$$\Gamma_b(E) = \frac{g_b m_b E s_{\text{inv}}(E)}{p^2 \hbar^2} \frac{r_{\text{B}}(E_f^*)}{r_C(E^*)}$$

where μ is the spin degeneracy of the emitted fragment, μ is the reduced mass of the emitting system, and $\text{E}s_{\text{inv}}(E)$ is the inverse cross section of the compound nucleus formation from the constituents of the exit channel. The decay width depends on the level density ratio of the final $r_{\mathrm{B}}\left(\mathrm{E}_f^*\right)$ to the initial $r_C\left(\mathrm{E}^*\right)$ state. The energy of the final state is $\mathrm{E}_f^* = E^* - S_b - E - \in$, where S_b is the separation energy of the fragment b from the compound nucleus. The same expression may be used for fragments emitted in excited-bound states, taking into account the intrinsic excitation energy $\in$ of the fragment.

For fragments emitted in the particle-unbound energy region (continuum), the corresponding expression for the decay width involves the folding integral of the total level densities of the emitted fragment b and the residual nucleus B [31, 32]:

$$\Gamma\left(E^*,E\right) = \frac{ES_{\mathrm{inv}}(E)m\int_0^{E^*} r_b(\in) r_B\left(E^* - \in\right)d\in}{p^2\hbar^2 \, r_C\left(E^*\right)}$$

Gamma-ray emission competes strongly in the last stages of deexcitation, i.e., excitation energies close to below the particle separation energy. For the gamma-decay emission width, similar relations hold. For example, assuming E1 emission

$$\Gamma_g(E,\in) = \frac{1}{p^2(\hbar c)^2} S_g\left(\in_g\right) \frac{r\left(E^* - \in_g\right)}{r\left(E^*\right)} \in_g^2$$

where $S_g\left(\in_g\right)$ is the photo-absorption cross section given by

$$S_g\left(\in_g\right) = \frac{S_0 \in_g^2 \Gamma_R^2}{\left(\in_g^2 - E_G^2\right)^2 + \Gamma_R^2 \in_g^2}$$

A parametrization of the empirical parameters of the giant dipole resonance S_0, E_G, and Γ_R has been reported by Iljinov *et al.* [33].

Having established expressions for the decay widths of all exit channels, we may express the energy spectrum of mode β as

$$S_{\mathrm{CN}} \frac{\Gamma_b}{\sum_b \Gamma_b} dE = S_{\mathrm{CN}} \frac{g_b m_b E S_{\mathrm{inv}}(E) r_B\left(E_f^*\right) dE_f^*}{\sum_b \int_0^{E_b^{\max}} g_b m_b E S_{\mathrm{inv}}(E) r_B\left(E_f^*\right) dE_f^*}$$

This expression is commonly referred to as the *evaporation equation*, in analogy with classical evaporation models. The energy spectra have typical Maxwellian shapes, shifted up in energy according to the respective emission (Coulomb) barriers.

Integrating over all emission energies, we obtain the *Weisskopf–Ewing formula* for the angle-integrated cross sections

$$S_{ab} = S_{\mathrm{CN}} \frac{\int_0^{E_b^{\max}} g_b m_b E S_{\mathrm{inv}}(E) r_B(E_f^*) dE_f^*}{\sum_b \int_0^{E_b^{\max}} g_b m_b E S_{\mathrm{inv}}(E) r_B(E_f^*) dE_f^*}$$

The previous formulation may be considered a generalized Weisskopf model [34]. The model does not involve any cluster pre-formation probabilities. By using the principle of detailed balance, we obtained absolute decay widths. An alternative approach based on the reciprocity theorem allows us to determine the decay probability, without reference to the decay widths. The reciprocity theorem is based on the proportionality between the cross section of the process $S_{a \to b}$ and the available phase space volume of the final states. Thus, $S_{a \to b}$ can be related to the cross section of its time-reversed process $S_{a \to b}$.

A detailed theory that takes angular momentum coupling into account has been introduced by Wolfenstein [35] and by Hauser and Feshbach [36]. This approach takes into account the coupling between the spin of the participating nuclei and orbital angular momentum removed by the emitted particle. Thus, it enables an accurate calculation of angular distributions and cross sections to discrete and continuum states of the residual nuclei. It is customarily referred to as the *Hauser–Feshbach theory* and is outlined below.

Let the projectile *a* have spin $\vec{i}$, energy E_a, and angular momentum $\ell\hbar$ relative to the target *A* with spin $\vec{I}$. The compound nucleus *C* is formed at a state with an excitation energy E^* and spin $\vec{J}$, denoted by $(E^*, \vec{J})$. It decays with the emission of particle *b*, which has spin $\vec{i}'$, energy E_b, and angular momentum $\ell'\hbar$ relative to the residual nucleus *B* formed at the state $(E_B^*, \vec{I}')$. In the entrance (as well as in the exit) channel, we define the channel spin $\vec{S} = \vec{i} + \vec{I}$ and the total angular momentum as $\vec{J} = \vec{S} + \vec{\ell}$. The Bohr independence hypothesis applied to each compound nuclear state of spin $S_{ab}^J = S_{\mathrm{CN}}^J G_b^J$ gives $S_{ab}^J = S_{\mathrm{CN}}^J G_b^J$, or

$$S_{ab}^J = \frac{p}{k_a^2} \frac{2J+1}{(2i+1)(2I+1)} \sum_{S,\ell} T_\ell G_b^J$$

The cross section of the process $\alpha \rightarrow \beta$ involves a summation over all the contributing values of *J*. Thus, we obtain the *Hauser–Feshbach formula* for the angle-integrated cross sections.

$$S_{ab} = \frac{p}{k_a^2} \sum_J \frac{2J+1}{(2i+1)(2I+1)} \sum_{s,\ell} T_\ell(a) \frac{\sum_{s',\ell'} T_{\ell'}(b)}{\sum_{g,s'',\ell''} T_{\ell''}(g)}$$

The energy spectra of the emitted particles are predicted to have Maxwellian shapes. More complicated expressions describe the angular distributions of emitted particles. They are symmetric about 90° in the center of mass system, as expected for the decay of a compound nucleus. The Hauser–Feshbach theory has been elaborated by many authors (see, e.g., Ref. [37]).

The required transmission coefficients are obtained from an optical model calculation for the inverse (capture) process or a barrier penetration calculation. At high excitation energies, the calculation of the above sums is facilitated with the introduction of level densities. Then, we end up with a formalism that would better be called *semi-classical* [28]. It is used extensively in statistical model evaporation codes. The Hauser–Feshbach theory reduces to the Weisskopf–Ewing theory for small values of the compound nucleus spin *J* and angular momentum of the emitted particles or the spin cutoff parameter in the level densities goes to infinity [38, 39].

Proton-induced reactions at high energies impart a negligible amount of angular momentum to the target nucleus. For these reactions, the Weisskopf formalism of compound nucleus decay seems reasonable. In reactions involving heavy-ion projectiles, such as fragmentation reactions, angular momentum effects are important and should be treated with the Hauser–Feshbach formalism.

The generalized Weiskopf evaporation formalism (GWEF) treats gamma decay, nucleon and cluster or IMF emission on an equal footing. The decay widths for IMF emission depend on the available phase space at the scission point of the decaying system, involving the level densities of the two fragments at infinite separation. This is in contrast to the transition-state approach of fission, which is based on the number of states at the more compact saddle-point configuration. The validity of the GWEF is limited to light compound nuclei for which the effective fissility parameter at zero spin ($x = Z^2/(50.14A)$) is below the Businaro–

Gallone point [40]. This corresponds to a value of $x_{BG} = 0.396$ according to the liquid-drop model. For nuclei along the valley of stability, this value points to the nucleus $^{107}_{46}Pd$. For nuclei with $x < x_{BG}$, the saddle and scission point configurations are expected to be very close and little damping is expected as the system proceeds between the two [40]. Then, nucleon emission as well as heavier massive divisions up to symmetry should be determined by the available phase space at scission rather than the saddle-point configuration. For heavier compound nuclei, GWEF is expected to work for the most asymmetric mass divisions.

The two basic ingredients of the model are the nuclear level densities and the transmission coefficients. They describe the available phase space for nuclear decay and the access to this phase space, respectively.

16.4.1.1 Level densities

Reliable calculations of the decay products of an excited nucleus require an accurate description of the level density at low excitation and their extrapolation into regions of high excitation energy, angular momentum, and nuclear shape, for which we have little experimental knowledge. Thus, the characterization of the level density has to be in large part phenomenological. Phenomenological treatments are based on the Fermi gas model. The subject of nuclear level densities has been discussed in the classic reviews by Ericson [41], Huizenga and Moretto [42], and more recently by Koning *et al.* [43].

16.4.1.2 Transmission coefficients

Transmission coefficients (T_L) for particle emission are related through the principle of detailed balance to the ones of the inverse process, namely the capture of the particle by the excited daughter nucleus. Proper use of optical model T_L-s should exclude possible contributions from direct and pre-equilibrium processes. Furthermore, the use of a T_L appropriate for capture by a cold nucleus to describe decay from an excited nucleus may be questioned [41]. Alexander, Magda, and Landowne [44] have reviewed the logical basis of using the optical model T_L-s in statistical model calculations. They have argued that T_L-s derived from an ingoing-wave boundary condition calculation may be more appropriate. In this context, several studies have addressed the deviations of the experimental observables from the statistical model predictions as originating from non-fusion components in the optical model T_L-s (see, e.g., Refs. [45, 46]). Modifications in the optical model T_L-s for particle emission have been considered in statistical model calculations [47, 48]. In spallation reactions, highly excited nuclear species are produced with wide distributions in A, Z, and excitation energy. From this point of view, the sensitivity of spallation data in the details of penetration factors may be difficult to assess.

16.4.2 Fission

Fission is a large-scale collective phenomenon, in which a massive nucleus splits into two massive fragments. Advances in this exciting research field have been summarized in a number of books and review articles [49–51]. In nuclear reactions, fission may become a strongly competing decay mode.

A universally accepted model for nuclear fission is not available. In most statistical model codes, the treatment of the fission decay mode is phenomenological. Bohr and Wheeler [52] have obtained an expression for the fission decay width in the framework of the transition-state theory. It is expressed in terms of transition states at the saddle-point configuration over the initial density of states

$$\Gamma_f^{BW}\left(E^*,J\right)=\frac{1}{2pr_g(I)}\int_0^{E^*-B_f} r_s\left(E^*-B_f-\in\right)d\in$$

where ρ_s and ρ_g are the densities of states at the saddle-point and the ground-state deformation, respectively, B^f is the fission barrier height, and $\in$ is the kinetic energy associated with the motion of the fission fragments at the saddle point. Fission barriers may be calculated with the finite-range liquid-drop model [53]. At low excitation energies, shell effects are responsible for creating a double-humped fission barrier structure (see Ref. [51] and references therein).

The above expression has to be modified by including an appropriate Hill–Wheeler penetrability factor to account for quantum mechanical tunneling through the fission barrier [50]. This standard theory of fission underestimates the measured pre-fission neutron multiplicities. A fission delay is needed to account for the slowing effects of nuclear dissipation as the system proceeds adiabatically from saddle to scission [50, 54]. These considerations are sufficient for the calculation of total fission cross sections. Going one step further, the role of mass asymmetry in the transition-state model was formulated by Moretto [55] and elaborated by Benlliure [56]. This facilitates the calculation of the mass and charge distributions of the fission products.

The transition-state calculation of fission widths treats both symmetric and asymmetric mass divisions. It is argued that it should converge to the Weisskopf calculation of decay widths for the most asymmetric mass divisions [30, 57].

The liquid-drop model provides an argument for the expected symmetry of mass distributions in the fission process. The major contributions to the nuclear potential energy are the surface and Coulomb energy. For a given nucleus, we may calculate a conditional saddle point under the constraint of a fixed mass asymmetry. The shape of the resulting *ridge line* depends on whether the effective fissility parameter lies above or below the Bussinaro–Gallone (BG) point [55, 58]. For a light system, the surface dominates over the Coulomb energy and produces low barriers favoring asymmetric fission. For a heavy system, the Coulomb energy dominates and favors symmetric fission. Therefore, the mass distributions are predicted asymmetric for a light and symmetric for a heavy mass system, respectively.

In contrast to the transition-state theory of fission, scission point models have also been developed. These models assume strong coupling of degrees of freedom from saddle to scission [59, 60]. The fission probability is proportional to the density of states at the scission configuration. Therefore, they require a delicate description of the scission configuration, where shell effects, pairing, deformation, and Coulomb energies greatly influence the outcome of the fission process. Scission point models are often criticized for neglecting the fact that fission is a dynamic process in which the adiabatic development from saddle to scission has important consequences.

16.4.3 Pre-equilibrium Decay

Pre-equilibrium particle emission may occur before the excited pre-fragment attains complete statistical equilibrium. This mechanism is described with either one of two semi-classical models: the exciton model of Griffin [61] and the hybrid model [62]. Quantum approaches comprise the multi-step compound model (MSC) and the multi-step direct model (MSD) of Nishioka *et al.* [63, 64].

16.4.4 Breakup of Light Nuclei

The deexcitation of very light systems with $A < 16$ in high-energy hadronic collisions needs a special treatment. Fermi proposed a simultaneous breakup mechanism into the available multiparticle exit channels. A partitioning of the excited light nucleus is considered. It is assumed that the energy released by the breakup is shared by all possible states, which occur with frequencies proportional to their statistical weights [65].

16.4.5 Multifragmentation

At excitation energies E^*/A greater than 2–3 MeV/A, the nucleus may become unstable to a simultaneous breakup involving multiple production of intermediate mass fragments—a process called multifragmentation [66]. The degree of equilibration indicated by the decay properties has suggested the use of statistical multifragmentation models (SMM) for handling the decays of this reaction channel. Many versions of the SMM have been proposed over the years. The comprehensive review of Bondorf *et al.* [67] discusses the formulation, numerical simulations, and the implementation of the model in intermediate energy hadron–nucleus and heavy-ion reactions at intermediate energies. Carlson *et al.* [68] examined the consistency of FBM with SMM. It is argued that a generalized FBM, in which densities of excited states are taken into account, fulfills the hypothesis of statistical equilibrium better and is equivalent to SMM.

16.5 Statistical Model Codes in Spallation Studies

Statistical model codes provide a useful tool for the interpretation of experiments involving the compound nucleus decay [30]. On the basis of the numerical treatment, they are based on deterministic grid search algorithms (GR) or the Monte-Carlo (MC) method

Table 16.1 Characteristics of statistical model codes.

Code	Type	g	EF	IMF	F	PE	FB	MF	Ref.
ABLA	MC	Y	WE	WE	Y	N	N	Y	[69]
ALICE	GR	N	WE	N	TS	Y	Y	N	[70, 71]
DRESNER	MC	Y	WE	N	TS	N	N	N	[72]
GEM	MC	N	WE	WE	PAR	N	N	N	[73]
GEMINI	MC	N	HF	TS	TS	N	N	N	[48]
MECO	MC	Y	WE	WE	TS	N	N	N	[32]
PACE2	MC	Y	HF	N	TS	N	N	N	[74]
SMM	MC	N	WE	WE	PAR	N	Y	Y	[75]
TALYS	GR	Y	HF	N	Y	Y	N	N	[76]

Abbreviations: EF (evaporation formalism), F (fission), PE (pre-equilibrium), FB (Fermi breakup), MF (multifragmentation), MC (Monte-Carlo), GR (Grid), WE (Weisskopf–Ewing), HF (Hauser–Feshbach), TS (transition state), PAR (parametrization)

[30]. They may differ in the treatment of the compound nucleus decay, i.e., the Weisskopf–Ewing (WE) or the Hauser–Feshbach (HF) formalism. Other differences may appear in the treatment of IMF emission, fission (F), pre-equilibrium emission (PE), Fermi breakup (FB), and multifragmentation (MF).

Table 16.1 describes the characteristics of statistical model codes commonly used in spallation reaction studies. It is an extension of the corresponding table given in Refs. [1, 30]. In the table, we included the sequential binary decay code MECO [32] developed some years ago and applied in spallation reaction studies just recently.

16.6 Two-Stage Model Calculation

In the following, we present a comparison of experimental spallation residue cross sections with the results of a two-stage calculation. We consider the reverse-kinematics reaction ^{56}Fe+p at 750 MeV studied at GSI in Darmstadt, Germany [77].

ISABEL produces the isotopic distribution shown in Fig. 16.1a. On this plot, the radii of circles are proportional to the corresponding cross sections. A wide distribution of excited

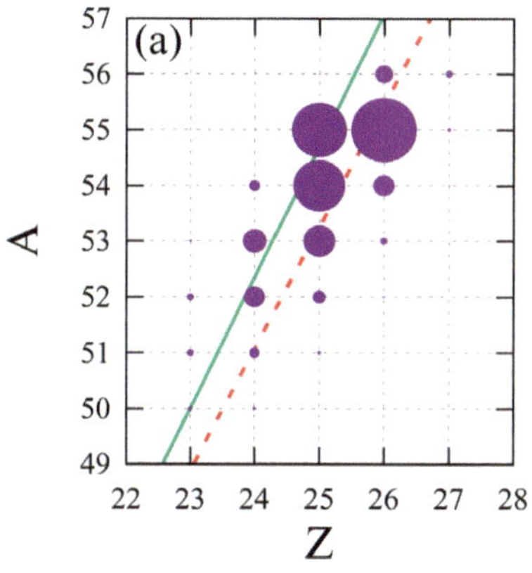

Figure 16.1 (a) Isotopic distribution of excited pre-fragments produced in spallation reactions of ^{56}Fe + p at 750 MeV, according to the code ISABEL. Radii of circles are proportional to the production cross sections. The dashed and solid lines mark the location of the evaporation attractor line and the line of beta stability, respectively. The deexcitation of these reaction events via the code MECO produces the spallation residue distribution shown in (b).

pre-fragments in A and Z is predicted. The distribution involves nuclear species starting with A and Z slightly greater than the target (due to pion interactions) and extends down in A and Z by 7 and 4 units, respectively. The distribution is located far from the line of beta stability, shown by the solid line. The dashed line shows the location of the evaporation attractor line, where the subsequent decay is expected to bring the final residues [78].

For the deexcitation, we employ the sequential binary decay code MECO [32]. We use level densities described with the composite level density formula of Gilbert and Cameron [43]. In the Fermi gas excitation energy region, we use an energy-independent level density parameter $a = A/8.0$ for all nuclei involved. Transmission coefficients for the emission of neutrons, protons, and IMFs with $Z < 5$ were calculated with the optical model, using global parameters. For the emission of IMFs with $Z \geq 5$, we use barrier penetration calculations with the parabolic model approximation of the barrier. The nuclear potential was generated with the parameters of Christensen and Winther [79].

The symbols in Fig. 16.2 show the experimental mass and charge distributions [77] of evaporation residues in 750 MeV p + ^{56}Fe spallation reactions. The experimental data are compared with calculations performed with the code ISABEL followed by MECO. Curves marked with number 1 refer to a calculation with MECO considering the emission of gamma rays, neutrons, protons, and alpha particles. This calculation agrees with the experimental data for $A > 40$ and $Z > 20$ but fails to describe the data at lower A and Z values. A better overall agreement with the data is obtained if we include in the calculation the emission of IMFs with A up to 20 and Z up to 9, all emitted in their ground states. This amounts to a total of 42 decay channels. In this case, we obtain the curves marked with number 2. There is a significant improvement in the description of the data, in the whole range of A and Z. Finally, we examine the role of emitted fragments in excited-bound and excited-unbound states, i.e., in the discrete and the continuum parts of their energy spectrum. This amounts to a total number of 181 decay channels. We obtain the curves marked with number 3. The agreement with the experimental A- and Z-distributions is very good. This was obtained with simple assumptions concerning level densities and transmission coefficients. However, the description of details, such as the odd–even structure of the Z-distributions, needs a more refined treatment.

The distribution of spallation residue cross sections in A and Z is shown in Fig. 16.1b. These cross sections are distributed on either side of the evaporation attractor line [78], shown with the dashed line. This line marks the location where the excited pre-

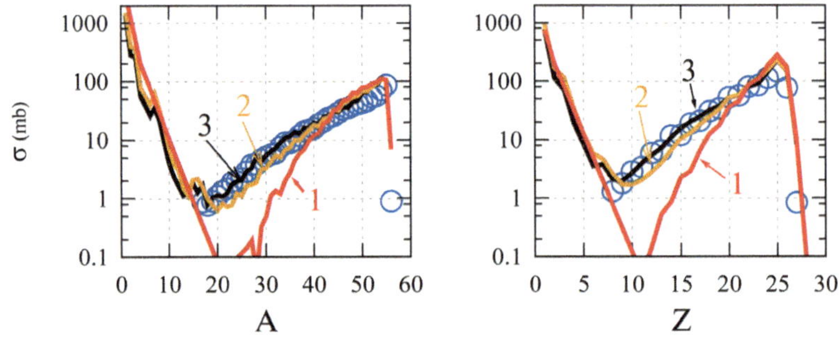

Figure 16.2 Mass and charge distributions of evaporation residues and IMFs in spallation reactions of 750 MeV ^{56}Fe + p. Experimental data (symbols) are compared with calculations performed with the codes ISABEL–MECO (solid curves) described in the text. Data are from Ref. [77].

fragments are expected to end up asymptotically. The evaporation attractor line is determined at the limit of low excitation energies from the equality between neutron and proton decay widths. The location of the line of stability, shown with the solid line in Fig. 16.1, is determined from the equality of the neutron and proton Fermi-level distances from zero. Along the line of stability, the Coulomb barrier inhibits proton emission and neutron emission may be favorable. Thus, the location of the evaporation attractor line is distinct from the line of stability.

16.7 IAEA Benchmark of Spallation Models

The International Atomic Energy Agency (IAEA) promotes a comprehensive benchmark of spallation models [80]. The specifications of the benchmark, including a set of selected experimental data to be compared to models, have been fixed in expert meetings. Experimental data consist of (a) excitation functions for isotope production, (b) double differential cross sections of emitted particles and nucleon clusters, and (c) average multiplicities and multiplicity distributions of emitted particles and nucleon clusters, including IMFs. The physics involved in commonly used physics models has been described in depth. Inter-comparisons of experimental results to model calculations and figures of merit have been reported. This facilitates the ranking of models considered in ongoing studies [81].

The nuclear reactions community greatly benefits from systematic comparisons involving the predictions of reliable models with good-quality experimental data. It is a great step toward the clarification of the underlying physics, enrichment of reaction models, and understanding of the spallation process.

Acknowledgments

The author acknowledges Prof. George Souliotis, Laboratory of Physical Chemistry, The University of Athens, Greece, for providing him with the event file of the ISABEL calculation and fruitful discussions.

References

1. Benlliure, J. (2006). Spallation reactions in applied and fundamental research, *Lect. Notes Phys.* **700**, pp. 191–238.
2. https://iaea.org/topics/nuclear-science/nuclear research/accelerators/spallation-sources
3. Goldenbaum, F. (2004). *The Physics of Spallation Process: Theory, Experiments and Applications* (ISBN: 978-3-89336-342-1).
4. David, J. C. (2015). Spallation reactions: A successful interplay between modeling and applications, *Eur. Phys. J. A* **51**, p. 68.
5. Metropolis, N. *et al.* (1958). Monte-Carlo calculations on intranuclear cascades. I. Low-energy studies, *Phys. Rev.* **110**, pp. 185.
6. Rudstam, G. (1966). Systematics of spallation yields, *Naturforschung Zeitschrift A* **21**, p. 1027.
7. Silberberg, R. and Tsao, C. H. (1990). Spallation processes and nuclear interaction products of cosmic rays, *Phys. Rep.* **191**, p. 351.
8. Silberberg, R., Tsao, C. H., and Barghouty, A. F. (1998). Updated partial cross sections of proton-nucleus reactions, *Astrophys. J.* **501**, p. 911.
9. Sümmerer, K. (2012). Improved empirical parametrization of fragmentation cross sections, *Phys. Rev. C* **86**, p. 014601.
10. Schmitt, C., Schmidt, K.-H., and Kelic-Heil, A. (2014). SPACS: A semi-empirical parameterization for isotopic spallation cross sections, *Phys. Rev. C* **90**, p. 064605.
11. Serber, R. (1947). Nuclear reactions at high energies, *Phys. Rev.* **72**, p. 1114.
12. Abdel-Waged, K. *et al.* (2010). Isospin effects in a covariant transport approach to spallation reactions: Analysis of p+Fe and p+Pb reactions at 0.8, 1.2, and 1.6 GeV, *Phys. Rev. C* **81**, p. 014605.
13. Napolitani, P. and Colonna, M. (2015). Fission dynamics of intermediate-fissility systems: A study within a stochastic three-dimensional approach, *Phys. Rev. C* **92**, p. 034607.
14. Papa, M. *et al.* (2001). Constrained molecular dynamics approach to fermionic systems, *Phys. Rev. C* **64**, p. 024612.
15. Jun, S., Long, Z., and Guo, C. (2018). Influence of the nuclear level density on the odd–even staggering in 56Fe+p spallation at energies from 300 to 1500 MeV/nucleon, *Phys. Rev. C* **97**, p. 054604.
16. Aichelin, J. (1991). Quantum molecular dynamics: A dynamical microscopic n-body approach to investigate fragment formation and the nuclear equation of state in heavy ion collisions, *Phys. Rep.* **202**, p. 233.

17. Hartnack, C. *et al.* (1989). Quantum molecular dynamics a microscopic model from UNILAC to CERN energies, *Nucl. Phys. A* **495**, p. 303.
18. Bertini, H. W. (1963). Low-energy intranuclear cascade calculation, *Phys. Rev.* **131**, p. 1801.
19. Yariv, Y. and Fraenkel, Z. (1979). Intranuclear cascade calculation of high-energy heavy-ion interactions, *Phys. Rev. C* **20**, p. 2227.
20. Cugnon, J. (1987). Proton–nucleus interaction at high energy, *Nucl. Phys. A* **462**, p. 751.
21. Budard, A., Cugnon, J., David, J.-C., Leray, S., and Mancusi, D. (2013). New potentialities of the Liège intranuclear cascade model for reactions induced by nucleons and light charged particles, *Phys. Rev. C* **87**, p. 014606.
22. Mancuci, D., Boudard, A., Cugnon, J., David, J.-C., and Kaitaniemi, P. (2014). Extension of the Liège intranuclear-cascade model to reactions induced by light nuclei, *Phys. Rev. C* **90**, p. 054602.
23. Bohr, N. (1936). Neutron capture and nuclear constitution, *Nature* **137**, p. 344.
24. Bohr, N. (1937). Transmutations of atomic nuclei, *Science* **86**, p. 161.
25. Bohr, N., Kalckar, F., and Danske, V. (1937). The penetration of atomic particles through matter, *Selsk. Math.-Phys. Medd.* **14**, p. 10.
26. Bethe, H. (1937). Nuclear physics B. Nuclear dynamics, theoretical, *Rev. Mod. Phys.* **9**, p. 69.
27. Weisskopf, V. F. and Ewing, D. H. (1940). On the yield of nuclear reactions with heavy elements, *Phys. Rev.* **57**, p. 472.
28. Thomas, T. D. (1968). Compound nuclear reactions induced by heavy ions, *Ann. Rev. Nucl. Sci.* **18**, p. 343.
29. Cole, A. J. (2000). *Statistical Models for Nuclear Decay: From Evaporation to Vaporization*, IOP Publishing Ltd, ISBN: 0 7503 0512 6.
30. Stokstad, R. G. (1984). *Treatise on Heavy Ion Science*, edited by D. A. Bromley, Vol. III, Plenum, New York.
31. Preston, M. A. (1962). *Physics of the Nucleus, Addisson*, Wesley Publishing Company, Reading Massachusetts, Palo Alto, London.
32. Nicolis, N. G. (2008). Sequential binary decay of highly excited nuclei, *Int. J. Mod. Phys. E* **17**, p. 1541.
33. Iljinov, A. S. *et al.* (1992). Phenomenological statistical analysis of level densities, decay widths and time lives of excited nuclei, *Nucl. Phys. A* **543**, p. 517.
34. Gross, D. H. E. (1990). Statistical decay of very hot nuclei: The production of large clusters, *Rep. Prog. Phys.* **53**, p. 605.
35. Wolfenstein, L. (1951). Conservation of angular momentum in the statistical theory of nuclear reactions, *Phys. Rev.* **82**, p. 690.

36. Feshbach, H. (1952). The inelastic scattering of neutrons, *Phys. Rev.* **87**, p. 366.
37. Douglas, A. C. and McDonald, M. (1959). Compound nucleus processes in medium mass nuclei, *Nucl. Phys.* **13**, p. 382.
38. Bass, R. (1980). *Nuclear Reactions with Heavy Ions*, Springer-Verlag, Berlin Heidelberg, New York.
39. Hodgson, P. E. (1987). Compound nucleus reactions, *Rep. Prog. Phys.* **50**, p. 1171.
40. Nicolis, N. G., Sarantites, D. G., Sobotka, L. G., and Charity, R. J. (1992). Statistical emission of deuterons and tritons from highly excited compound nuclei, *Phys. Rev. C* **45**, p. 2393.
41. Ericson, T. (1960). The statistical model and nuclear level densities, *Adv. Phys.* **9**, p. 425.
42. Moretto, L. G. and Wosniak, G. J. (1988). The role of the compound nucleus in complex fragment emission at low and intermediate energies, *Progr. Part. Nucl. Phys.* **21**, p. 401.
43. Koning, A. J., Hilaire, S., and Goriely, S. (2008). Global and local level density models, *Nucl. Phys. A* **810**, p. 13.
44. Alexander, J. M., Magda, M. T., and Landowne, S. (1990). Inverse reactions and the statistical evaporation model: Ingoing-wave boundary-condition and optical models, *Phys. Rev. C* **42**, p. 1092.
45. Emery, G. (1973). Perturbation of nuclear decay rates, *Ann. Rev. Nucl. Sci.* **22**, p. 378.
46. Czakanski, J. *et al.* (1992). Critical angular momentum of cluster evaporation: 16O (16O, α) 28Si and 16O (16O, 8Be) 24 Mg, *Nucl. Phys. A* **542**, p. 278.
47. Hodgson, P. E. and Betak, E. (2003). Cluster emission, transfer and capture in nuclear reactions, *Phys. Rep.* **374**, p. 1.
48. Charity, R. J. *et al.* (1988). Systematic of complex fragment emission in niobium-induced reactions, *Nucl. Phys. A* **483**, p. 371.
49. Vandebosch, R. and Huizenga, J. R. (1973). *Nuclear Fission*, Academic Press, New York.
50. Krappe, H. J. and Pomorski, K. (2012). *Theory of Nuclear Fission: A Textbook*, Springer-Verlag, Berlin Heidelberg.
51. Schmidt, K.-H. and Jurado, B. (2018). Review on the progress in nuclear fission: Experimental methods and theoretical descriptions, *Rep. Prog. Phys.* **81**, p. 106301.
52. Bohr, N. and Wheeler, J. A. (1939). The mechanism of nuclear fission, *Phys. Rev.* **56**, p. 426.
53. Sierk, A. J. (1986). Macroscopic model of rotating nuclei, *Phys. Rev. C* **33**, p. 2039.
54. Kramers, H. A. (1940). Brownian motion in a field of force and the diffusion model of chemical reactions, *Physica VII* **4**, p. 284.

55. Moretto, L. G. (1975). Statistical emission of large fragments: A general theoretical approach, *Nucl. Phys. A* **247**, p. 211.
56. Benlliure, J. *et al.* (1998). Yields in relativistic collisions of fissile nuclei, *Nucl. Phys. A* **628**, p. 458.
57. Swiatecki, W. J., Siwek-Wilczynska, K., and Wilczynski, J. (2008). Ratios of disintegration rates for distinct decay modes of an excited nucleus, *Phys. Rev. C* **78**, p. 054604.
58. Businaro, U. L. and Gallone, S. (1955). Asymmetric equilibrium shapes in the liquid drop model, *Nuovo Cimento* **1**, p. 1277.
59. Fong, P. (1969). Statistical theory of nuclear fission: A collection of papers, *Gordon and Breach*, **58**, p. 1.
60. Wilkins, B. D., Steinberg, E. P., and Chasman, R. R. (1976). Scission-point model of nuclear fission based on deformed-shell effects, *Phys. Rev. C* **14**, p. 1832.
61. Griffin, J. (1966). Statistical model of intermediate structure, *Phys. Lett.* **17**, p. 478.
62. Kalbach-Cline, C. (1973). Residual two-body matrix elements for pre-equilibrium calculations, *Nucl. Phys. A* **210**, p. 590.
63. Nishioka, H. *et al.* (1986). Statistical theory of precompound reactions: The multistep compound process, *Ann. Phys.* **172**, p. 67.
64. Nishioka, H., Weidenmuller, H. A., and Yoshida, S. (1988). Statistical theory of precompound reactions: The multistep direct process, *Ann. Phys.* **183**, p. 166.
65. Fermi, E. (1950). High energy nuclear events, *Progr. Theoret. Phys.* **5**, pp. 70.
66. Trautmann, W. (1997). Multifragmentation in relativistic heavy-ion reactions. In: *Correlations and Clustering Phenomena in Subatomic Physics*, edited by Harakeh, M. N., Koch, J. H., and Scholten, O., NATO ASI Series (Series B: Physics), Vol 359, Springer, Boston, MA.
67. Bondorf, J. P., Botvina, A. S., Iljinov, A. S., Mishustin, I. N., and Sneppen, K. (1995). Statistical multifragmentation of nuclei, *Phys. Rep.*, **257**, p. 133.
68. Carlson, B. V. *et al.* (2012). Fermi breakup and the statistical multifragmentation model, *Nucl. Phys. A* **876**, p. 77.
69. Junghans, A. R. *et al.* (1998). Projectile-fragment yields as a probe for the collective enhancement in the nuclear level density, *Nucl. Phys. A* **629**, p. 635.
70. Blann, M. (1966). Evidence for the influence of shell structure on level densities in the continuum, *Nucl. Phys.* **80**, p. 223.
71. Plasil, F. (1978). Deexcitation of compound nuclei with high angular momenta, *Phys. Rev. C* **17**, p. 823.
72. Dostrovsky, I. *et al.* (1960). Monte-Carlo calculations of nuclear evaporation processes. IV. Spectra of neutrons and charged particles from nuclear reactions, *Phys. Rev.* **118**, p. 781.

73. Furihata, S. (2000). *Nucl. Instr. Methods B* **171**, p. 251.
74. Gavron, A. (1980). Statistical model calculations in heavy ion reactions, *Phys. Rev. C* **21**, p. 230.
75. Bondorf, J. P. *et al.* (1995). Statistical multifragmentation of nuclei, *Phys. Rep.* **257**, p. 133.
76. http://www.talys.eu/home/
77. Villagrasa-Canton, C. *et al.* (2007). Spallation residues in the reaction 56Fe+p at 0.3 A, 0.5 A, 0.75 A, 1.0 A and 1.5 A GeV, *Phys. Rev. C* **75**, p. 044603.
78. Charity, R. J. (1998). N-Z distributions of secondary fragments and the evaporation attractor line, *Phys. Rev. C* **58**, p. 1073.
79. Christensen, P. R. and Winther, A. (1976). The evidence of the ion–ion potentials from heavy ion elastic scattering, *Phys. Lett. B* **65**, p. 19.
80. http://www-nds.iaea.org/spallations/
81. Singh, U. *et al.* (2018). The predictive power of spallation models for isotopic cross sections, *Eur. Phys. J. A* **54**, p. 109.

Index